An Introduction to Genetic Engineering

An Introduction to Genetic Engineering

Edited by David Irvine

Syrawood Publishing House
New York

Published by Syrawood Publishing House,
750 Third Avenue, 9th Floor,
New York, NY 10017, USA
www.syrawoodpublishinghouse.com

An Introduction to Genetic Engineering
Edited by David Irvine

International Standard Book Number: 978-1-68286-592-7 (Hardback)

Cataloging-in-Publication Data

An introduction to genetic engineering / edited by David Irvine.
p. cm.
Includes bibliographical references and index.
ISBN 978-1-68286-592-7
1. Genetic engineering. 2. Biotechnology. I. Irvine, David.
TP248.6 .I58 2018
660.65--dc23

Printed in China.

TABLE OF CONTENTS

PREFACE

Genetic engineering is the study of the process of transforming genomes. It uses the elements and techniques of biotechnology to manipulate the genes and transfer them across organisms to develop better organisms and crops. The processes used in genetic engineering are gene isolation, molecular cloning, genetic transformation, etc. This book is a compilation of chapters that discuss the most vital concepts in the field of genetic engineering. It presents topics which include various methods and theories used in this vast field. Different approaches, evaluations and methodologies have been included in it. This textbook attempts to assist those with a goal of delving into the field of genetic engineering.

To facilitate a deeper understanding of the contents of this book a short introduction of every chapter is written below:

Chapter 1- Genetic engineering is the science of altering the genome of an organism using biotechnology. Genome comprises the genes, non-coding DNA, and either mitochondria or chloroplast and consists the complete set of genetic information of an organism. This is an introductory chapter which will introduce briefly all the significant aspects of genetic engineering.

Chapter 2- A restriction enzyme is a kind of enzyme that cuts a DNA at a specific or random location known as restriction sites. There are two sorts of restriction enzymes, exonucleases and endonucleases. Types of enzymes discussed in this chapter are polynucleotide phosphorylase, phosphatases, methylases, etc. The chapter strategically encompasses and incorporates genetic engineering, providing a complete understanding.

Chapter 3- Nucleic acids are polymers that store DNA. It is classified into types, DNA and RNA. The nucleic acids are essential in researches for which they need to be isolated. There are many methods to detect nucleic acids. A few of them are fluorometric quantification, UV spectroscopy, silver staining, ethidium bromide staining, etc. The aspects elucidated in this chapter are of vital importance, and provide a better understanding of genetic engineering.

Chapter 4- For gene manipulation, nucleic acid needs to be isolated and separated. Isolation goes through three steps. First, the cell membrane is broken to free cellular components and DNA. Then, the nucleic acid is separated from other organelles. Finally, the nucleic acid is purified. Organic extraction and enzymatic digestion, ion-exchange chromatography, DNA samples concentration are few methods of purifying nucleic acid. This chapter will provide an integrated understanding of genetic engineering.

Chapter 5- Genes can be transferred using biological, chemical and mechanical methods. Biological methods include bactofection and transduction; chemical methods include calcium phosphate, cationic lipid, etc. A few mechanical methods are microinjection, sonoporation, bead transfection, electroporation, etc. In order to completely understand genetic engineering, it is necessary to understand the processes related to it. The following chapter elucidates the varied processes and methods associated with this area of study.

Chapter 6- Transgenic science is used for both animals and plants. In plants, genes are introduced normally to strengthen them against diseases, pests and stress. Transgenic animals, on the other hand, are used for clinical research. Knockout and knockin animals are two types of transgenic animals. The aspects elucidated in this chapter are of vital importance, and provide a better understanding of the advancements made in genetic engineering.

I owe the completion of this book to the never-ending support of my family, who supported me throughout the project.

Editor

1

An Overview of Genetic Engineering

Genetic engineering is the science of altering the genome of an organism using biotechnology. Genome comprises the genes, non-coding DNA, and either mitochondria or chloroplast and consists the complete set of genetic information of an organism. This is an introductory chapter which will introduce briefly all the significant aspects of genetic engineering.

Function of Gene

The main distinction between a living and non-living entity is the ability to replicate and reproduce similar offsprings. Nucleic acid molecules (DNA and RNA) present in a living organism acts as a genetic template to pass the hereditary information from one generation to the next. Nucleic acid molecules are organised as genes which code for a particular phenotype via specific proteins and the expression of a gene is regulated by both external and internal factors which aid the developmental process of an organism. This relation between genes and proteins forms the "central dogma of life".

Gene

A gene can be defined as the region of DNA (or RNA in case of virus) that controls a discrete hereditary characteristic, usually corresponding to a single protein or RNA. This includes the entire functional unit, encompassing coding (exons) and noncoding sequences (introns and regulatory sequences).

- Exons and introns which represent the coding and noncoding regions are present in a eukaryotic gene. Introns are absent in prokaryotes.
- The introns are removed by splicing and the exons are translated in tandem to yield the functional polypeptide that further undergoes post translational modification to become functional. These functional polypeptides (proteins) are targeted to various organelles in the cell or exported out of the cell for carrying out various intracellular and extracellular processes respectively.

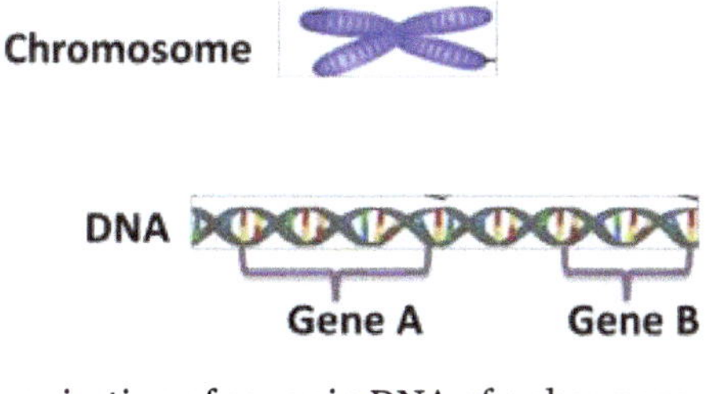

Organization of genes in DNA of a chromosome

Genome

Genome is the complete set of genetic information of a cell or an organism; in particular, the complete sequence of DNA/RNA that carries this information. In diploid organisms, it refers to the haploid set of chromosomes present in a cell. Depending on its localization, genome may be nuclear or organellar. Organellar genomes are again of two types: mitochondrial and chloroplast genome. Genome size of organisms differs significantly between different species. The size of the genome governs the size and complexity of an organism. However, many small sized organisms, in fact have bigger genomes than their larger counterparts.

Various organisms have different sized genome as can be seen in the table below:

Species	Organism	Genome Size (Mb)
Triticum aestivum	Plant	16000
Homo sapiens	Mammal	3200
Arabidopsis thaliana	Plant	125
Drosophila melanogaster	Insect	180
Caenorhabditis elegans	Nematode worm	97
Saccharomyces cerevisiae	Yeast	12.1
Escherichia coli	Bacterium	4.64
Haemophilus influenzae	Bacterium	1.83
Mycoplasma genitalium	Bacterium	0.58

- The genome contains all the genes present in the nucleus of a cell. Gene varies in size from a few hundred DNA/RNA bases to more than few thousand bases.
- The haploid set of chromosome contains the total genome of the organism.
- The bacterium *Mycoplasma genitalium* has a small genome size of 0.58Mb and the plant *Triticum aestivim* has a large genome size of 16000Mb. The genome size in human is 3200Mb.

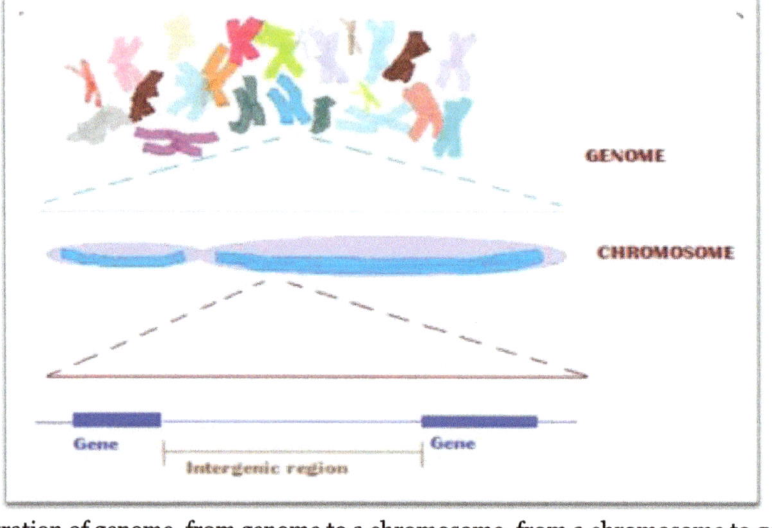

Illustration of genome, from genome to a chromosome, from a chromosome to gene

Role of Genes Within Cells

- Genes contain the instructions for each cell to make proteins and RNAs. Genes are made up of DNA fragments.
- Within the cell the DNA performs two tasks:
  - Act as information repository including instructions in making the component molecules of the cells.
  - Pass on the information to the next generation.
- The mere presence of DNA does not implicate a cell to be alive and functional. Mammalian red blood cells (RBCs) discard nucleus during developmental process and thus lacks DNA in mature state.
- Genes are transcribed to RNA which are processed to various forms like mRNA, tRNA, rRNA etc. mRNA are translated to proteins depending on the regulatory signals. tRNA and rRNA serve as the components of translational machinery.
- New functions of RNA are also being discovered like regulatory (miRNA, siRNA etc) and catalytic (ribozymes) functions.
- Proteins are structural components of cells, enzymes, hormones, various signalling molecules, receptors and other factors which are involved in performing the chemistry of life and are essential for the normal body function, for example, sugar conversion to energy and metabolite (small molecules) production in cell.

Genetic Code

The genetic code is the set of instructions that translates the information encoded in genetic material (mRNA or DNA sequences) into proteins (amino acid sequences) by living cells.

- The genetic code is a triplet code (i.e. a group of three adjacent nucleotides) called codon. This three nucleotide codon in nucleic acid sequence specifies a single amino acid. However, genetic code in human mitochondria differs from the standard nuclear genetic code.
- Epigenetic effects, however is not stored using the genetic code. Besides all organisms have DNA that contains regulatory sequences, intergenic segments, chromosomal structural areas and other non-coding DNA that can contribute greatly to phenotype.
- These codons are always written with the 5′-terminal nucleotide to the left.
- The code is unambiguous i.e. each triplet specifies only a single amino acid.
- The genetic code is degenerate i.e. more than one triplet codon can code for a single amino acid. (61 codons code for 20 amino acids)
- Three codons do not specify any amino acid but acts as terminal sites (stop codons), signalling the end of protein coding sequence. They are namely UAA (Ochre), UAG (Amber) and UGA (Opal).

- AUG is an initiation codon that signals the initiation of translation and also codes for methionine. In some mRNA, GUG and UUG also act as initiation codon.
- The genetic codes for each 20 amino acids were defined by pioneering works of Marshall W. Nirenberg and Har Gobind Khorana in 1964.

Second letter→

First letter↓	U		C		A		G		Third letter↓
	U		C		A		G		
U	UUU	Phe	UCU	Ser	UAU	Tyr	UGU	Cys	U
	UUC	Phe	UCC	Ser	UAC	Tyr	UGC	Cys	C
	UUA	Leu	UCA	Ser	UAA	Stop	UGA	Stop	A
	UUG	Leu	UCG	Ser	UAG	Stop	UGG	Irp	G
C	CUU	Leu	CCU	Pro	CAU	His	CGU	Arg	U
	CUC	Leu	CCC	Pro	CAC	His	CGC	Arg	C
	CUA	Leu	CCA	Pro	CAA	Gln	CGA	Arg	A
	CUG	Leu	CCG	Pro	CAG	Gln	CGG	Arg	G
A	AUU	Ile	ACU	Thr	AAU	Asn	AGU	Ser	U
	AUC	Ile	ACC	Thr	AAC	Asn	AGC	Ser	C
	AUA	Ile	ACA	Thr	AAA	Lys	AGA	Arg	A
	AUG	Met	ACG	Thr	AAG	Lys	AGG	Arg	G
G	GUU	Val	GCU	Ala	GAU	Asp	GGU	Gly	U
	GUC	Val	GCC	Ala	GAC	Asp	GGC	Gly	C
	GUA	Val	GCA	Ala	GAA	Glu	GGA	Gly	A
	GUG	Val	GCG	Ala	GAG	Glu	GGG	Gly	G

Table: Genetic code

Gene Expression

Let us now understand the central dogma of life. You now know that most of the genes contain the information needed to make functional molecules called proteins. However, some genes produce other molecule (tRNA, rRNA, microRNA etc) that assist or regulates the protein expression and assembly. These complex events within each cell consist of two main steps: transcription and translation. In prokaryotes where there is no nuclear membrane, both transcription and translation occur in the cytoplasm whereas in eukaryotes, transcription occurs inside the nucleus and the translation occurs in the cytoplasm.

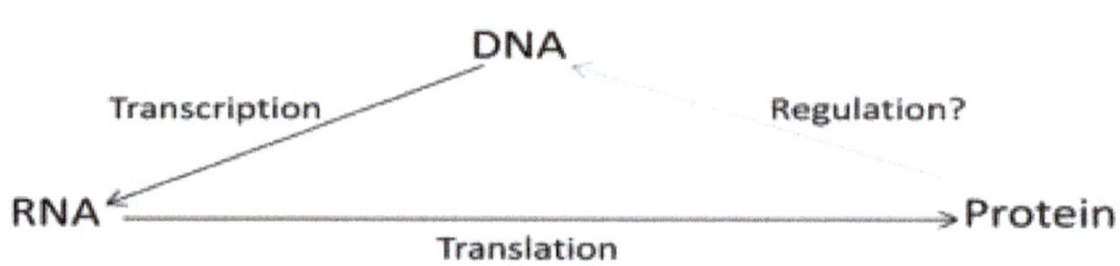

Central dogma of life

Events Involved in Gene Expression

The events involved in genome expression in higher organisms with respect to protein coding genes are discussed below. The genes that give rise to non-coding RNAs are transcribed and processed but are never translated.

- Access to gene: Genes are inaccessible as they are buried deep within the highly packaged chromosomes. The initial step involves a preparative process that opens the chromatin structure and positions of the nucleosome in the region of genome containing active genes.
- Formation of transcription initiation complex involves the assembly of a set of proteins into a complex that copy DNA into RNA. This is a highly regulated process as the transcription initiation complex must be constructed at the precise position in the genome, adjacent to active genes to form a RNA copy.

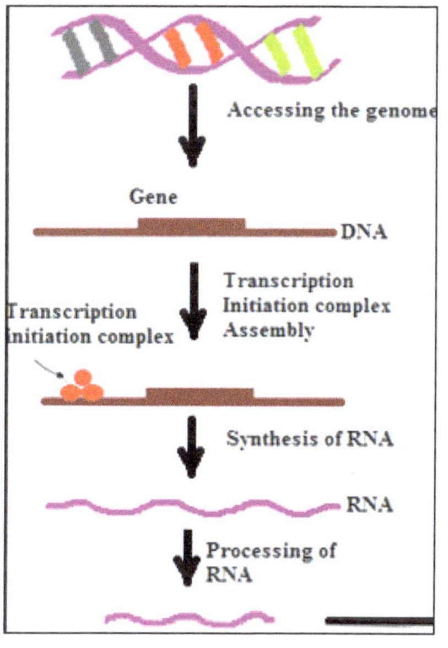

Steps involved in Gene Expression

- RNA Synthesis involves the transcription of a gene into RNA molecule and it occurs in the nucleus.
- RNA Processing comprises of post transcriptional modification/alterations of the RNA molecule and its chemical structure required for the RNA to be translated into protein or non-coding RNA (rRNA, tRNA, miRNA). RNA splicing (deletion of introns and combination of exons), 5' capping, polyadenylation etc are commonly occurred RNA processing steps in eukaryotes. However, prokaryotic organisms do not have a well developed RNA processing mechinary.
- Degradation of RNA is the controlled turnover of RNA molecules and should not be viewed simply as a mean of getting rid of unwanted RNAs. It determines the makeup of the transcriptome and is considered as an important step in genome expression. Different ribonucleases (RNases) plays the prime role in this process and multiple cofactors like small RNA (siRNA, miRNA etc), molecular cheparons (Lsm1-7, Lsm2-8, Hfq etc) regulate this process.
- Protein synthesis is initiated after the assembly of the translation initiation complex near the 5' termini of a mRNA molecule. It involves translation of RNA molecules into proteins.
- Protein folding and protein processing may occur together after protein synthesis. Post translation events like folding involve the protein attaining its correct three dimentional

configuration. Processing (phosphorylation, glycosylation, carboxylation etc) involves the modification of the protein by addition of chemical groups and removal of one or more functional units of the protein.

Types of Gene Expression

- Constitutive expression: Housekeeping genes are essential and necessary for sustaining life, and are therefore continuously expressed. *gapdh* (glyceraldehydes 3 phosphate dehydrogenase), *sdha* (succinate dehydrogenase) etc are human housekeeping genes which are expressed throughout the development.
- Induction and repression: The expression levels of some genes fluctuate in response to external signals. Also, under a certain situation, some genes show higher expression level, while others show lower expression levels. The former is called induced expression and the latter is called repressed expression.

Control of Gene Expression

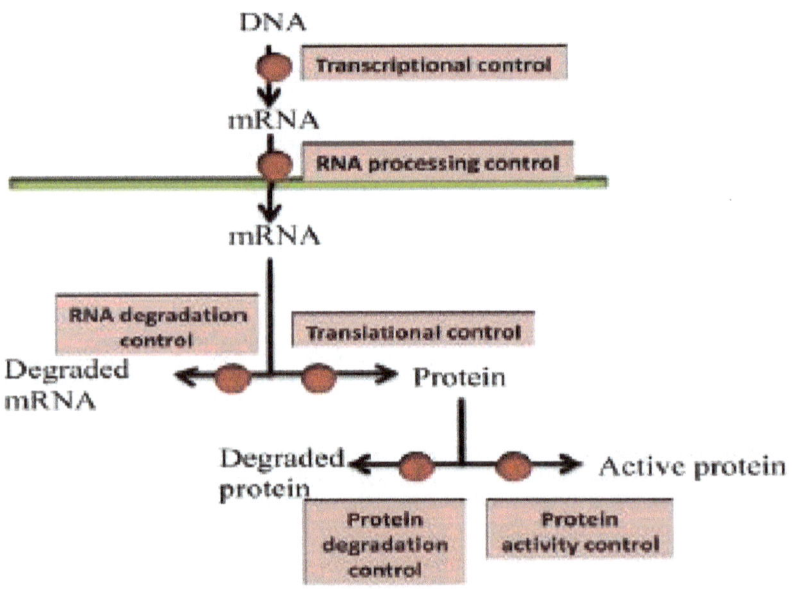

Points of control of gene expression

The eukaryotic gene expressions in a cell are controlled at about six stages as shown in the figure.

- Transcriptional control decides when and how frequently a given gene be transcribed. It enables the cell to check any discrepancy in the quantity of transcriptomes generated.
- RNA processing regulates how the RNA transcript is spliced or otherwise processed. (applicable only to eukaryotes), which have split genes consisting of exons and introns.
- Translational control decides which mRNAs in the cytoplasm are to be translated by ribosomes.
- mRNA degradation control selectively destabilize some mRNA molecules in the cytoplasm.

- Protein activity control selectively activates, inactivates, degrades, or compartmentalizes specific protein molecules after they have been synthesized. (applicable only to eukaryotes)

Regulatory Elements

Gene expression is a complex multi-step process. Amongst various steps involved, transcription initiation is a vital key point in controlling gene expression.

Fundamental elements that regulate the process of transcription are:

- cis-acting elements comprising of special DNA sequences.
- trans-acting elements comprising of regulatory proteins.
- DNA –protein interaction.
- Protein-protein interaction
- RNA polymerase

Cis-acting Elements

In Latin "cis", means "same side as". Cis acting elements are thus a region of DNA or RNA that regulates the expression of genes located on the same molecule. The cis-regulatory elements are often binding sites for one or more trans-acting factors. Cis-element may be located upstream to the coding sequence of the gene it controls (in the promoter region or further upstream 5'), in an intron, or 3' to the gene's coding sequence, either in the untranslated or untranscribed region.

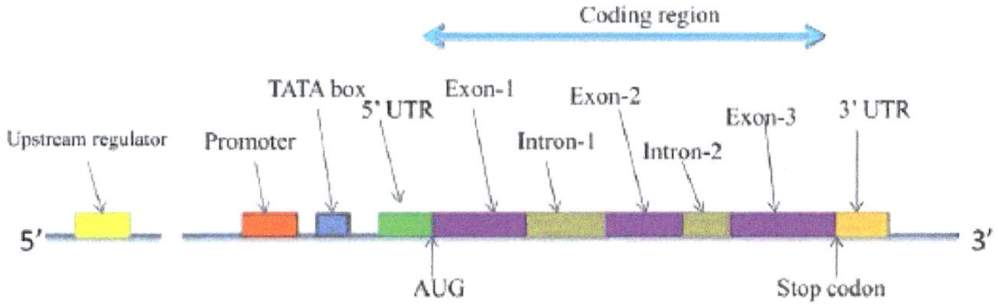

Structure of eukaryotic gene with upstream regulatory elements

Prokaryotic Systems

- Promoter is a DNA sequence where RNA polymerase binds to initiate transcription. There are two promoter sequences in prokaryotic systems known as -10 (Pribnow box or Pribnow-Schaller box) and -35 sequences.
- Operator is regulatory sequence of DNA located immediately upstream of the structural gene that controls transcription of an operon.
- Inducers are located upstream of promoter region.

- Downstream regulatory sequences comprise of GC-rich inverted repeats followed by four adenine (A) residues signal the termination of transcription.

Trans-acting Elements

The protein factors which regulate the expression of gene by binding to cis acting DNA sequence are termed as trans-acting elements. Trans-acting molecules generally have two domains: DNA binding domain (which binds to cis elements) and protein binding domain (required for activation or suppression of transcription). Transcription initiation is a tightly regulated process controlled by trans-acting elements both in prokaryotes as well as eukaryotes.

DNA – Protein Interactions

Gene regulatory proteins and the transcription factors are capable of binding to the DNA based on the interaction of amino acids of the protein with the nucleotides of the DNA. The regulation is implemented through various interactions between cis-acting elements and trans-acting factors. Examples of some DNA binding proteins are:

- Eukaryotic TATA-binding protein
- σ subunit of bacterial RNA polymerase etc.

There are four types of structures of DNA binding proteins:

- Zinc finger proteins
- Helix loop Helix protein
- Leucine zipper proteins
- Homeodomain proteins

Protein – Protein Interactions

Protein-protein interaction is present in both prokaryotes and eukaryotes. The external signals affect the gene expression with the help of such interaction. Proteins interact with each other to form a homo or hetero-dimers before binding to DNA molecules. Eg. transcription initiation factors interacts with TATA box binding proteins (TBP) to activate transcription of a gene.

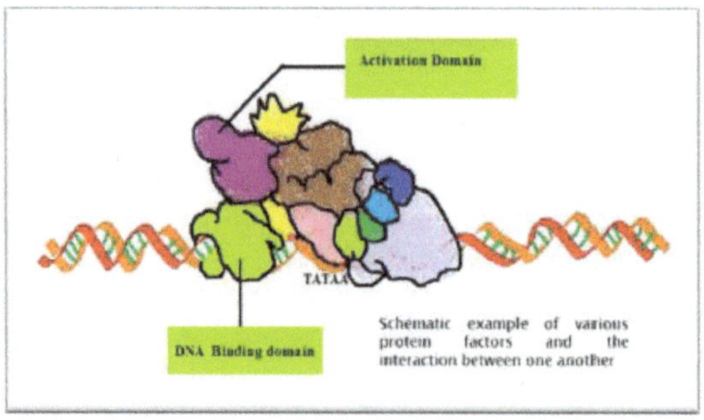

Interaction between various protein factors

RNA Polymerase

- A single RNA polymerase is responsible for transcribing all types of RNA in prokaryotic system.
- However, eukaryotes have three different RNA polymerases, which have been found to specialize in the synthesis of various types of RNA:
  1. RNA polymerase I (Pol I) - transcribes rRNA (ribosomal RNA) genes.
  2. RNA polymerase II (Pol II) - transcribes protein-coding genes or mRNA (messenger RNA).
  3. RNA polymerase III (Pol III) - transcribes other functional RNA genes (e.g., tRNA).
- In eukaryotes, transcription occurs inside the nucleus. All the enzymes responsible for translation are present in the cytosol therefore the transcripts formed then move out of the nucleus through nuclear pores into the cytosol (the liquid phase of the cytoplasm), where translation occurs. Organelle genomes (like the mitochondrial genome) are transcribed within the organelle (the mitochondria) and translation is also within the organelle (the mitochondria).
- Since prokaryotes have no nucleus, the step involving the movement of transcripts from nucleus to cytoplasm does not take place, and translation can take place immediately in the cytoplasm, directly on the growing transcript.

Vector

DNA molecule used for carrying an exogenous DNA into a host organism and facilitates stable integration and replication inside the host system is termed as Vector. Molecular cloning involves series of sequential steps which includes restriction digestion of DNA fragments both target DNA and vector, ligation of the target DNA with the vector and introduction into a host organism for multiplication. Then the fragments resulted after digestion with restriction enzymes are ligated to other DNA molecules that serve as vectors.

In general, vectors should have following characteristics:

- Capable of replicating inside the host.
- Have compatible restriction site for insertion of DNA molecule (insert).
- Capable of autonomous replication inside the host (*ori* site).
- Smaller in size and able to incorporate larger insert size.
- Have a selectable marker for screening of recombinant organism.

Plasmid

A plasmid is a small DNA molecule within a cell that is physically separated from a chromosomal DNA and can replicate independently. They are most commonly found in bacteria as small circular, double-stranded DNA molecules; however, plasmids are sometimes present in archaea and eukaryotic organisms. In nature, plasmids often carry genes that may benefit the survival of the organism, for example antibiotic resistance. While the chromosomes are big and contain all the essential genetic information for living under normal conditions, plasmids usually are very small and contain only additional genes that may be useful to the organism under certain situations or particular conditions. Artificial plasmids are widely used as vectors in molecular cloning, serving to drive the replication of recombinant DNA sequences within host organisms.

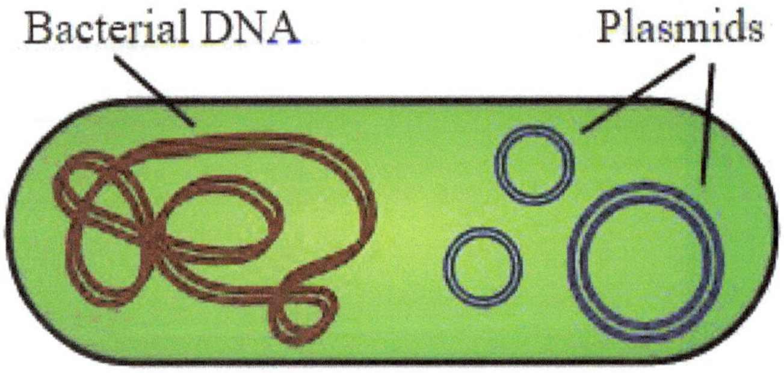

Illustration of a bacterium showing chromosomal DNA and plasmids. Not to scale.

Plasmids are considered *replicons*, a unit of DNA capable of replicating autonomously within a suitable host. However, plasmids, like viruses, are not generally classified as life. Plasmids can be transmitted from one bacterium to another (even of another species) via three main mechanisms: transformation, transduction, and conjugation. This host-to-host transfer of genetic material is called horizontal gene transfer, and plasmids can be considered part of the mobilome. Unlike viruses (which encase their genetic material in a protective protein coat called a capsid), plasmids are "naked" DNA and do not encode genes necessary to encase the genetic material for transfer to a new host. However, some classes of plasmids encode the conjugative "sex" pilus necessary for their own transfer. The size of the plasmid varies from 1 to over 200 kbp, and the number of identical plasmids in a single cell can range anywhere from one to thousands under some circumstances.

The relationship between microbes and plasmid DNA is neither parasitic nor mutualistic, because each implies the presence of an independent species living in a detrimental or commensal state with the host organism. Rather, plasmids provide a mechanism for horizontal gene transfer within a population of microbes and typically provide a selective advantage under a given environmental state. Plasmids may carry genes that provide resistance to naturally occurring antibiotics in a competitive environmental niche, or the proteins produced may act as toxins under similar circumstances, or allow the organism to utilize particular organic compounds that would be advantageous when nutrients are scarce.

Properties and Characteristics

The American molecular biologist Joshua Lederberg first introduced the term *plasmid* in 1952 - originally to describe any bacterial genetic material that exists in an extrachromosomal state for at least part of its replication cycle. Later in 1968, it was decided that the term plasmid should be adopted as the term for extrachromosomal genetic element, and to distinguish it from viruses, the definition was narrowed to genetic elements that exist exclusively or predominantly outside of the chromosome and can replicate autonomously.

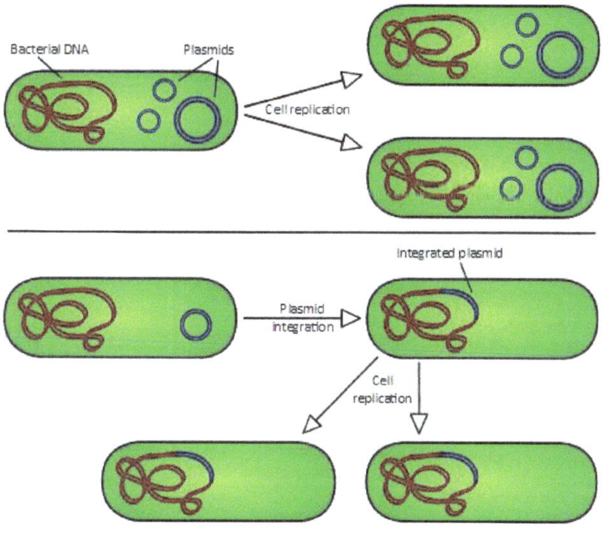

There are two types of plasmid integration into a host bacteria: Non-integrating plasmids replicate as with the top instance, whereas episomes, the lower example, can integrate into the host chromosome.

In order for plasmids to replicate independently within a cell, they must possess a stretch of DNA that can act as an origin of replication. The self-replicating unit, in this case the plasmid, is called a replicon. A typical bacterial replicon may consist of a number of elements, such as the gene for plasmid-specific replication initiation protein (Rep), repeating units called iterons, DnaA boxes, and an adjacent AT-rich region. Smaller plasmids make use of the host replicative enzymes to make copies of themselves, while larger plasmids may carry genes specific for the replication of those plasmids. A few types of plasmids can also insert into the host chromosome, and these integrative plasmids are sometimes referred to as episomes in prokaryotes.

Plasmids almost always carry at least one gene. Many of the genes carried by a plasmid are beneficial for the host cells, for example: enabling the host cell to survive in an environment that would otherwise be lethal or restrictive for growth. Some of these genes encode traits for antibiotic resistance or resistance to heavy metal, while others may produce virulence factors that enable a bacterium to colonize a host and overcome its defences, or have specific metabolic functions that allow the bacterium to utilize a particular nutrient, including the ability to degrade recalcitrant or toxic organic compounds. Plasmids can also provide bacteria with the ability to fix nitrogen. Some plasmids, however, have no observable effect on the phenotype of the host cell or its benefit to the host cells cannot be determined, and these plasmids are called cryptic plasmids.

Naturally occurring plasmids vary greatly in their physical properties. Their size can range from very small mini-plasmids of less than a 1 kilobase pairs (Kbp), to very large megaplasmids of several megabase pairs (Mbp). At the upper end, little can differentiate between a megaplasmid and a minichromosome. Plasmids are generally circular, however examples of linear plasmids are also known. These linear plasmids require specialized mechanisms to replicate their ends.

Plasmids may be present in an individual cell in varying number, ranging from one to several hundreds. The normal number of copies of plasmid that may be found in a single cell is called the copy number, and is determined by how the replication initiation is regulated and the size of the molecule. Larger plasmids tend to have lower copy numbers. Low-copy-number plasmids that exist only as one or a few copies in each bacterium are, upon cell division, in danger of being lost in one of the segregating bacteria. Such single-copy plasmids have systems that attempt to actively distribute a copy to both daughter cells. These systems, which include the parABS system and parMRC system, are often referred to as the partition system or partition function of a plasmid.

Classifications and Types

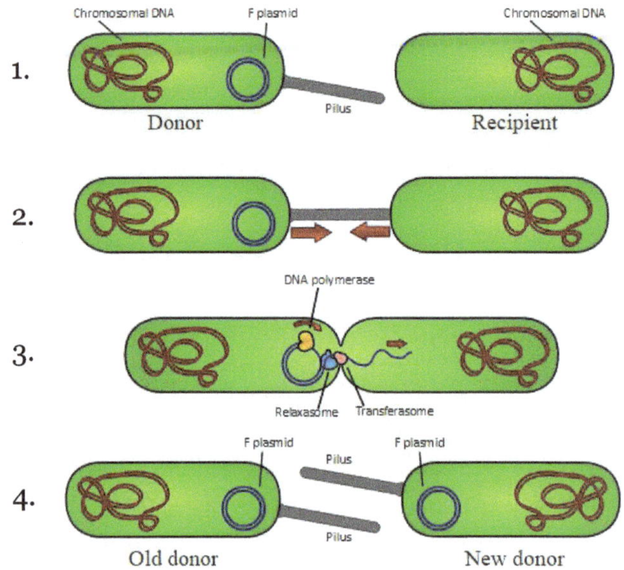

Overview of bacterial conjugation

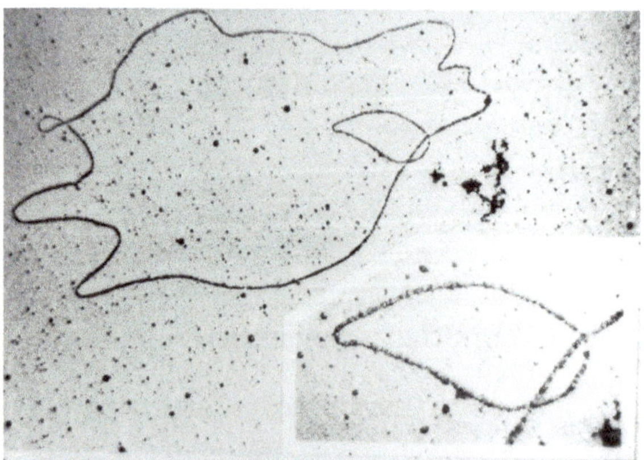
Electron micrograph of a DNA fiber bundle, presumably of a single bacterial chromosome loop.

Plasmids may be classified in a number of ways. Plasmids can be broadly classified into conjugative plasmids and non-conjugative plasmids. Conjugative plasmids contain a set of transfer or *tra* genes which promote sexual conjugation between different cells. In the complex process of conjugation, plasmid may be transferred from one bacterium to another via sex pili encoded by some of the *tra* genes. Non-conjugative plasmids are incapable of initiating conjugation, hence they can be transferred only with the assistance of conjugative plasmids. An intermediate class of plasmids are mobilizable, and carry only a subset of the genes required for transfer. They can parasitize a conjugative plasmid, transferring at high frequency only in its presence.

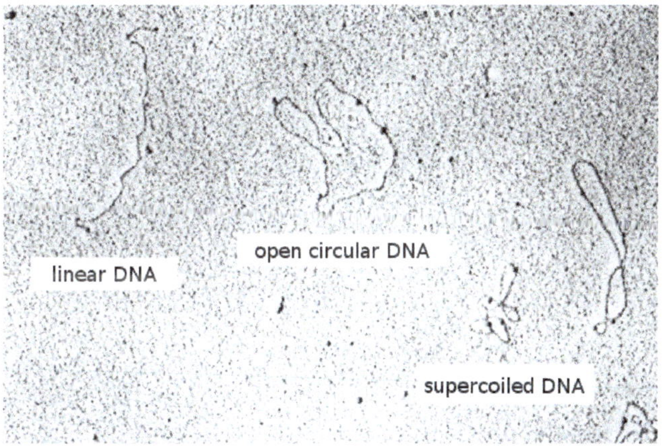

Electron micrograph of a bacterial DNA plasmid (chromosome fragment).

Plasmids can also be classified into incompatibility groups. A microbe can harbour different types of plasmids, however, different plasmids can only exist in a single bacterial cell if they are compatible. If two plasmids are not compatible, one or the other will be rapidly lost from the cell. Different plasmids may therefore be assigned to different incompatibility groups depending on whether they can coexist together. Incompatible plasmids normally share the same replication or partition mechanisms.

Another way to classify plasmids is by function. There are five main classes:

- Fertility F-plasmids, which contain *tra* genes. They are capable of conjugation and result in the expression of sex pili.
- Resistance (R) plasmids, which contain genes that provide resistance against antibiotics or poisons. Historically known as R-factors, before the nature of plasmids was understood.
- Col plasmids, which contain genes that code for bacteriocins, proteins that can kill other bacteria.
- Degradative plasmids, which enable the digestion of unusual substances, e.g. toluene and salicylic acid.
- Virulence plasmids, which turn the bacterium into a pathogen.

Plasmids can belong to more than one of these functional groups.

Vectors

Artificially constructed plasmids may be used as vectors in genetic engineering. These plasmids serve as important tools in genetics and biotechnology labs, where they are commonly used to clone and amplify (make many copies of) or express particular genes. A wide variety of plasmids are commercially available for such uses. The gene to be replicated is normally inserted into a plasmid that typically contains a number of features for their use. These include a gene that confers resistance to particular antibiotics (ampicillin is most frequently used for bacterial strains), an origin of replication to allow the bacterial cells to replicate the plasmid DNA, and a suitable site for cloning.

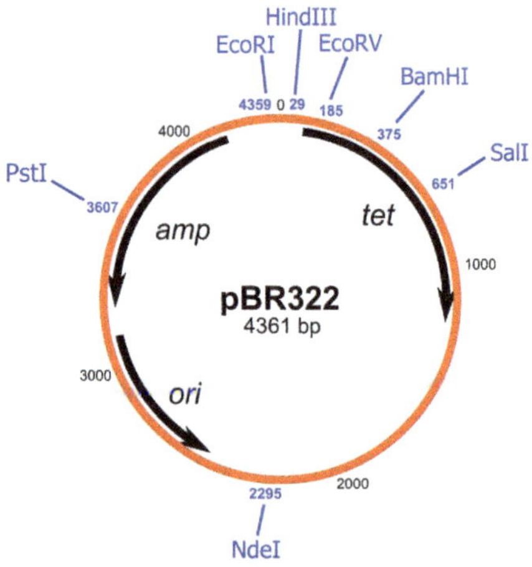

A schematic representation of the pBR322 plasmid, one of the first plasmids to be used widely as a cloning vector. Shown on the plasmid diagram are the genes encoded (*amp* and *tet* for ampicillin and tetracycline resistance respectively), its origin of replication (*ori*), and various restriction sites (indicated in blue).

Cloning

Plasmids are the most-commonly used bacterial cloning vectors. These cloning vectors contain a site that allows DNA fragments to be inserted, for example a multiple cloning site or polylinker which has several commonly used restriction sites to which DNA fragments may be ligated. After the gene of interest is inserted, the plasmids are introduced into bacteria by a process called transformation. These plasmids contain a selectable marker, usually an antibiotic resistance gene, which confer on the bacteria an ability to survive and proliferate in a selective growth medium containing the particular antibiotics. The cells after transformation are exposed to the selective media, and only cells containing the plasmid may survive. In this way, the antibiotics act as a filter to select only the bacteria containing the plasmid DNA. The vector may also contain other marker genes or reporter genes to facilitate selection of plasmid with cloned insert. Bacteria containing the plasmid can then be grown in large amounts, harvested, and the plasmid of interest may then be isolated using various methods of plasmid preparation.

A plasmid cloning vector is typically used to clone DNA fragments of up to 15 kbp. To clone longer lengths of DNA, lambda phage with lysogeny genes deleted, cosmids, bacterial artificial chromosomes, or yeast artificial chromosomes are used.

Protein Production

Another major use of plasmids is to make large amounts of proteins. In this case, researchers grow bacteria containing a plasmid harboring the gene of interest. Just as the bacterium produces proteins to confer its antibiotic resistance, it can also be induced to produce large amounts of proteins from the inserted gene. This is a cheap and easy way of mass-producing the protein the gene codes for, for example, insulin.

Gene Therapy

Plasmid may also be used for gene transfer into human cells as potential treatment in gene therapy so that it may express the protein that is lacking in the cells. Some strategies of gene therapy require the insertion of therapeutic genes at pre-selected chromosomal target sites within the human genome. Plasmid vectors are one of many approaches that could be used for this purpose. Zinc finger nucleases (ZFNs) offer a way to cause a site-specific double-strand break to the DNA genome and cause homologous recombination. Plasmids encoding ZFN could help deliver a therapeutic gene to a specific site so that cell damage, cancer-causing mutations, or an immune response is avoided.

Disease Models

Plasmids were historically used to genetically engineer the embryonic stem cells of rats in order to create rat genetic disease models. The limited efficiency of plasmid-based techniques precluded their use in the creation of more accurate human cell models. However, developments in Adeno-associated virus recombination techniques, and Zinc finger nucleases, have enabled the creation of a new generation of isogenic human disease models.

Episomes

The term episome was proposed by François Jacob and Élie Wollman in 1958 to describe extra-chromosomal genetic material that may replicate autonomously or become integrated into the chromosome. The use of the term, however, has diverged since it was first coined as plasmid became the preferred word for autonomously replicating extrachromosomal DNA as proposed in the symposium in 1968 – it was suggested by some that the use of the term episome be abandoned, although others continued to use the term with a shift in meaning. In prokaryotes, episome is now used by some to refer to plasmid that is capable of integrating into the chromosome. The integrative plasmids may be replicated and stably maintained in a cell through multiple generations, but always at some stage they exist as an independent plasmid molecule.

In eukaryotes, episomes are used to mean non-integrated extrachromosomal closed circular DNA molecule that may be replicated in the nucleus. Viruses are the most common examples of this, such as herpesviruses, adenoviruses, and polyomaviruses, but some are plasmids. Other examples include aberrant chromosomal fragments, such as double minute chromosomes, that can arise during artificial gene amplifications or in pathologic processes (e.g., cancer cell transformation). Episomes in eukaryotes behave similarly to plasmids in prokaryotes in that the DNA is stably maintained and replicated with the host cell. Cytoplasmic viral episomes (as in poxvirus infections) can also occur. Some episomes, such as herpesviruses, replicate in a rolling circle mechanism, similar to bacterial phage viruses. Others replicate through a bidirectional replication mechanism (*Theta type* plasmids). In either case, episomes remain physically separate from host cell chromosomes. Several cancer viruses, including Epstein-Barr virus and Kaposi's sarcoma-associated herpesvirus, are maintained as latent, chromosomally distinct episomes in cancer cells, where the viruses express oncogenes that promote cancer cell proliferation. In cancers, these episomes passively replicate together with host chromosomes when the cell divides. When these viral episomes initiate lytic replication to generate multiple virus particles, they in general activate cellular innate immunity defense mechanisms that kill the host cell.

Plasmid Maintenance

Some plasmids or microbial hosts include an addiction system or postsegregational killing system (PSK), such as the hok/sok (host killing/suppressor of killing) system of plasmid R1 in *Escherichia coli*. This variant produces both a long-lived poison and a short-lived antidote. Several types of plasmid addiction systems (toxin/ antitoxin, metabolism-based, ORT systems) were described in the literature and used in biotechnical (fermentation) or biomedical (vaccine therapy) applications. Daughter cells that retain a copy of the plasmid survive, while a daughter cell that fails to inherit the plasmid dies or suffers a reduced growth-rate because of the lingering poison from the parent cell. Finally, the overall productivity could be enhanced.

In contrast, virtually all biotechnologically used plasmids (such as pUC18, pBR322 and derived vectors) do not contain toxin-antitoxin addiction systems and thus need to be kept under antibiotic pressure to avoid plasmid loss.

Yeast Plasmids

Yeast are organisms that naturally harbour plasmids. Notable plasmids are 2 μm plasmids - small circular plasmids often used for genetic engineering of yeast, and linear pGKL plasmids from *Kluyveromyces lactis*, that are responsible for killer phenotypes.

Other types of plasmids are often related to yeast cloning vectors that include:

- *Yeast integrative plasmid (YIp)*, yeast vectors that rely on integration into the host chromosome for survival and replication, and are usually used when studying the functionality of a solo gene or when the gene is toxic. Also connected with the gene URA3, that codes an enzyme related to the biosynthesis of pyrimidine nucleotides (T, C);
- *Yeast Replicative Plasmid (YRp)*, which transport a sequence of chromosomal DNA that includes an origin of replication. These plasmids are less stable, as they can *get lost* during the budding.

Plasmid DNA Extraction

As alluded to above, plasmids are often used to purify a specific sequence, since they can easily be purified away from the rest of the genome. For their use as vectors, and for molecular cloning, plasmids often need to be isolated.

There are several methods to isolate plasmid DNA from bacteria, the archetypes of which are the miniprep and the maxiprep/bulkprep. The former can be used to quickly find out whether the plasmid is correct in any of several bacterial clones. The yield is a small amount of impure plasmid DNA, which is sufficient for analysis by restriction digest and for some cloning techniques.

In the latter, much larger volumes of bacterial suspension are grown from which a maxi-prep can be performed. In essence, this is a scaled-up miniprep followed by additional purification. This results in relatively large amounts (several hundreds micrograms) of very pure plasmid DNA.

In recent times, many commercial kits have been created to perform plasmid extraction at various scales, purity, and levels of automation. Commercial services can prepare plasmid DNA at quoted prices below $300/mg in milligram quantities and $15/mg in gram quantities (early 2007).

Conformations

Plasmid DNA may appear in one of five conformations, which (for a given size) run at different speeds in a gel during electrophoresis. The conformations are listed below in order of electrophoretic mobility (speed for a given applied voltage) from slowest to fastest:

- *Nicked open-circular* DNA has one strand cut.
- *Relaxed circular* DNA is fully intact with both strands uncut, but has been enzymatically *relaxed* (supercoils removed). This can be modeled by letting a twisted extension cord unwind and relax and then plugging it into itself.
- *Linear* DNA has free ends, either because both strands have been cut or because the DNA was linear *in vivo*. This can be modeled with an electrical extension cord that is not plugged into itself.
- *Supercoiled* (or *covalently closed-circular*) DNA is fully intact with both strands uncut, and with an integral twist, resulting in a compact form. This can be modeled by twisting an extension cord and then plugging it into itself.
- *Supercoiled denatured* DNA is like *supercoiled DNA*, but has unpaired regions that make it slightly less compact; this can result from excessive alkalinity during plasmid preparation.

The rate of migration for small linear fragments is directly proportional to the voltage applied at low voltages. At higher voltages, larger fragments migrate at continuously increasing yet different rates. Thus, the resolution of a gel decreases with increased voltage.

At a specified, low voltage, the migration rate of small linear DNA fragments is a function of their length. Large linear fragments (over 20 kb or so) migrate at a certain fixed rate regardless of length. This is because the molecules 'resperate', with the bulk of the molecule following the leading end through the gel matrix. Restriction digests are frequently used to analyse purified plasmids. These enzymes specifically break the DNA at certain short sequences. The resulting linear fragments form 'bands' after gel electrophoresis. It is possible to purify certain fragments by cutting the bands out of the gel and dissolving the gel to release the DNA fragments.

Because of its tight conformation, supercoiled DNA migrates faster through a gel than linear or open-circular DNA.

Software for Bioinformatics and Design

The use of plasmids as a technique in molecular biology is supported by bioinformatics software. These programs record the DNA sequence of plasmid vectors, help to predict cut sites of restriction enzymes, and to plan manipulations. Examples of software packages that handle plasmid maps are ApE, Clone Manager, GeneConstructionKit, Geneious, Genome Compiler, LabGenius, Lasergene, MacVector, pDraw32, Serial Cloner, VectorFriends, Vector NTI, and WebDSV. These software help conduct entire experiments in silico before doing wet experiments.

Expression Vector

An expression vector, otherwise known as an expression construct, is usually a plasmid or virus designed for gene expression in cells. The vector is used to introduce a specific gene into a target cell, and can commandeer the cell's mechanism for protein synthesis to produce the protein encoded by the gene. Expression vectors are the basic tools in biotechnology for the production of proteins.

The vector is engineered to contain regulatory sequences that act as enhancer and promoter regions and lead to efficient transcription of the gene carried on the expression vector. The goal of a well-designed expression vector is the efficient production of protein, and this may be achieved by the production of significant amount of stable messenger RNA, which can then be translated into protein. The expression of a protein may be tightly controlled, and the protein is only produced in significant quantity when necessary through the use of an inducer, in some systems however the protein may be expressed constitutively. *Escherichia coli* is commonly used as the host for protein production, but other cell types may also be used. An example of the use of expression vector is the production of insulin, which is used for medical treatments of diabetes.

Elements of Expression Vectors

An expression vector has features that any vector may have, such as an origin of replication, a selectable marker, and a suitable site for the insertion of a gene such as the multiple cloning site. The cloned gene may be transferred from a specialized cloning vector to an expression vector, although it is possible to clone directly into an expression vector. The cloning process is normally performed in *Escherichia coli*, and vectors used for protein production in organisms other than *E.coli* may have, in addition to a suitable origin of replication for its propagation in *E. coli*, elements that allow them to be maintained in another organism, and these vectors are called shuttle vectors.

Elements for Expression

An expression vector must have elements necessary for gene expression. These may include a promoter, the correct translation initiation sequence such as a ribosomal binding site and start codon, a termination codon, and a transcription termination sequence. There are differences in the machinery for protein synthesis between prokaryotes and eukaryotes, therefore the expression vectors must have the elements for expression that is appropriate for the chosen host. For example, prokaryotes expression vectors would have a Shine-Dalgarno sequence at its translation initiation site for the binding of ribosomes, while eukaryotes expression vectors would contain the Kozak consensus sequence.

The promoter initiates the transcription and is therefore the point of control for the expression of the cloned gene. The promoters used in expression vector are normally inducible, meaning that protein synthesis is only initiated when required by the introduction of an inducer such as IPTG. Gene expression however may also be constitutive (i.e. protein is constantly expressed) in some expression vectors. Low level of constitutive protein synthesis may occur even in expression vectors with tightly controlled promoters.

Protein Tags

After the expression of the gene product, it is usually necessary to purify the expressed protein; however, separating the protein of interest from the great majority of proteins of the host cell can be a protracted process. To make this purification process easier, a purification tag may be added to the cloned gene. This tag could be histidine (His) tag, other marker peptides, or a fusion partners such as glutathione S-transferase or maltose-binding protein. Some of these fusion partners may also help to increase the solubility of some expressed proteins. Other fusion proteins such as green fluorescent protein may act as a reporter gene for the identification of successful cloned genes.

Others

The expression vector is transformed or transfected into the host cell for protein synthesis. Some expression vectors may have elements for transformation or the insertion of DNA into the host chromosome, for example the *vir* genes for plant transformation, and integrase sites for chromosomal insertion.

Some vectors may include targeting sequence that may target the expressed protein to a specific location such as the periplasmic space of bacteria.

Expression/Production Systems

Different organisms may be used to express a gene's target protein, and the expression vector used will therefore have elements specific for use in the particular organism. The most commonly used organism for protein production is the bacterium *Escherichia coli*. However, not all proteins can be successfully expressed in *E. coli*, or be expressed with the correct form of post-translational modifications such as glycosylations, and other systems may therefore be used.

Bacterial

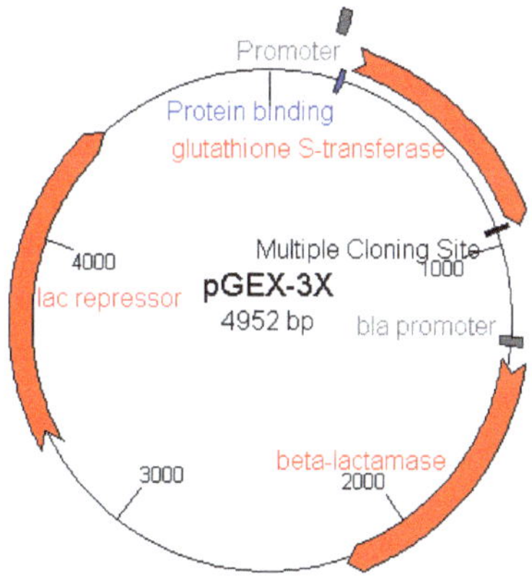

An example of a bacterial expression vector is the pGEX-3x plasmid

The expression host of choice for the expression of many proteins is *Escherichia coli* as the production of heterologous protein in *E. coli* is relatively simple and convenient, as well as being rapid and cheap. A large number of *E. coli* expression plasmids are also available suitable for a wide variety of needs. Other bacteria used for protein production include *Bacillus subtilis*.

Most heterologous proteins are expressed in the cytoplasm of *E. coli*. However, not all proteins formed may be soluble in the cytoplasm, and incorrectly folded proteins formed in cytoplasm can form insoluble aggregates called inclusion bodies. Such insoluble proteins will require refolding, which can be an involved process and may not necessarily produce high yield. Proteins which have disulphide bonds are often not able to fold correctly due to the oxidizing environment in the cytoplasm which prevents such bond formation, and a possible solution is to target the protein to the periplasmic space by the use of an N-terminal signal sequence. Other more sophisticated systems are also being developed; such systems may allow for the expression of proteins previously thought impossible in *E. coli*, such as glycosylated proteins.

The promoters used for these vector are usually based on the promoter of the *lac* operon or the T7 promoter, and they are normally regulated by the *lac* operator. These promoters may also be hybrids of different promoters, for example, the Tac-Promoter is a hybrid of *trp* and *lac* promoters. Note that most commonly used *lac* or *lac*-derived promoters are based on the *lac*UV5 mutant which is insensitive to catabolite repression. This mutant allows for expression of protein under the control of the *lac* promoter when the growth medium contains glucose since glucose would inhibit gene expression if wild-type *lac* promoter is used. Presence of glucose nevertheless may still be used to reduce background expression through residual inhibition in some systems.

Examples of *E. coli* expression vectors are the pGEX series of vectors where glutathione S-transferase is used as a fusion partner and gene expression is under the control of the tac promoter, and the pET series of vectors which uses a T7 promoter.

It is possible to simultaneously express two or more different proteins in *E. coli* using different plasmids. However, when 2 or more plasmids are used, each plasmid needs to use a different antibiotic selection as well as a different origin of replication, otherwise the plasmids may not be stably maintained. Many commonly used plasmids are based on the ColE1 replicon and are therefore incompatible with each other; in order for a ColE1-based plasmid to coexist with another in the same cell, the other would need to be of a different replicon, e.g. a p15A replicon-based plasmid such as the pACYC series of plasmids. Another approach would be to use a single two-cistron vector or design the coding sequences in tandem as a bi- or poly-cistronic construct.

Yeast

A yeast commonly used for protein production is *Pichia pastoris*. Examples of yeast expression vector in *Pichia* are the pPIC series of vectors, and these vectors use the AOX1 promoter which is inducible with methanol. The plasmids may contain elements for insertion of foreign DNA into the yeast genome and signal sequence for the secretion of expressed protein. Proteins with disulphide bonds and glycosylation can be efficiently produced in yeast. Another yeast used for protein production is *Kluyveromyces lactis* and the gene is expressed, driven by a variant of the strong lactase LAC4 promoter.

Saccharomyces cerevisiae is particularly widely used for gene expression studies in yeast, for example in yeast two-hybrid system for the study of protein-protein interaction. The vectors used in yeast two-hybrid system contain fusion partners for two cloned genes that allow the transcription of a reporter gene when there is interaction between the two proteins expressed from the cloned genes.

Baculovirus

Baculovirus, a rod-shaped virus which infect insect cells, is used as the expression vector in this system. Insect cell lines derived from Lepidopterans (moths and butterflies), such as *Spodoptera frugiperda*, are used as host. The shuttle vector is called bacmid, and gene expression is under the control of a strong promoter pPolh. Baculovirus has also been used with mammalian cell lines in the BacMam system.

Baculovirus is normally used for production of glycoproteins, although the glycosylations may be different from those found in vertebrates. In general, it is safer to use than mammalian virus as it has a limited host range and does not infect vertebrates without modifications.

Plant

Many plant expression vectors are based on the Ti plasmid of *Agrobacterium tumefaciens*. In these expression vectors, DNA to be inserted into plant is cloned into the T-DNA, a stretch of DNA flanked by a 25-bp direct repeat sequence at either end, and which can integrate into the plant genome. The T-DNA also contains the selectable marker. The *Agrobacterium* provides a mechanism for transformation, integration of into the plant genome, and the promoters for its *vir* genes may also be used for the cloned genes. Concerns over the transfer of bacterial or viral genetic material into the plant however have led to the development of vectors called intragenic vectors whereby functional equivalents of plant genome are used so that there is no transfer of genetic material from an alien species into the plant.

Plant viruses may be used as vectors since the *Agrobacterium* method does not work for all plants. Examples of plant virus used are the tobacco mosaic virus (TMV), potato virus X, and cowpea mosaic virus. The protein may be expressed as a fusion to the coat protein of the virus and is displayed on the surface of assembled viral particles, or as an unfused protein that accumulates within the plant. Expression in plant using plant vectors is often constitutive, and a commonly used constitutive promoter in plant expression vectors is the cauliflower mosaic virus (CaMV) 35S promoter.

Mammalian

Mammalian expression vectors offer considerable advantages for the expression of mammalian proteins over bacterial expression systems - proper folding, post-translational modifications, and relevant enzymatic activity. It may also be more desirable than other eukaryotic non-mammalian systems whereby the proteins expressed may not contain the correct glycosylations. It is of particular use in producing membrane-associating proteins that require chaperones for proper folding and stability as well as containing numerous post-translational modifications. The downside, however, is the low yield of product in comparison to prokaryotic vectors as well as the costly nature of the techniques involved. Its complicated technology, and potential contamination with animal viruses of mammalian cell expression have also placed a constraint on its use in large-scale industrial production.

Cultured mammalian cell lines such as the Chinese hamster ovary (CHO), COS, including human cell lines such as HEK and HeLa may be used to produce protein. Vectors are transfected into the cells and the DNA may be integrated into the genome by homologous recombination in the case of stable transfection, or the cells may be transiently transfected. Examples of mammalian expression vectors include the adenoviral vectors, the pSV and the pCMV series of plasmid vectors, vaccinia and retroviral vectors, as well as baculovirus. The promoters for cytomegalovirus (CMV) and SV40 are commonly used in mammalian expression vectors to drive gene expression. Non-viral promoter, such as the elongation factor (EF)-1 promoter, is also known.

Cell-free Systems

E. coli cell lysate containing the cellular components required for transcription and translation are used in this *in vitro* method of protein production. The advantage of such system is that protein may be produced much faster than those produced *in vivo* since it does not require time to culture the cells, but it is also more expensive. Vectors used for *E. coli* expression can be used in this system although specifically designed vectors for this system are also available. Eukaryotic cell extracts may also be used in other cell-free systems, for example, the wheat germ cell-free expression systems. Mammalian cell-free systems have also been produced.

Applications

Laboratory use

Expression vector in an expression host is now the usual method used in laboratories to produce proteins for research. Most proteins are produced in *E. coli*, but for glycosylated proteins and those with disulphide bonds, yeast, baculovirus and mammalian systems may be used.

Production of Peptide and Protein Pharmaceuticals

Most protein pharmaceuticals are now produced through recombinant DNA technology using expression vectors. These peptide and protein pharmaceuticals may be hormones, vaccines, antibiotics, antibodies, and enzymes. The first human recombinant protein used for disease management, insulin, was introduced in 1982. Biotechnology allows these peptide and protein pharmaceuticals, some of which were previously rare or difficult to obtain, to be produced in large quantity. It also reduces the risks of contaminants such as host viruses, toxins and prions. Examples from the past include prion contamination in growth hormone extracted from pituitary glands harvested from human cadavers, which caused Creutzfeldt–Jakob disease in patients receiving treatment for dwarfism, and viral contaminants in clotting factor VIII isolated from human blood that resulted in the transmission of viral diseases such as hepatitis and AIDS. Such risk is reduced or removed completely when the proteins are produced in non-human host cells.

Transgenic Plant and Animals

In recent years, expression vectors have been used to introduce specific genes into plants and animals to produce transgenic organisms, for example in agriculture it is used to produce transgenic plants. Expression vectors have been used to introduce a vitamin A precursor, beta-carotene, into rice plants. This product is called golden rice. This process has also been used to introduce a gene into plants that produces an insecticide, called Bacillus thuringiensis toxin or Bt toxin which reduces the need for farmers to apply insecticides since it is produced by the modified organism. In addition expression vectors are used to extend the ripeness of tomatoes by altering the plant so that it produces less of the chemical that causes the tomatoes to rot. There have been controversies over using expression vectors to modify crops due to the fact that there might be unknown health risks, possibilities of companies patenting certain genetically modified food crops, and ethical concerns. Nevertheless, this technique is still being used and heavily researched.

Genetically engineered fluorescent GloFish

Transgenic animals have also been produced to study animal biochemical processes and human diseases, or used to produce pharmaceuticals and other proteins. They may also be engineered to have advantageous or useful traits. Green fluorescent protein is sometimes used as tags which results in animal that can fluoresce, and this have been exploited commercially to produce the fluorescent GloFish.

Gene Therapy

Gene therapy is a promising treatment for a number of diseases where a "normal" gene carried by the vector is inserted into the genome, to replace an "abnormal" gene or supplement the expression of particular gene. Viral vectors are generally used but other nonviral methods of delivery are being developed. The treatment is still a risky option due to the viral vector used which can cause ill-effects, for example giving rise to insertional mutation that can result in cancer. However, there have been promising results.

Viral Vector

In recombinant molecular biology, virus particles has been modified to use as a carrier of nucleic acid into a cell, termed as viral vectors. Viral vectors are highly efficient in transferring target DNA/RNA segment to the host cells with high specificity. Wild type virus are modified by deleting the non-essential genes and incorporating exogenous nucleic acid segments to construct a viral vector. Viral vectors have wide application in gene therapy and targeted drug delivery systems. Main advantages of viral vectors are-high transfer efficiency and high cell specificity. Although, there are certain safety issues associated with viral vectors and careful handing is essential during the experimental procedure.

Commonly used viral vectors are- Adenovirus, retrovirus, lentivirus, adeno associated virus (AAV), herpes simplex virus (HSV) etc. The properties of different viral vectors are summarized in the table below.

Viral vector	Insert type	Insert size	Immunogenicity	Host genome integration
Adeno virus	DNA	2-8 kb	Very high	Non integrating
Retro virus	RNA	2-8 kb	Low	Integrating
Lentivirus	RNA	7-18 kb	Low	Integrating
Adeno associated virus (AAV)	DNA	4.5 kb	Low	Non integrating
Herpes simplex virus (HSV)	DNA	>30kb	Low	Non integrating

Table: Types of viral vectors

Phage Vectors

To insert DNA fragments of more than 10 kb, normally plasmids are not the suitable vehicles, as large inserts may trigger plasmid rearrangement or affect plasmid replication. This leads to development of a new class of vectors based on bacteriophages. Amongst various bacteriophages available such as λ, T4, T5, and T7 phages; the λ phage gained favourable attention due to its unique life cycle.

λ phage

Bacteriophage λ contains ~49kb of DNA and has a very efficient mechanism for delivering its genome into a bacterium. Two key features contribute to its utility as a vector to clone larger DNA fragments:

1. One-third λ genome is nonessential and could be replaced with foreign DNA. Approximately 24.6kb of λ genome can be deleted, hence maximum insert size could be upto 26 kb.
2. Packing of DNA in phage could only take place if the size is between 40 and 52 kb long, a constraint that can be used to ensure packaging.

Two problems had to be addressed before λ -based cloning vectors could be developed:

- The size limitation of the insert is determined by the genome size of phage λ(distance between the cos sites). The size range of the modified genome size should be within the range between 78-105% of the genome size for proper packaging. If >2.4kb is inserted to full length λ vector, the packaging efficiency is reduced. Hence λ vector should be smaller in size than wild type λ genome.

- The large λ genome has a few unique recognition sequences for bacterial restriction endonuclease. Bacterial restriction digestion system may target the modified vector and cleave the λ DNA molecule. This limitation can be overcome by replacing or mutating the restriction sites.

Two types of vector have been developed using λ genome:

Insertion Vectors

- Foreign DNA sequence is inserted into the λ genome without any significant change of the wild type genome.
- Smaller insert size (upto ~10kb).
- They may contain a multiple cloning site inserted in *lacZ* system for screening of recombinant bacterial colonies.
- Can be used to clone smaller DNA molecule.
- Eg: λ ZAP, λ gt etc.

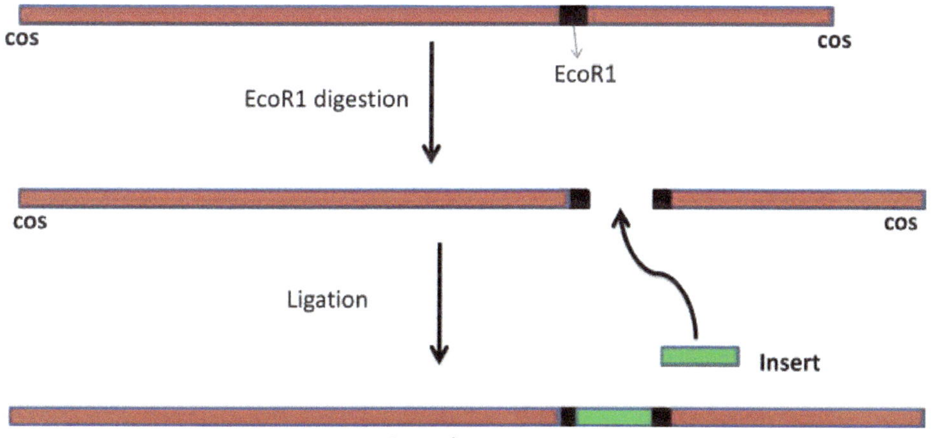

Insertion vector

Replacement Vectors

- Full length λ molecule having two identical restriction sites flanked by "stuffer fragment".
- Stuffer fragment is replaced by foreign DNA during restriction cloning.
- The vector without the foreign insert cannot be packaged due to the size limitation (smaller than the required).
- Insert size ranges between 10-23 kb.
- Eg. λ EMBL 3, λ EMBL 4, λ DASH etc.

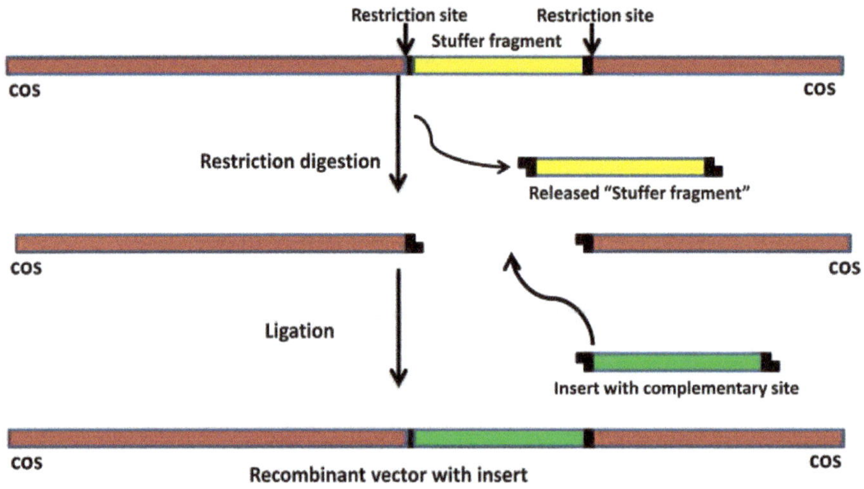

Replacement vector

Features of λ Phage Vectors

The λ genome is linear, but in the ends has 12-nucleotides overhangs, termed as *cos* sites, which are complementary to each other. A λ cloning vector can be circularized using *cos* site which can be manipulated and replicated inside *E. coli* via the process of transfection.

Alternatively, a more efficient uptake system called in vitro packaging can be utilized. Treatment with the appropriate restriction endonuclease followed by ligation in presence of insert DNA, produces ' left arm-new DNA-right arm' concatemers that are then added to an *in vitro* packaging mix to form λ phage particles. These phages are then co-incubated with *E. coli* cells, and the infection process naturally transports the vector plus new DNA into the bacteria. Bacteria that are infected with the packaged cloning vector die within about 20 minutes and several rounds of phage replication and bacterial lysis forms a zone of clearing, called a plaque.

M13 Phage Vectors

M13 phage is filamentous phage that infects *E. coli* via F-pilus . The genome is a single stranded circular DNA of size ~6.4kb surrounded by a proteinaceous coat. The DNA strand present in phage is called plus (+) strand. After entering to *E. coli* host, it converts into double stranded DNA molecule called replicative form (RF) by utilizing bacterial machinery. M13 phage as cloning vector can be obtained in both single stranded as well as double stranded form. Replicative form double stranded vector are modified and replicated inside *E. coli* host similar to a plasmid vector. Single stranded vectors can be isolated by collecting M13 phage. M13 vectors have useful application in following areas:

- DNA sequencing
- Mutagenesis study
- probe generation
- Phage display

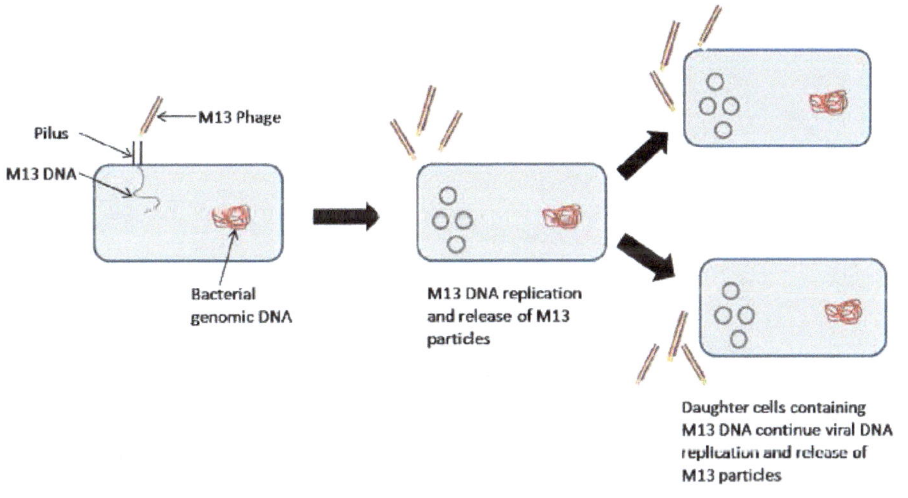

M13 Phage Vectors

Cosmid

- A cosmid, first described by Collins and Hohn in 1978, is a type of hybrid plasmid with a bacterial "ori" sequence and a "*cos*" sequences derived from the lambda phage.
- *Cos* site is the sequence required by a DNA molecule in order to be recognized as a 'λ genome› by the proteins that package DNA into λ phage particles.
- Cosmid DNA containing particles are as transmittable as real λ phages, but once inside the cell, the cosmid cannot control synthesis of new phage particles and instead replicates as a plasmid.
- Recombinant DNA is therefore obtained from colonies rather than plaques. They frequently also contain a gene for selection such as antibiotic resistance.
- They are able to load 37 to 52 kb of DNA, while normal plasmids are able to carry only 1–26 kb.
- Sometimes helper phage is used to assist in packaging of cosmid inside phage. Helper phage provides the essential proteins required for packaging which are lacked by cosmid vector.
- For packaging into a phage, concatemer formation is required (cosmid-insert-cosmid). This is generated by using two cos sites flanked by the insertion site for foreign DNA. Providing the inserted DNA in the right size, *in vitro* packaging cleaves the *cos* sites and replaces the recombinant cosmids in mature phage particles.
- Recombinant λ phages are used to infect an *E. coli* culture. Infected cells are plated on a selective medium and antibiotic-resistant colonies are grown. All colonies are recombinants, as non-recombinant linear cosmids are too small to be packaged into λ heads.
- Cosmids are widely exploited to build genomic libraries. The upper limit for the length of the cloned DNA is set by the space available within the λ phage particle. New DNA

insert of size upto 44 kb can be inserted before the packaging of the λ phage particle is reached.

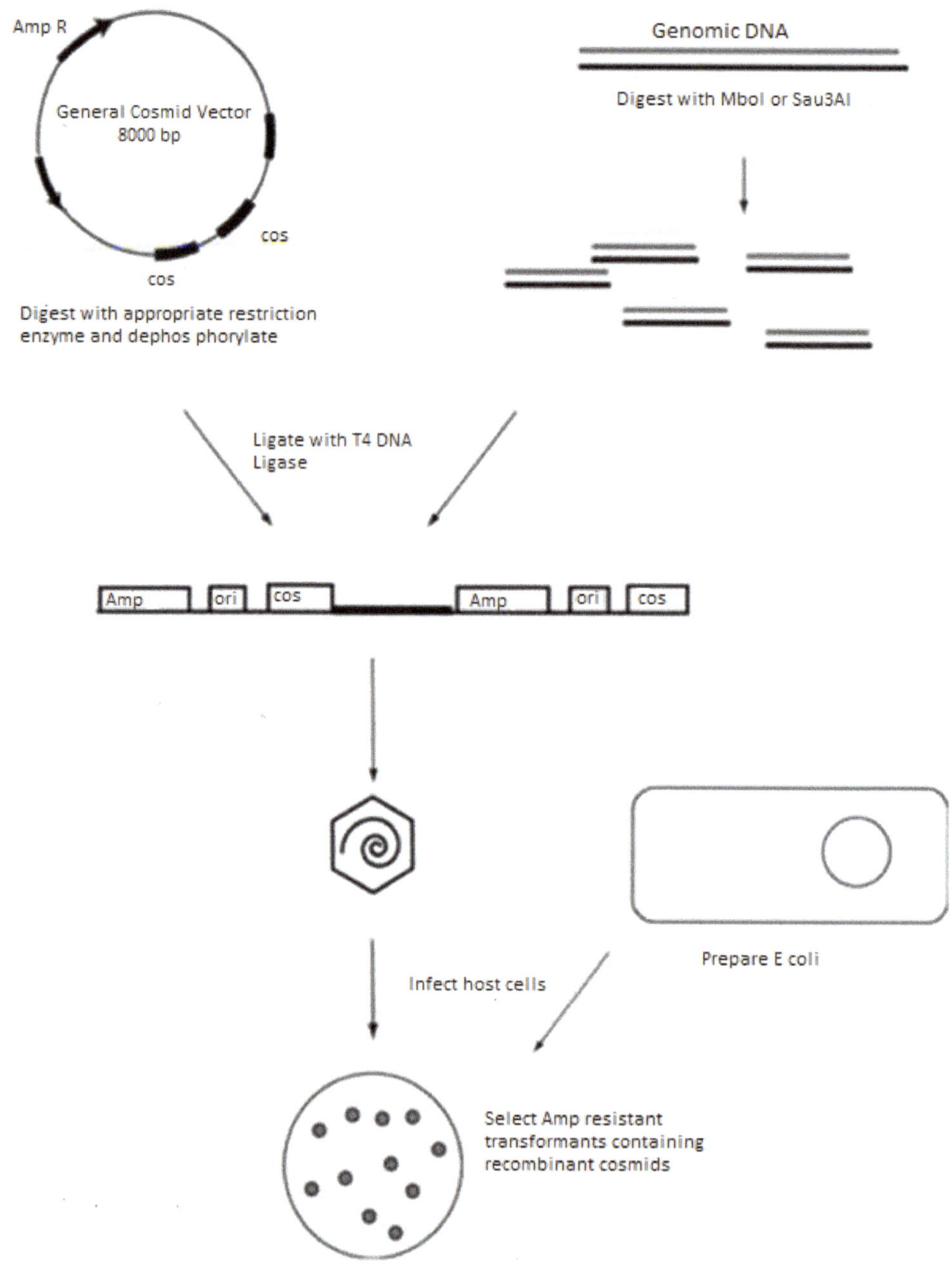

Schematics for Cosmid Library Construction

Limitation of Cosmid Vector

- Slower replication
- Higher frequency of recombination inside bacterial host.
- Unstable inside *E.coli* host and thus easy to lose vector.

Example

pJB8 is 5.4 kb in size and carries the ampicillin-resistance gene (amp^R), a segment of λ DNA containing the *cos* site, and an *Escherichia coli* origin of replication (ori).

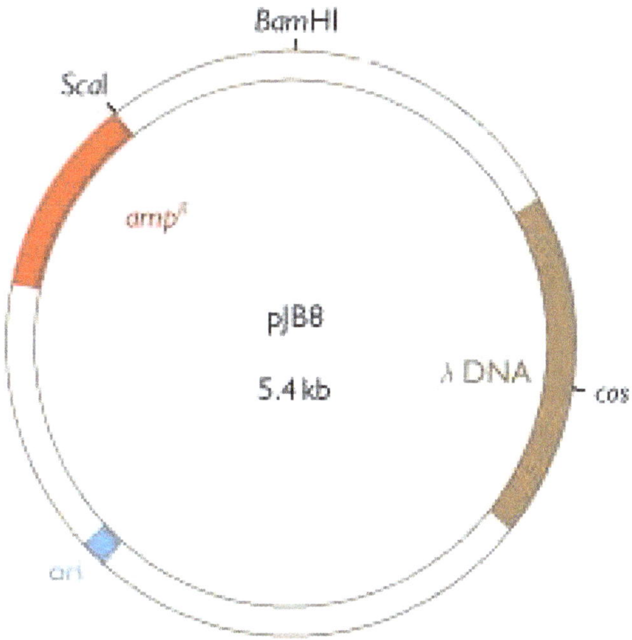

Schematic representation of pJB8

Fosmid

- Fosmids are similar to cosmids, however they are primarily based on bacterial F-plasmid.
- Simon and co-workers, in the year 1992, first developed F-factor based vector named as pFOS for stable propagation of cosmid sized human genomic DNA inserts.
- They carry the F plasmid origin of replication and a λ *cos* site.
- Fosmids can carry up to 40 kb of insert DNA.
- The cloning vector is limited, as a host (usually *E. coli*) can only contain one fosmid molecule. Low copy number offers higher stability as compared to high copy number cosmids.
- Fosmids have high structural stability and have been found to maintain human DNA effectively even after 100 generations of growth. It is ideal to use a fosmid vectors for constructing genomic and meta-genomic libraries.

Fosmids contain several functional elements as discussed below:

- OriT (Origin of Transfer): The sequence which marks the starting point of conjugative transfer.
- OriV (Origin of Replication): The sequence starting with which the plasmid-DNA will be replicated in the recipient cell.

- *tra* -region (transfer genes): Genes coding the F-Pilus and DNA transfer process.
- IS (Insertion Elements): so-called "selfish genes" (sequence fragments which can integrate copies of themselves at different locations).

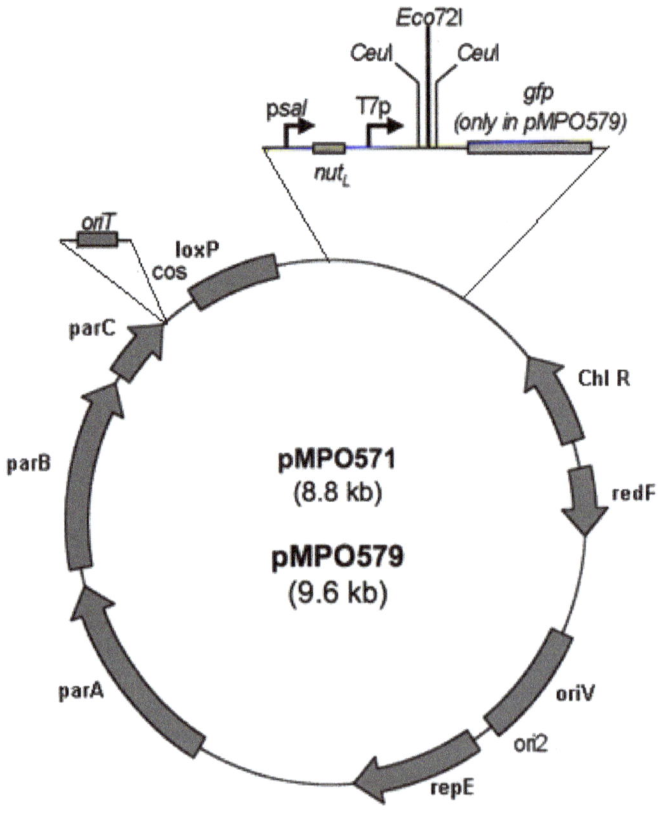

Schematic diagram of fosmids derived from pCC1FOS- *Ceu* I

Phagemid

Although M13 vectors are very useful for the production of single-stranded recombinant genes, they have certain disadvantages. There is a limit to the size of DNA fragment that can be cloned in an M13vector, with 1.5 kb being the ideal capacity, although fragments up to 3 kb have occasionally been cloned. To overcome this limitation, phagemid vectors were developed by combining a part of the M13 genome with plasmid DNA.

- Phagemids are cloning vectors developed as a hybrid of the filamentous phage M13 and plasmids to produce a vector that can get packed as a phage particle but also can propagate as a plasmid.
- They contain an origin of replication (ori) for double stranded replication inside *E. coli* host, as well as an " f1 ori" to enable single stranded replication and packaging into phage particles. Many commonly used plasmids contain an f1 ori and are thus phagemids.
- Phagemid generally encode no or only one of the capsid proteins of virus. Other structural and functional proteins necessary for phage lifecycle are provided by the helper phage.

- The components present in a phagemid vector are:
    - Origin of replication (ori) of a plasmid.
    - Intergenic region (IG region) which contains the packaging signal for the phage particle and also has replication origin inside phage.
    - A gene encoding phage coat protein.
    - A selection marker.
    - Restriction enzyme recognition sites.

Phagemid vectors are commonly used for "phage display technology" by which a broad range of proteins and peptides can be expressed as fusions to phage coat proteins and displayed on the viral surface. The advantage of phagemid vectors is that double stranded phagemid vectors can be converted into single stranded vectors and packaged into virion particles by infecting the cells with helper phage.

- Phagemid has certain advantages over phage vectors:
    - The carrying capacity of phagemid is higher than phage vectors.
    - Phagemid has higher efficiency in transformation than phage vectors.
    - Phagemids are genetically more stable than recombinant phage vectors.
    - Phagemids can be exploited to generate single stranded DNA template for sequencing purposes.
    - Single stranded phagemid vectors inside the phage can be targeted for site-directed mutagenesis.
    - Single stranded vectors can be used to generate hybridization probes for mRNA or cDNA.

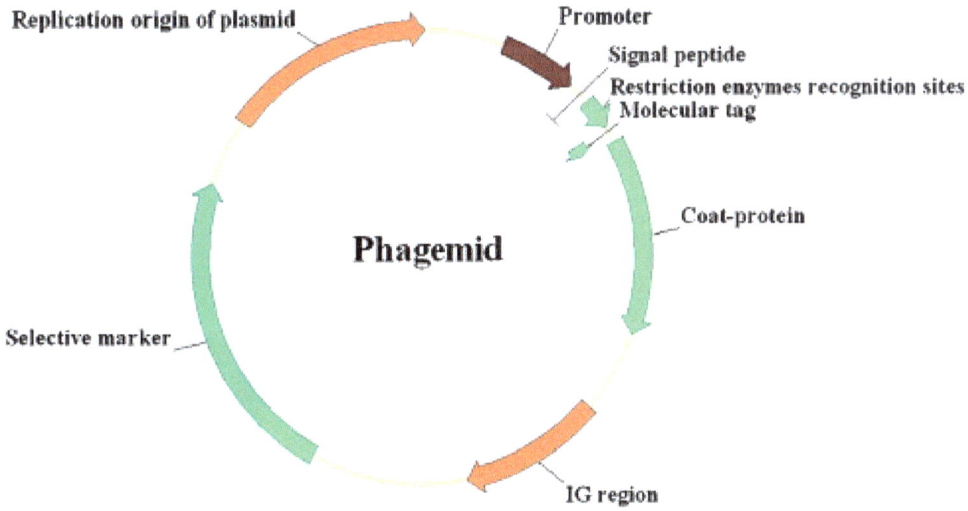

Phagemid vector for phage display

Examples of Phagemid

A pEMBL

- One of the first hybrid phagmid vectors was pEMBL constructed in 1983. They are characterized by the presence of –

    1. The *bla* gene as selectable marker for ampicillin resistance.
    2. A short segment coding for the alpha-peptide of beta-galactosidase (lacZ) and containing a MCS.
    3. The intragenic (IG) region of phage F1.

- These vectors have been used successfully for DNA sequencing with the dideoxy method, and can be used for other purposes for which M13 derivatives are used. However, the pEMBL plasmids have the advantage of being smaller than M13 vectors, and the purification of DNA is simpler. In addition, long inserts have a higher stability in pEMBL plasmids than M13 vectors.

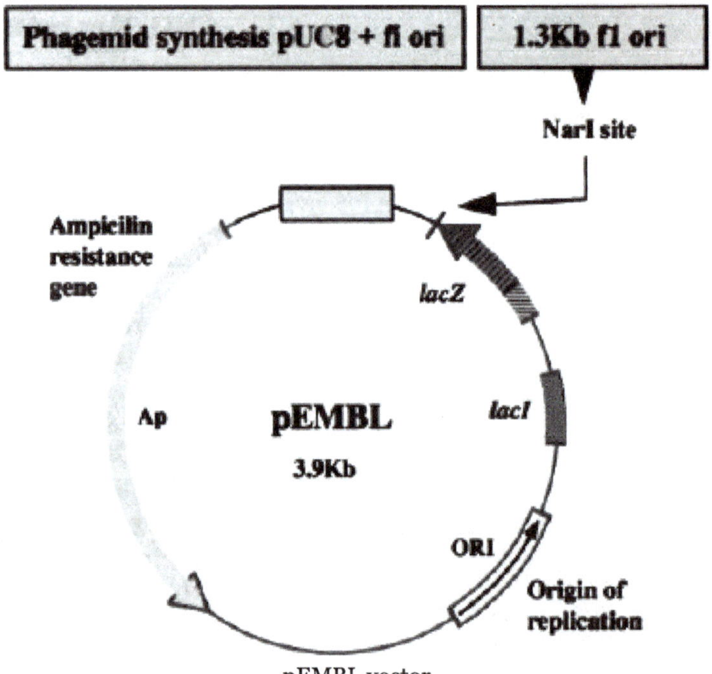

pEMBL vector

Artificial Chromosomes

Artificial chromosomes are DNA molecules assembled *in vitro* from defined constituents that can function like natural chromosomes.

Types of artificial chromosomes:

i. BACs: Bacterial artificial chromosomes

ii. YACs: Yeast artificial chromosomes

iii. MACs: Mammalian artificial chromosomes

iv. HACs: Human artificial chromosomes

v. PACs: P1-derived artificial chromosomes

Some of the artificial chromosomes are discussed below-

Bacterial Artificial Chromosomes: BAC

Bacterial artificial chromosomes (BACs) are designed for the cloning of large DNA insert (typically 100 to 300 kb) in *E. coli* host. BAC vectors contain a single copy F-plasmid origin of replication (ori). The F (fertility) plasmid is relatively large and vectors derived from it have a higher capacity than normal plasmid vectors. F-plasmid has F (fertility) factor which controls the replication and maintain low copy number. Also conjugation can take place between F+ bacteria (male) and F- bacteria (female) to transfer F-plasmid via pilus.

Common gene components of a bacterial artificial chromosome are:

1. oriS, repE – F for plasmid replication and regulation of copy number.
2. parA and parB for maintaining low copy number and avoiding two F plasmids in a single cell during cell division.
3. A selectable marker for antibiotic resistance; some BACs also have *lacZ* at the cloning site for blue/white selection.
4. T7 and Sp6 phage promoters for transcription of inserted genes.

The *par* genes, derived from F plasmid assist in the even distribution of plasmids to daughter cells during cell division and increase the likelihood of each daughter cell carrying one copy of the plasmid, even when few copies are present. The low number of copies is useful in cloning large fragments of DNA because it limits the opportunities for unwanted recombination reactions that can unpredictably alter large cloned DNA over time.

The first BAC vector, pBAC108L, did not contain a selectable marker for recombinants. Thus, positive recombinants had to be identified by colony hybridization. Two widely used BAC vectors, pBeloBAC11 and pECBAC1, are derivatives of pBAC108L in which the original cloning site is replaced with a *lac* Z gene carrying a multiple cloning site. PBeloBAC11 has two *Eco* RI sites, one in the *lac* Z gene and one in the CMR gene, whereas pECBAC1 has only one *Eco* RI site in the *lac* Z gene. Further improvements to BACs have been made by replacing the *lac* Z gene with the *sac* B gene which is a negative selection marker. The product of *sacB* gene is levansucrase which can convert sucrose present in the media into levan, a toxin for the bacteria. Hence the colonies without insert would have intact *sacB* gene and thus cells die before forming colonies.

The F plasmid is relatively large and vectors constructed on it have a higher capacity for accepting inserted DNA. A similar cloning vector called a P1-derived artificial chromosome or PAC has also been produced from the bacterial P1 bacteriophage DNA. Both BACs and PACs can be used to clone fragments of 300kb and longer. They are often used to sequence the genome of organisms in genome projects.

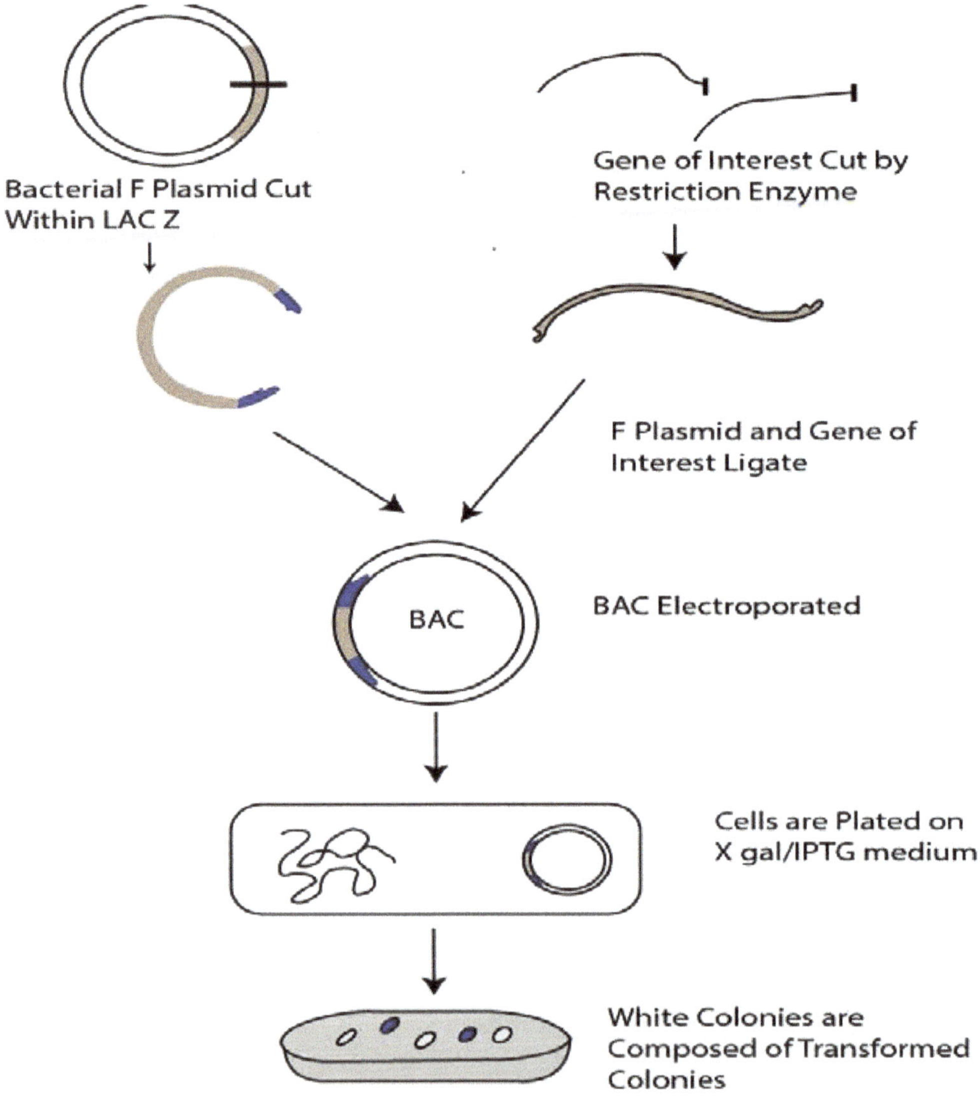

Transforming a Bacterium Using a BAC Vector

Yeast Artificial Chromosomes: YAC

- First described in 1983 by Murray and Szostak, a yeast artificial chromosome has sequences to exist inside E. coli as a circular plasmid and contains sequences to maintain as linear nuclear chromosome in yeast.
- As YAC vectors can accommodate 100-500 kb of insert DNA. The number of clones in a genomic library can be greatly reduced.
- YAC vectors have following elements:
    - E. coli origin of replication

- Yeast origin of replication
- Elements of eukaryotic yeast chromosome (centromere and telomere region)
- Selection markers for both the host.

- YAC vector is initially propagated as circular plasmid inside bacterial host utilizing bacterial ori sequence. Circular plasmid is cut at specific site using restriction enzymes to generate a linear chromosome with two telomere sites at terminals. The linear chromosome is again digested at specific site with two arms with different selection marker. Genomic insert is then ligated into YAC vector using DNA ligase enzyme. The recombinant vectors are transformed into yeast cells and screened for the selection markers to obtain recombinant colonies.
- Yeast expression vectors, such as YACs, YIPs (yeast integrating plasmids), and YEPs (yeast episomal plasmids), have advantageous over bacterial artificial chromosomes (BACs). They can be used to express eukaryotic proteins that require post-translational modification. However, YACs have been found to be less stable than BACs.

Some recombinant plasmids have the ability to incorporate multiple replication origins and other elements that allow them to be used in more than one species (for example, yeast or E. coli).

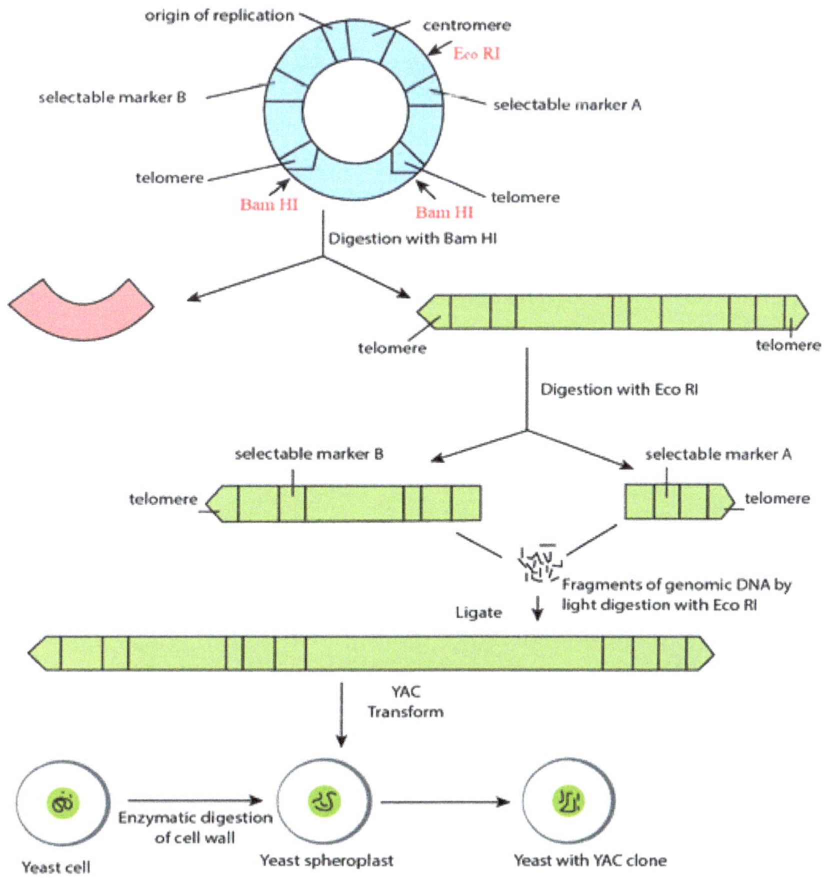

YAC vector

Mammalian Artificial Chromosomes: MAC

MACs or mammalian artificial chromosomes, like YACs, rely on the presence of centromeric and telomeric sequences and origin of DNA replication. They involve autonomous replication and segregation in mammalian cells, as opposed to random integration into chromosomes (as for other vectors). They can be modified for their use as expression systems of large genes, including not only the coding region but can contain control elements. Two principal procedures exist for the generation of MACs.

1) In one method, telomere-directed fragmentation of natural chromosomes is used. For example, a human artificial chromosome (HAC) has been derived from chromosome 21 using this method.

2) Another method involves de novo assembly of cloned centromeric, telomeric, and replication origins in vitro .

MAC vectors are difficult to assemble as compared to YAC vectors. Mammalian DNA has higher degree of repetition and larger centromere and telomere regions. Also the sequences necessary for chromosome replication in mammalian system are not well defined till now. MAC vectors have application in the field of gene therapy and eukaryotic protein expression and production.

Biosafety Guidelines

During early era of life sciences, biosafety principles and guidelines were mostly applied in the field of microbiology and medical practices. In recent time, biosafety guidelines are designed and applied to research involving recombinant DNA (rDNA) techniques. Handling, production, storing and transportation of genetically modified organism (GMOs) involve different biosafety issues under different category. Biosafety practices deal with the application of standard safety principles handling hazardous material/agents to minimize potential harmful effect on human health and environment.

The definition of biosafety corresponds to recombinant DNA technology can be described as: "Application of safety principles to laboratory practices in which potentially hazardous materials or organism are manipulated or handled."

National Institutes of Health (NIH) developed certain guidelines on rDNA research in May, 1976. Similarly, Department of Biotechnology (Govt. of India) has also come out with its set of guidelines. Scope of the guidelines suggested by DBT encompasses research, large scale operations and environmental risks. The guidelines prescribe specific actions that include establishing safety procedures for rDNA research, production and release to the environment and setting up containment conditions for certain experiments.

Biosafety regulatory principles and protocols regulates the potential risk and allow access to the benefits of rDNA technology. Risk assessment and risk management are two important component of biosafety.

Biosafety Levels (BSL):

All the facilities handling microorganisms and materials containing recombinant DNA molecules have risk assessment program. Depending on the risk possessed by the samples, four biosafety levels have been assigned to rDNA research facilities. Each BSL facility has requirement of unique design features and safety equipments.

Biosafety Level-I (BSL-I):

- Agents: Characterized strains of microorganisms known to cause no disease in healthy adults. eg. *E. coli, S. cerevesiae, B. subtilis* etc.
- Recombinant DNA based research activities involving non-pathogenic micro-organisms for expression of genes using plasmid vectors or low risk viral vectors.
- Work practice: Standard aseptic microbiological techniques.
- Safety equipment requirement: Lab coats and eye protection recommended.
- Facilities: Bench top, sink etc.

Biosafety Level-II (BSL-II):

- Agents: Handling of micro-organisms which possess moderate hazard to personal and environment.
- rDNA based research activities in micro-organisms using non-viral or viral vectors.
- Work practice: Standard BSL-I practices with addition of limited access, biohazard sign, defined procedure for disposal of "Regulated Medical Waste", proper training to lab personal and medical surveillance.
- Safety equipment: Class-II biological safety cabinet, lab coats, gloves, eye/face protection, physical containment equipment to reduce infectious aerosol exposure or splashes.
- Facility: BSL-I facility with addition of autoclave, decontamination facility and proper airflow.

Biosafety Level-III (BSL-III):

- Agents: Handling of micro-organisms which are designated as hazardous or potentially lethal agents to personal and environment.
- Laboratory personnel must have specific training in handling infectious micro-organisms and should be supervised by scientist competent in handling infectious agents.
- Work practices: BSL-2 practices, with the addition of: controlled access, on-site decontamination of all waste and lab clothing and medical surveillance.
- Safety equipment: Class-III biological safety cabinet, lab coats, gloves, eye/face protection, respiratory protection, physical containment equipment to reduce infectious aerosol exposure or splashes.

- Facility: BSL-III facility has specific criteria to meet. Lab should have double door entry with physical separation of working area from the access corridors, directional airflow in lab, and no recirculation of exhaust air in the lab, sufficient decontamination facility, in lab autoclave etc.

Biosafety Level-IV (BSL-IV):

- Agents: Hazardous and potentially lethal organisms that posses high individual risk of laboratory transmitted disease for which there is no vaccine or treatment, or a related agent with unknown risk of transmission.
- Laboratory personnel must have specialized training in handling BSL-IV agents and should be supervised by scientist competent in handling infectious agents.
- Safety equipment: Class-IV biological safety cabinet, lab coats, gloves, eye/face protection, respiratory protection, physical and containment equipment to reduce infectious aerosol exposure or splashes.
- Facility: BSL-IV facility requires specialized design to minimize the exposure to risk and only the authorized entry should be permitted in laboratory area in BSL-IV labs.

Risk Analysis

The foundation of any safety program is the use of control measures appropriate for the risk posed by the activities and the agents in use. The process of analyzing and determining the risk associated with recombinant DNA work is called as Risk analysis. The principle behind biosafety regulations is to minimize the risk to human health and safety, and the conservation of environment including safe handling of hazardous material. Risk analysis consists of three components: risk assessment, risk management and risk communication.

Risk Assessment: Estimation and determination of risk associated with the handling and production of a recombinant DNA molecule.

Risk Management: The process of analyzing possible prevention measures to minimize the risk and designing policies accordingly including implementation of them.

Risk Communication: The exchange of information and opinions on risk management between academic parties, industry, consumers and policy makers.

Risk Assessment

The biosafety level is determined based on the risk associated with the work. The principle investigator is responsible for implementing the necessary safety requirements in his/her laboratory. Risk assessment process accounts the following criteria to determine biosafety level:

i. Pathogenicity – The ability of an organism to cause disease in human system.

ii. Virulence – The severity of the disease (lethal/non lethal, availability of cure etc) in a healthy adult.

iii. Proliferation – the subsequent multiplication, genetic reconstruction, growth, transport, modification and die-off of these micro-organisms in the environment, including possible transfer of genetic material to other micro-organisms.

iv. Transmission route – The possible route of transmission (mucous membrane, inhalation etc) to establish the disease in human or other organism.

v. Infectious dose (ID) – The amount of infectious agent required to cause disease in healthy human.

vi. Antibiotic/disinfectant resistance – The resistance acquired by the infectious agent to available antibiotic/disinfectant.

The risk associated with recombinant DNA technology can be categorized under different headings based on their implication on different platforms.

General Scientific Considerations

A. Characteristics of Donor and Recipient Organisms

- Taxonomy, identification, source and culture
- Genetic characteristics of donor and recipient organisms
- Pathogenic and physiological traits of donor and recipient organisms

B. Properties of the modified/engineered organism

C. Description of (a) modification, (b) nature, function and source of the insert, (c) vector construction, (d) transfer into host, (e) stability of insert, (f) frequency of mobilization, (g) rate and level of expression, and (h) Influence of the recipient organism on the activity of the foreign protein.

Human Health Considerations

A. Characteristics of the modified/engineered organism.

- Comparison of the recombinant organism to the wildtype organism regarding pathogenicity.
- Transmission route to human.
- Pathogenicity to humans (or to animals if appropriate).

B. Health considerations generally associated with the presence of non-viable organisms or with the products of rDNA processes.

C. Management of personnel exposure, including biological measures and physical and organizational measures.

Environmental and Agricultural Considerations

A. Ecological traits relating to the donor and recipient environment.

B. Properties of environment where the engineered organism are being applied.

C. Survival, multiplication and dissemination of the engineered organism in the environment.

D. Interactions of engineered organism(s) with biological systems (target and non-target populations, stability, and routes of dissemination).

E. Potential environmental impacts (Effect on target and non-target organisms and ecosystems).

Risk Management

Risk management in biosafety issues is related to the target site where the practice is conducting (laboratory, industry, agriculture field etc.).

Recommendations: General

i. Harmonization of approaches to rDNA techniques can be facilitated by exchanging principles or guidelines for national regulations; developments in risk analysis; and practical experience in risk management. Therefore, information should be shared as freely as possible.

ii. There is no scientific basis for specific legislation for the implementation of rDNA techniques and applications. Member countries should examine their existing oversight and review mechanisms to ensure that adequate review and control may be applied while avoiding any undue burdens that may hamper technological developments in this field.

iii. Any approach to implement guidelines should not impede future developments in rDNA techniques. International harmonization should recognize this need.

iv. To facilitate data exchange and minimize trade barriers between countries, further developments such as testing methods, equipment design, and knowledge of microbial taxonomy should be considered at both national and international levels. Due account should be taken of ongoing work on standards within international organizations.

v. Special efforts should be made to improve public understanding of the various aspects of rDNA techniques.

vi. For rDNA applications in industry, agriculture and the environment, it will be important for member countries to watch the development of these techniques. For certain industrial applications and for environmental and agricultural applications of rDNA organisms, some countries may wish to have a notification scheme.

vii. Recognizing the need for innovation, it is important to consider appropriate means to protect intellectual property and confidentiality interests while assuring safety.

Recommendations: Specific for Industry

i. The large-scale industrial application of rDNA techniques wherever possible should utilize microorganisms that are intrinsically of low risk. Such microorganisms can be handled under conditions of Good Industrial Large-Scale Practice (GILSP).

ii. If a recombinant microorganism cannot be handled merely by GILSP, measures of containment corresponding to the risk assessment should be used in addition to GILSP.

iii. Further, research to improve techniques for monitoring and controlling non-intentional release of rDNA organisms should be encouraged in large-scale industrial applications requiring physical containment.

Containment Levels

Biosafety containment levels have to be designated for a facility depending on the level of risk associated with the biological and chemical agents used and released from it. Following NIH (National Institute of Health, USA) and DBT (Department of Biotechnology, India) guidelines, different facilities for biological research have been classified under three containment levels.

Containment Category 1:

- Viable organisms should be handled in a production system which physically separates the process from the environment;
- Exhaust gases should be treated to minimize (i.e. to reduce to the lowest practicable level consistent with safety) the release of viable organisms;
- Sample collection, addition of materials to the system and the transfer of viable organisms to another system should be done in a manner which minimizes release;
- Bulk quantities of culture fluids should not be removed from the system unless the viable organisms have been inactivated by validated means;
- Effluent from the production facility should be inactivated by validated means prior to discharge.

Containment Category 2:

- Viable organisms should be handled in a production system which physically separates the process from the environment;
- Exhaust gases should be treated to prevent the release of viable organisms;
- Sample collection, addition of materials to a closed system and the transfer of viable organisms to another closed system should be done in a manner which prevents release;
- Culture fluids should not be removed from the closed system unless the viable organisms have been inactivated by validated chemical or physical means;
- Seals should be designed to prevent leakage or should be fully enclosed in ventilated housings;
- Closed systems should be located in an area controlled according to the requirements;

- Effluent from the production facility should be inactivated by validated chemical or physical means prior to discharge.

Containment Category 3:

- Viable organisms should be handled in a production system which physically separates the process from the environment;
- Exhaust gases should be treated to prevent the release of viable organisms;
- Sample collection, addition of materials to a closed system and the transfer of viable organisms to another closed system should be done in a manner which prevents release;
- Culture fluids should not be removed from the closed system unless the viable organisms have been inactivated by validated chemical or physical means;
- Seals should be designed to prevent leakage or should be fully enclosed in ventilated housings;
- Production systems should be located within a purpose built controlled area according to the requirements;
- Entry should be restricted in the laboratory area and only persons with appropriate authority should be allowed access to the working area.

Effluent from the production facility should be inactivated by validated chemical or physical means prior to discharge.

Different containment levels have been assigned for rDNA GILSP (Good industrial large scale practice) micro-organisms. Examples of containment approaches for recombinant organisms are discussed below:

S. No	Specifications	Containment categories		
		1	2	3
1.	Viable organisms should be handled in a system which physically separates the process from the environment (closed system)	Yes	Yes	Yes
2.	Exhaust gases from the closed system should be treated so as to:	Minimum release	Prevent release	Prevent release
3.	Sample collection, addition of materials to closed system and transfer of viable organisms to another closed system, should be performed so as to:	Minimum release	Prevent release	Prevent release
4.	Bulk culture fluids should not be removed from the closed system unless the viable organisms have been:	Inactivated by validated means	Inactivated by validated chemical or physical means	Inactivated by validated chemical or physical means
5.	Seale should be designed so as to:	Minimum release	Prevent release	Prevent release
6.	Closed system should be located within a controlled area	Optional	Optional	Yes, and purpose-built

a) Biohazard sings should	Optional	Yes	Yes
b) Access should be restricted to nominated personnel only	Optional	Yes	Yes, via an air-lock
c) Personnel should wear protective clothing	Yes, work clothing	Yes	Yes, Specially designed for such facilities to providc protection
d) Decontamination and washing facility should be provided for personnel	Yes	Yes	Yes
e) personnel should shower before leaving the controlled area	No	Optional	Yes
f) Effluent from sinks and showers should be collected and inactivated before release	No	Optional	Yes
g) The controlled area should be adequately ventilated to minimize are contamination	Optional	Optional	Yes
h) The controlled area should be maintained at an air pressure negative to atmosphere	No	Optional	Yes
i) Input air and extract air to the controlled area should be HEPA filtered	No	Optional	Yes
j) The controlled area should be designed to contain spillage of the entire contents of the closed to system	No	Optional	Yes
k) The controlled area should be sealable to permit fumigation	No	Optional	Yes
7. Effluent treatment before final discharge	Inactivated by validated means	Inactivated by validated or physical means	Inactivated by validated or physical means

Table: Containment levels for different facilities.

National Institute of Health (NIH) has developed certain management procedures that can be adopted by research workers themselves to contain minor spills involving small quantities of biological materials/ experimental samples related to recombinant DNA. However, management of decontamination and cleanup of large spill or a spill involving a highly infectious agent are best left to the biosafety officer of the laboratory. It is the responsibility of the principal investigator to maintain an enough supply of a chemical disinfectant effective against the investigated microorganisms.

A spill may happen under various circumstances as below-

- in a biological safety cabinet,
- in the open laboratory,
- in a centrifuge,
- biological spill on a person and
- spill involving radioactive materials.

In each case, before starting the procedure and during the procedure, protective gloves, lab coat or gown, and eye protection should be worn. Researchers in the vicinity and the principal investigator should also be warned immediately. Decontamination with a recommended disinfectant procedure should be followed. Similarly, rDNA and transgenic organisms must be treated the same as medical or infectious waste before disposal. All wastes resulting from the procedure are to be disposed in the designated biowaste container.

Spill in a Biological Safety Cabinet

A spill that is confined within a biological safety cabinet generally presents little or no hazard to personnel in the area. Chemical disinfection procedures are to be initiated at once while the cabinet continues to operate. The disinfectant should be active against the organisms of potential hazard.

i. Spray or wipe the walls, work surfaces, and equipment with the chosen disinfectant. Disinfectants with detergent have the advantage of detergent activity that will help clean the surfaces by removing both dirt and microorganisms.

ii. Minimize the generation of aerosols and use sufficient disinfectant to ensure that drain pans and catch basins below the work surface contain disinfectant. The front exhaust shall also be wiped and the disinfectant drained into a container.

iii. Lift the front exhaust grill and tray and wipe all surfaces. Wipe the catch basin and drain the disinfectant into a container.

This procedure will not disinfect the filters, fans, air ducts, and other interior parts of the cabinet. Contact bio safety officer if the interior cabinet needs to be disinfected.

Spill in the Open Laboratory

For a spill in the open laboratory outside a biological safety cabinet, the spill response depends on the size of the spill and hazard of the material.

i. A minimally hazardous material spilled without generating appreciable aerosols can be cleaned with a paper towel soaked in a chemical disinfectant.

ii. A spill of a larger volume of hazardous material with aerosol generation requires evacuation of the room, waiting for aerosol reduction, donning personal protective gear (including appropriate respiratory protection), selecting a disinfectant effective against the organisms involved, and cleaning as described above. Following cleanup, responsible personnel shall wash or shower the infected area with a disinfectant soap.

iii. If clothing gets contaminated, it should be carefully removed and the contaminated area of the garment folded inward and kept for autoclaving.

iv. Wash arms, face, and hands with disinfectant.

Spill in a Centrifuge

A biological spill in a centrifuge has the potential for producing large volumes of aerosols.

i. On becoming aware that a spill may have occurred within a centrifuge or other piece of equipment, turn off the equipment, allow aerosols to settle, and then decontaminate it.

ii. Place contaminated equipment in a leak proof bag and move it to a biological safety cabinet, if possible for decontamination.

Biological Spill on a Person

If a biological material is spilled on a person, emergency response is based on the hazardous level of the biological agent spilled, the amount of material spilled, and whether significant aerosols were generated.

i. If aerosol formation is believed to have been associated with the spill, a contaminated person shall leave the contaminated area immediately. If possible, he/she should go to another laboratory area so that hallways and other public areas do not become contaminated.

ii. Contaminated clothes should be removed and placed in red or orange biohazard bags for disinfection.Contaminated skin shall be flushed with water followed by washing with a disinfectant soap. Showering can also be done, depending on the extent of the spill. A designated shower room should be earmarked near the laboratory.

Spill Involving Radioactive Materials

Spills involving radioactive materials should be strictly handled by safety officer in charge. Following safety precautions should be taken under such situation:

i. Other laboratory personals should be warned immediately and if possible the place should be vacated.

ii. The principal investigator should be informed about the spill.

iii. The institute biosafety officer should be contacted immediately.

iv. One should not try to handle anything by himself.

Mechanism of Implementation of Biosafety Guidelines

The guidelines suggest compliance of the safeguards through voluntary as well as regulatory approach. The implementation is enforced through the development of institutional frame work of advisory and regulatory bodies to deal with the specific and discretionary actions on the following:

i. Self regulation and control in the form of guidelines on recombinant research activities; and

ii. Regulation of large scale use of engineered organisms in production activity and release of organisms in environmental applications under statutory provisions.

The institutional mechanism as proposed for implementation of guidelines is shown in the Figure.

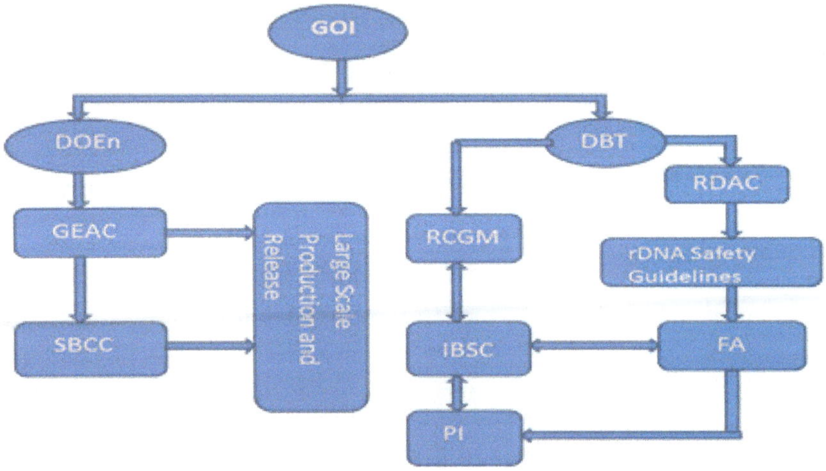

Implementation of Biosafety guidelines

(GOI: Govt. of India, DBT: Department of Biotechnology; RDAC: Recombinant DNA Advisory Committee; IBSC: Institutional Biosafety Committee; RCGM: Review Committee on Genetic Manipulation; DOEn: Department of Environment; GEAC: Genetic Engineering Approval Committee; SBCC: State Biotechnology Coordination Committee; PI: Principle Investigator (R&D/Industry/Others); FA: Funding Agency (Govt./Private or Public Institution))

Mainly it consists of the following:

i. Recombinant DNA Advisory Committee (RDAC)

ii. Institutional Biosafety Committee (IBSC)

iii. Review Committee on Genetic Manipulation (RCGM)

iv. Genetic Engineering Approval Committee (GEAC)

Recombinant DNA Advisory Committee (RDAC)

The Committee should take note of development and advances at national and international levels in Biotechnology towards the correctness of the safety regulation for India on recombinant research use and applications. The committee should be aware of recent developments and advances in safety regulation in recombinant research at both national and international levels. It should meet once in 6 months or sooner for this purpose.

The specific terms of reference for Recombinant Advisory Committee include the following:

i. To evolve long term policy for research and development in Recombinant DNA research.

ii. To formulate the safety guidelines for Recombinant DNA Research to be followed in India.

iii. To recommend type of training program for technicians and research fellows for making them adequately aware of hazards and risks involved in recombinant DNA research and methods for avoiding it.

Institutional Biosafety Committee (IBSC)

Institutional Biosafety Committee (IBSC) is to be constituted in all centres engaged in genetic engineering research and production activities. The Institutional Biosafety Committee shall be the nodal point for interaction within institution for implementation of the guidelines. Any research project which is likely to have biohazard potential (as envisaged by the guidelines) during the execution stage or which involve the production of either microorganisms or biologically active molecules that might cause bio-hazard should be notified to IBSC.

The biosafety functions and activity include the following:

i. Registration of Bio-safety Committee membership composition with RCGM and submission of reports: IBSC will provide half yearly report on the ongoing projects to RCGM regarding the observance of the safety guidelines on accidents, risks and on deviations if any. A computerized Central Registry for collation of periodic report on approved projects will be set up with RCGM to monitor compliance on safeguards as stipulated in the guidelines.

ii. Review and clearance of project proposals falling under restricted category that meets the requirements under the guidelines: IBSC should make efforts to issue clearance quickly on receiving the research proposals from investigators.

iii. Tailoring biosafety program to the level of risk assessment.

iv. Training of personnel on biosafety.

v. Instituting health monitoring program for laboratory personnel.

vi. Adopting emergency plans.

Review Committee on Genetic Manipulation (RCGM)

The RCGM will have the following functions:

i. To establish procedural guidance manual - for regulatory process involving genetically engineered organisms in research, production and applications related to environmental safety.

ii. To review the reports in all approved ongoing research projects involving high risk category and controlled field experiments and ensures that safeguards are maintained as per guidelines.

iii. To recommend the type of containment facility and the special containment conditions to be followed for experimental trials and for certain experiments.

iv. To advise customs authorities on import of biologically active material, genetically engineered substances or products and on excisable items to Central Revenue and Excise.

v. To assist Department of Industrial Development, Banks towards the clearance of applications in setting up industries based on genetically engineered organisms.

vi. To assist the Bureau of Indian Standards to evolve standards for biologics produced by recombinant DNA (rDNA) technology.

vii. To advise on intellectual property rights with respect to rDNA technology on patents.

Genetic Engineering Approval Committee (GEAC)

Genetic Engineering Approval Committee (GEAC) will function under the Department of Environment (DOEn) as statutory body for review and approval of activities involving large scale use of genetically engineered organisms and their products in research and development, industrial production, environmental release and field applications.

The functions include giving approval from environmental view on:

i. Import, export, transport, manufacture, process, selling of any microorganisms or genetically engineered substances or cells including food stuffs and additives that contains products derived by Gene Therapy.

ii. Discharge of Genetically engineered/classified organisms/cells from Laboratory, hospitals and related areas into environment.

iii. Large scale use of genetically engineered organisms/classified microorganisms in industrial production and applications. (Production shall not be commenced without approval).

iv. Deliberate release of genetically engineered organisms. The approval will be for a period of 4 years.

The funding agency will be responsible for approval and clearing of research proposals for grants in aid in respect of rDNA research activities. The funding agency at the centre and state level will be advised to ensure that the guidelines are taken into account for compliance while supporting grants on research projects. Investigators will be required to submit as part of the project application an evaluation of biohazards that may arise and also the requirement on the type of containment facility, certified by IBSC. The funding agency should state clearly that support on approved projects will be withdrawn in case of deliberate violation or avoidable negligence of the rDNA guidelines. The investigators will also be asked to make a declaration in their publications that the work was carried out following the national guidelines. The funding agency will annually submit to RCGM the list of approved projects that come under high risk categories.

References

- Morange M (2009). "What history tells us XIX. The notion of the episome" (PDF). Journal of Biosciences. 34 (6): 845–8. PMID 20093737. doi:10.1007/s12038-009-0098-z
- Thomas, Christopher M; Summers, David (2008). "Bacterial Plasmids". Encyclopedia of Life Sciences. ISBN 0470016175. doi:10.1002/9780470015902.a0000468.pub2
- Burgess RR (2009). "Refolding solubilized inclusion body proteins". Methods in Enzymology. 463: 259–82. PMID 19892177. doi:10.1016/S0076-6879(09)63017-2
- Gunge, N; Murata, K; Sakaguchi, K (July 1982). "Transformation of Saccharomyces cerevisiae with linear DNA killer plasmids from Kluyveromyces lactis". Journal of Bacteriology. 151 (1): 462–4. PMC 220260. PMID 7045080

- Wolfgang Schumann (2008). "Chapter 1 - Escherichia coli Cloning and Expression Vectors". In Georg Lipps. Plasmids: Current Research and Future Trends. Caister Academic Press. pp. 1–2. ISBN 978-1-904455-35-6
- Huang CJ, Lin H, Yang X (2012). "Industrial production of recombinant therapeutics in Escherichia coli and its recent advancements". J Ind Microbiol Biotechnol. 39 (3): 383–99. PMID 22252444. doi:10.1007/s10295-011-1082-9
- T. A. Brown (2010). "Chapter 2 - Vectors for Gene Cloning: Plasmids and Bacteriophages". Gene Cloning and DNA Analysis: An Introduction (6th ed.). Wiley-Blackwell. ISBN 978-1405181730
- Smith DB, Johnson KS (1988). "Single-step purification of polypeptides expressed in Escherichia coli as fusions with glutathione S-transferase". Gene. 67 (1): 31–40. PMID 3047011. doi:10.1016/0378-1119(88)90005-4
- Dubendorff JW, Studier FW (1991). "Controlling basal expression in an inducible T7 expression system by blocking the target T7 promoter with lac repressor". Journal of Molecular Biology. 219 (1): 45–59. PMID 1902522. doi:10.1016/0022-2836(91)90856-2
- David Summers (1996). "Chapter 1 - The Function and Organization of Plasmids". The Biology of Plasmids. Wiley-Blackwell; First Edition. pp. 21–22. ISBN 978-0632034369
- Kishwar Hayat Khan (2013). "Gene Expression in Mammalian Cells and its Applications". Adv Pharm Bull. 3 (2): 257–263. PMC 3848218. PMID 24312845. doi:10.5681/apb.2013.042
- Walden R, Schell J (1990). "Techniques in plant molecular biology--progress and problems". European Journal of Biochemistry. 192 (3): 563–76. PMID 2209611. doi:10.1111/j.1432-1033.1990.tb19262.x
- George Acquaah (16 August 2012). Principles of Plant Genetics and Breeding. John Wiley & Sons Inc. ISBN 9781118313695
- Fischer, A.; Hacein-Bey-Abina, S.; Cavazzana-Calvo, M. (2010). "20 years of gene therapy for SCID". Nature Immunology. 11 (6): 457–460. PMID 20485269. doi:10.1038/ni0610-457
- Alexander Dorozynski (2002). "Parents sue over contaminated human growth hormone". British Medical Journal. 324 (7349): 1294. PMC 1123268. PMID 12039815. doi:10.1136/bmj.324.7349.1294/b

2

Role of Enzymes in Genetic Engineering

A restriction enzyme is a kind of enzyme that cuts a DNA at a specific or random location known as restriction sites. There are two sorts of restriction enzymes, exonucleases and endonucleases. Types of enzymes discussed in this chapter are polynucleotide phosphorylase, phosphatases, methylases, etc. The chapter strategically encompasses and incorporates genetic engineering, providing a complete understanding.

Restriction Enzyme

A restriction enzyme or restriction endonuclease is an enzyme that cuts DNA at or near specific recognition nucleotide sequences known as restriction sites. Restriction enzymes are commonly classified into four types, which differ in their structure and whether they cut their DNA substrate at their recognition site, or if the recognition and cleavage sites are separate from one another. To cut DNA, all restriction enzymes make two incisions, once through each sugar-phosphate backbone (i.e. each strand) of the DNA double helix.

These enzymes are found in bacteria and archaea and provide a defense mechanism against invading viruses. Inside a prokaryote, the restriction enzymes selectively cut up *foreign* DNA in a process called *restriction*; meanwhile, host DNA is protected by a modification enzyme (a methyltransferase) that modifies the prokaryotic DNA and blocks cleavage. Together, these two processes form the restriction modification system.

Over 3000 restriction enzymes have been studied in detail, and more than 600 of these are available commercially. These enzymes are routinely used for DNA modification in laboratories, and are a vital tool in molecular cloning.

History

The term restriction enzyme originated from the studies of phage λ and the phenomenon of host-controlled restriction and modification of a bacterial virus. The phenomenon was first identified in work done in the laboratories of Salvador Luria and Giuseppe Bertani in early 1950s. It was found that, for a bacteriophage λ that can grow well in one strain of *Escherichia coli*, for example *E. coli* C, when grown in another strain, for example *E. coli* K, its yields can drop significantly, by as much as 3-5 orders of magnitude. The host cell, in this example *E. coli* K, is known as the restricting host and appears to have the ability to reduce the biological activity of the phage λ. If a phage becomes established in one strain, the ability of that phage to grow also becomes restricted in other strains. In the 1960s, it was shown in work done in the laboratories of Werner Arber and Matthew Meselson that the restriction is caused by an enzymatic cleavage of the phage DNA, and the enzyme involved was therefore termed a restriction enzyme.

The restriction enzymes studied by Arber and Meselson were type I restriction enzymes, which cleave DNA randomly away from the recognition site. In 1970, Hamilton O. Smith, Thomas Kelly and Kent Wilcox isolated and characterized the first type II restriction enzyme, *Hind*II, from the bacterium *Haemophilus influenzae*. Restriction enzymes of this type are more useful for laboratory work as they cleave DNA at the site of their recognition sequence. Later, Daniel Nathans and Kathleen Danna showed that cleavage of simian virus 40 (SV40) DNA by restriction enzymes yields specific fragments that can be separated using polyacrylamide gel electrophoresis, thus showing that restriction enzymes can also be used for mapping DNA. For their work in the discovery and characterization of restriction enzymes, the 1978 Nobel Prize for Physiology or Medicine was awarded to Werner Arber, Daniel Nathans, and Hamilton O. Smith. The discovery of restriction enzymes allows DNA to be manipulated, leading to the development of recombinant DNA technology that has many applications, for example, allowing the large scale production of proteins such as human insulin used by diabetics.

Restriction enzymes likely evolved from a common ancestor and became widespread via horizontal gene transfer. In addition, there is mounting evidence that restriction endonucleases evolved as a selfish genetic element.

Recognition Site

```
5'...GAT|ATC...3'
3'...CTA|TAG...5'
```

A palindromic recognition site reads the same on the reverse strand as it does on the forward strand when both are read in the same orientation

Restriction enzymes recognize a specific sequence of nucleotides and produce a double-stranded cut in the DNA. The recognition sequences can also be classified by the number of bases in its recognition site, usually between 4 and 8 bases, and the amount of bases in the sequence will determine how often the site will appear by chance in any given genome, e.g., a 4-base pair sequence would theoretically occur once every 4^4 or 256bp, 6 bases, 4^6 or 4,096bp, and 8 bases would be 4^8 or 65,536bp. Many of them are palindromic, meaning the base sequence reads the same backwards and forwards. In theory, there are two types of palindromic sequences that can be possible in DNA. The *mirror-like* palindrome is similar to those found in ordinary text, in which a sequence reads the same forward and backward on a single strand of DNA, as in GTAATG. The *inverted repeat* palindrome is also a sequence that reads the same forward and backward, but the forward and backward sequences are found in complementary DNA strands (i.e., of double-stranded DNA), as in GTATAC (GTATAC being complementary to CATATG). Inverted repeat palindromes are more common and have greater biological importance than mirror-like palindromes.

*Eco*RI digestion produces "sticky" ends,

```
G AATTC
CTTAA G
```

whereas SmaI restriction enzyme cleavage produces "blunt" ends:

CCC|GGG
GGG|CCC

Recognition sequences in DNA differ for each restriction enzyme, producing differences in the length, sequence and strand orientation (5' end or 3' end) of a sticky-end "overhang" of an enzyme restriction.

Different restriction enzymes that recognize the same sequence are known as neoschizomers. These often cleave in different locales of the sequence. Different enzymes that recognize and cleave in the same location are known as isoschizomers.

Types

Naturally occurring restriction endonucleases are categorized into four groups (Types I, II III, and IV) based on their composition and enzyme cofactor requirements, the nature of their target sequence, and the position of their DNA cleavage site relative to the target sequence. DNA sequence analyses of restriction enzymes however show great variations, indicating that there are more than four types. All types of enzymes recognize specific short DNA sequences and carry out the endonucleolytic cleavage of DNA to give specific fragments with terminal 5'-phosphates. They differ in their recognition sequence, subunit composition, cleavage position, and cofactor requirements, as summarised below:

- Type I enzymes (EC 3.1.21.3) cleave at sites remote from a recognition site; require both ATP and S-adenosyl-L-methionine to function; multifunctional protein with both restriction and methylase (EC 2.1.1.72) activities.
- Type II enzymes (EC 3.1.21.4) cleave within or at short specific distances from a recognition site; most require magnesium; single function (restriction) enzymes independent of methylase.
- Type III enzymes (EC 3.1.21.5) cleave at sites a short distance from a recognition site; require ATP (but do not hydrolyse it); S-adenosyl-L-methionine stimulates the reaction but is not required; exist as part of a complex with a modification methylase (EC 2.1.1.72).
- Type IV enzymes target modified DNA, e.g. methylated, hydroxymethylated and glucosyl-hydroxymethylated DNA.

Type l

Type I restriction enzymes were the first to be identified and were first identified in two different strains (K-12 and B) of *E. coli*. These enzymes cut at a site that differs, and is a random distance (at least 1000 bp) away, from their recognition site. Cleavage at these random sites follows a process of DNA translocation, which shows that these enzymes are also molecular motors. The recognition site is asymmetrical and is composed of two specific portions—one containing 3–4 nucleotides, and another containing 4–5 nucleotides—separated by a non-specific spacer of about 6–8 nucleotides. These enzymes are multifunctional and are capable of both restriction and modification activities, depending upon the methylation status of the target DNA. The cofactors S-Adenosyl methionine (AdoMet), hydrolyzed adenosine triphosphate (ATP), and magnesium (Mg^{2+}) ions, are required for their full activity. Type I restriction enzymes possess three subunits called HsdR, HsdM, and HsdS; HsdR is required for restriction; HsdM is necessary for adding methyl groups to host DNA (methyltransferase activity), and HsdS is important for specificity of the recognition (DNA-binding) site in addition to both restriction (DNA cleavage) and modification (DNA methyltransferase) activity.

Type II

Typical type II restriction enzymes differ from type I restriction enzymes in several ways. They form homodimers, with recognition sites that are usually undivided and palindromic and 4–8 nucleotides in length. They recognize and cleave DNA at the same site, and they do not use ATP or AdoMet for their activity—they usually require only Mg^{2+} as a cofactor. These are the most commonly available and used restriction enzymes. In the 1990s and early 2000s, new enzymes from this family were discovered that did not follow all the classical criteria of this enzyme class, and new subfamily nomenclature was developed to divide this large family into subcategories based on deviations from typical characteristics of type II enzymes. These subgroups are defined using a letter suffix.

Type IIB restriction enzymes (e.g., BcgI and BplI) are multimers, containing more than one subunit. They cleave DNA on both sides of their recognition to cut out the recognition site. They require both AdoMet and Mg^{2+} cofactors. Type IIE restriction endonucleases (e.g., NaeI) cleave DNA following interaction with two copies of their recognition sequence. One recognition site acts as the target for cleavage, while the other acts as an allosteric effector that speeds up or improves the efficiency of enzyme cleavage. Similar to type IIE enzymes, type IIF restriction endonucleases (e.g. NgoMIV) interact with two copies of their recognition sequence but cleave both sequences at the same time. Type IIG restriction endonucleases (e.g., Eco57I) do have a single subunit, like classical Type II restriction enzymes, but require the cofactor AdoMet to be active. Type IIM restriction endonucleases, such as DpnI, are able to recognize and cut methylated DNA. Type IIS restriction endonucleases (e.g., *Fok*I) cleave DNA at a defined distance from their non-palindromic asymmetric recognition sites; this characteristic is widely used to perform in-vitro cloning techniques such as Golden Gate cloning. These enzymes may function as dimers. Similarly, Type IIT restriction enzymes (e.g., Bpu10I and BslI) are composed of two different subunits. Some recognize palindromic sequences while others have asymmetric recognition sites.

Type III

Type III restriction enzymes (e.g., EcoP15) recognize two separate non-palindromic sequences that are inversely oriented. They cut DNA about 20–30 base pairs after the recognition site. These enzymes contain more than one subunit and require AdoMet and ATP cofactors for their roles in DNA methylation and restriction, respectively. They are components of prokaryotic DNA restriction-modification mechanisms that protect the organism against invading foreign DNA. Type III enzymes are hetero-oligomeric, multifunctional proteins composed of two subunits, Res and Mod. The Mod subunit recognises the DNA sequence specific for the system and is a modification methyltransferase; as such, it is functionally equivalent to the M and S subunits of type I restriction endonuclease. Res is required for restriction, although it has no enzymatic activity on its own. Type III enzymes recognise short 5–6 bp-long asymmetric DNA sequences and cleave 25–27 bp downstream to leave short, single-stranded 5' protrusions. They require the presence of two inversely oriented unmethylated recognition sites for restriction to occur. These enzymes methylate only one strand of the DNA, at the N-6 position of adenosyl residues, so newly replicated DNA will have only one strand methylated, which is sufficient to protect against restriction. Type III enzymes belong to the beta-subfamily of N6 adenine methyltransferases, containing the nine motifs that characterise this family, including motif I, the AdoMet binding pocket (FXGXG), and motif IV, the catalytic region (S/D/N (PP) Y/F).

Type IV

Type IV enzymes recognize modified, typically methylated DNA and are exemplified by the McrBC and Mrr systems of *E. coli.*

Type V

Type V restriction enzymes (e.g., the cas9-gRNA complex from CRISPRs) utilize guide RNAs to target specific non-palindromic sequences found on invading organisms. They can cut DNA of variable length, provided that a suitable guide RNA is provided. The flexibility and ease of use of these enzymes make them promising for future genetic engineering applications.

Artificial Restriction Enzymes

Artificial restriction enzymes can be generated by fusing a natural or engineered DNA binding domain to a nuclease domain (often the cleavage domain of the type IIS restriction enzyme *Fok*I). Such artificial restriction enzymes can target large DNA sites (up to 36 bp) and can be engineered to bind to desired DNA sequences. Zinc finger nucleases are the most commonly used artificial restriction enzymes and are generally used in genetic engineering applications, but can also be used for more standard gene cloning applications. Other artificial restriction enzymes are based on the DNA binding domain of TAL effectors.

In 2013, a new technology CRISPR-Cas9, based on a prokaryotic viral defense system, was engineered for editing the genome, and it was quickly adopted in laboratories. For more detail, read CRISPR (Clustered regularly interspaced short palindromic repeats).

In 2017 a group in Illinois announced using an Argonaute protein taken from Pyrococcus furiosus (PfAgo) along with guide DNA to edit DNA as artificial restriction enzymes.

Nomenclature

Derivation of the *Eco*RI name		
Abbreviation	Meaning	Description
E	*Escherichia*	genus
co	*coli*	specific epithet
R	RY13	strain
I	First identified	order of identification in the bacterium

Since their discovery in the 1970s, many restriction enzymes have been identified; for example, more than 3500 different Type II restriction enzymes have been characterized. Each enzyme is named after the bacterium from which it was isolated, using a naming system based on bacterial genus, species and strain. For example, the name of the *Eco*RI restriction enzyme was derived as shown in the box.

Applications

Isolated restriction enzymes are used to manipulate DNA for different scientific applications.

They are used to assist insertion of genes into plasmid vectors during gene cloning and protein production experiments. For optimal use, plasmids that are commonly used for gene cloning are modified to include a short *polylinker* sequence (called the multiple cloning site, or MCS) rich in restriction enzyme recognition sequences. This allows flexibility when inserting gene fragments into the plasmid vector; restriction sites contained naturally within genes influence the choice of endonuclease for digesting the DNA, since it is necessary to avoid restriction of wanted DNA while intentionally cutting the ends of the DNA. To clone a gene fragment into a vector, both plasmid DNA and gene insert are typically cut with the same restriction enzymes, and then glued together with the assistance of an enzyme known as a DNA ligase.

Restriction enzymes can also be used to distinguish gene alleles by specifically recognizing single base changes in DNA known as single nucleotide polymorphisms (SNPs). This is however only possible if a SNP alters the restriction site present in the allele. In this method, the restriction enzyme can be used to genotype a DNA sample without the need for expensive gene sequencing. The sample is first digested with the restriction enzyme to generate DNA fragments, and then the different sized fragments separated by gel electrophoresis. In general, alleles with correct restriction sites will generate two visible bands of DNA on the gel, and those with altered restriction sites will not be cut and will generate only a single band. A DNA map by restriction digest can also be generated that can give the relative positions of the genes. The different lengths of DNA generated by restriction digest also produce a specific pattern of bands after gel electrophoresis, and can be used for DNA fingerprinting.

In a similar manner, restriction enzymes are used to digest genomic DNA for gene analysis by Southern blot. This technique allows researchers to identify how many copies (or paralogues) of a gene are present in the genome of one individual, or how many gene mutations (polymorphisms) have occurred within a population. The latter example is called restriction fragment length polymorphism (RFLP).

Artificial restriction enzymes created by linking the *Fok*I DNA cleavage domain with an array of DNA binding proteins or zinc finger arrays, denoted zinc finger nucleases (ZFN), are a powerful tool for host genome editing due to their enhanced sequence specificity. ZFN work in pairs, their dimerization being mediated in-situ through the *Fok*I domain. Each zinc finger array (ZFA) is capable of recognizing 9–12 base pairs, making for 18–24 for the pair. A 5–7 bp spacer between the cleavage sites further enhances the specificity of ZFN, making them a safe and more precise tool that can be applied in humans. A recent Phase I clinical trial of ZFN for the targeted abolition of the CCR5 co-receptor for HIV-1 has been undertaken.

Others have proposed using the bacteria R-M system as a model for devising human anti-viral gene or genomic vaccines and therapies since the RM system serves an innate defense-role in bacteria by restricting tropism by bacteriophages. There is research on REases and ZFN that can cleave the DNA of various human viruses, including HSV-2, high-risk HPVs and HIV-1, with the ultimate goal of inducing target mutagenesis and aberrations of human-infecting viruses. Interestingly, the human genome already contains remnants of retroviral genomes that have been inactivated and harnessed for self-gain. Indeed, the mechanisms for silencing active L1 genomic retroelements by the three prime repair exonuclease 1 (TREX1) and excision repair cross complementing 1(ERCC) appear to mimic the action of RM-systems in bacteria, and the non-homologous end-joining (NHEJ) that follows the use of ZFN without a repair template.

Examples

Examples of restriction enzymes include:

Enzyme	Source	Recognition Sequence	Cut
*Eco*RI	*Escherichia coli*	5′GAATTC 3′CTTAAG	5′---G AATTC---3′ 3′---CTTAA G---5′
*Eco*RII	*Escherichia coli*	5′CCWGG 3′GGWCC	5′--- CCWGG---3′ 3′---GGWCC ---5′
*Bam*HI	*Bacillus amyloliquefaciens*	5′GGATCC 3′CCTAGG	5′---G GATCC---3′ 3′---CCTAG G---5′
*Hin*dIII	*Haemophilus influenzae*	5′AAGCTT 3′TTCGAA	5′---A AGCTT---3′ 3′---TTCGA A---5′
*Taq*I	*Thermus aquaticus*	5′TCGA 3′AGCT	5′---T CGA---3′ 3′---AGC T---5′
*Not*I	*Nocardia otitidis*	5′GCGGCCGC 3′CGCCGGCG	5′---GC GGCCGC---3′ 3′---CGCCGG CG---5′
*Hin*FI	*Haemophilus influenzae*	5′GANTC 3′CTNAG	5′---G ANTC---3′ 3′---CTNA G---5′
*Sau*3AI	*Staphylococcus aureus*	5′GATC 3′CTAG	5′--- GATC---3′ 3′---CTAG ---5′
*Pvu*II*	*Proteus vulgaris*	5′CAGCTG 3′GTCGAC	5′---CAG CTG---3′ 3′---GTC GAC---5′
*Sma*I*	*Serratia marcescens*	5′CCCGGG 3′GGGCCC	5′---CCC GGG---3′ 3′---GGG CCC---5′
*Hae*III*	*Haemophilus aegyptius*	5′GGCC 3′CCGG	5′---GG CC---3′ 3′---CC GG---5′
*Hga*I	*Haemophilus gallinarum*	5′GACGC 3′CTGCG	5′---NN NN---3′ 3′---NN NN---5′
*Alu*I*	*Arthrobacter luteus*	5′AGCT 3′TCGA	5′---AG CT---3′ 3′---TC GA---5′
*Eco*RV*	*Escherichia coli*	5′GATATC 3′CTATAG	5′---GAT ATC---3′ 3′---CTA TAG---5′
*Eco*P15I	*Escherichia coli*	5′CAGCAGN$_{25}$NN 3′GTCGTCN$_{25}$NN	5′---CAGCAGN$_{25}$ NN---3′ 3′---GTCGTCN$_{25}$NN ---5′
*Kpn*I	*Klebsiella pneumoniae*	5′GGTACC 3′CCATGG	5′---GGTAC C---3′ 3′---C CATGG---5′
*Pst*I	*Providencia stuartii*	5′CTGCAG 3′GACGTC	5′---CTGCA G---3′ 3′---G ACGTC---5′

SacI	*Streptomyces achromogenes*	5′GAGCTC 3′CTCGAG	5′---GAGCT C---3′ 3′---C TCGAG---5′
SalI	*Streptomyces albus*	5′GTCGAC 3′CAGCTG	5′---G TCGAC---3′ 3′---CAGCT G---5′
*ScaI**	*Streptomyces caespitosus*	5′AGTACT 3′TCATGA	5′---AGT ACT---3′ 3′---TCA TGA---5′
SpeI	*Sphaerotilus natans*	5′ACTAGT 3′TGATCA	5′---A CTACT---3′ 3′---TGATC A---5′
SphI	*Streptomyces phaeochromogenes*	5′GCATGC 3′CGTACG	5′---GCATG C---3′ 3′---C GTACG---5′
*StuI**	*Streptomyces tubercidicus*	5′AGGCCT 3′TCCGGA	5′---AGG CCT---3′ 3′---TCC GGA---5′
XbaI	*Xanthomonas badrii*	5′TCTAGA 3′AGATCT	5′---T CTAGA---3′ 3′---AGATC T---5′

Key: * = blunt ends
N = C or G or T or A
W = A or T

A restriction enzyme is a nuclease enzyme that cleaves DNA sequence at a random or specific recognition sites known as restriction sites. In bacteria, restriction enzymes form a combined system (restriction+ modification system) with modification enzymes that methylate the bacterial DNA. Methylation of bacterial DNA at the recognition sequence typically protects the own DNA of the bacteria from being cleaved by restriction enzyme.

There are two different kinds of restriction enzymes:

(1) Exonucleases catalyses hydrolysis of terminal nucleotides from the end of DNA or RNA molecule either 5'to 3' direction or 3' to 5' direction. Example: exonuclease I, exonuclease II etc.

(2) Endonucleases can recognize specific base sequence (restriction site) within DNA or RNA molecule and cleave internal phosphodiester bonds within a DNA molecule. Example: EcoRI, Hind III, BamHI etc.

Endonuclease

Endonucleases are enzymes that cleave the phosphodiester bond within a polynucleotide chain. Some, such as Deoxyribonuclease I, cut DNA relatively nonspecifically (without regard to sequence), while many, typically called restriction endonucleases or restriction enzymes, cleave only at very specific nucleotide sequences.

Restriction enzymes are endonucleases from eubacteria and archaea that recognize a specific DNA sequence. The nucleotide sequence recognized for cleavage by a restriction enzyme is called the restriction site. Typically, a restriction site will be a palindromic sequence about four to six nucleotides long. Most restriction endonucleases cleave the DNA strand unevenly, leaving complementary single-stranded ends. These ends can reconnect through hybridization and are termed "sticky ends". Once paired, the phosphodiester bonds of the fragments can be joined by DNA ligase. There are hundreds of restriction endonucleases known, each attacking a different restriction site. The DNA fragments cleaved by the same endonuclease can be joined together regardless of the origin of the DNA. Such DNA is called recombinant DNA; DNA formed by the joining of genes into new combinations. *Restriction endonucleases* (restriction enzymes) are divided into three categories, Type I, Type II, and Type III, according to their mechanism of action. These enzymes are often used in genetic engineering to make recombinant DNA for introduction into bacterial, plant, or animal cells, as well as in synthetic biology.

Categories

Ultimately, there are three categories of restriction endonucleases that relatively contribute to the cleavage of specific sequences. The types I and III are large multisubunit complexes that include both the endonucleases and methylase activities. Type I can cleave at random sites of about 1000 base pairs or more from the recognition sequence and it requires ATP as source of energy. Type II behaves slightly differently and was first isolated by Hamilton Smith in 1970. They are simpler versions of the endonucleases and require no ATP in their degradation processes. Some examples of type II restriction endonucleases include *BamHI, EcoRI, EcoRV,* and *Haelll.* Type III, however, cleaves the DNA at about 25 base pairs from the recognition sequence and also requires ATP in the process.

Notations

The commonly used notation for restriction endonucleases is of the form "*vwxyz*", where "*vwx*" names the life form (bacteria) where this restriction endonuclease may be found, "*y*" names the strain (and is optional), and "*z*" (in Roman numerals) indicates different restriction endonucleases in the same life form (bacteria). Thus for example, "EcoRI" means that the restriction endonuclease is found in *Escherichia coli* ("Eco"); strain RY13 ("R"), restriction endonuclease number "I". Another example: "HaeII" and "HaeIII" refer to bacterium *Haemophilus aegyptius*, number II and number III, respectively. The restriction enzymes used in molecular biology usually recognize short target sequences of about 4 – 8 base pairs. For instance, the *EcoRI* enzyme recognizes and cleaves the sequence 5' – GAATTC – 3'.

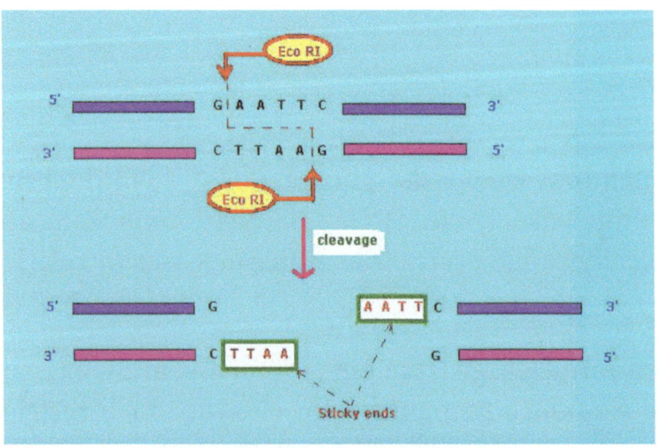

Restriction enzyme Eco RI

Restriction endonucleases come in several types. A restriction endonuclease typically requires a recognition site and a cleavage pattern (typically of nucleotide bases: A, C, G, T). If the recognition site is outside the region of the cleavage pattern, then the restriction endonuclease is referred to as Type I. If the recognition sequence overlaps with the cleavage sequence, then the restriction endonuclease restriction enzyme is Type II.

Further Discussion

Restriction endonucleases may be found that cleave standard dsDNA (double-stranded DNA), or ssDNA (single-stranded DNA), or even RNA. This discussion is restricted to dsDNA; however, the discussion can be extended to the following:

- Standard dsDNA
- Non-standard DNA
    1. Holliday junctions Holliday junction
    2. Triple-stranded DNA, quadruple-stranded DNA (G-quadruplex)
    3. Double-stranded hybrids of DNA and RNA (one strand is DNA, the other strand is RNA)
    4. Synthetic or artificial DNA (for example, containing bases other than A, C, G, T, refer to the work of Eric T. Kool). Research with synthetic codons, refer to the research by S. Benner, and enlarging the amino acid set in polypeptides, thus enlarging the proteome or proteomics.

In addition, research is now underway to construct synthetic or artificial restriction endonucleases, especially with recognition sites that are unique within a genome.

Restriction endonucleases or restriction enzymes typically cleave in two ways: blunt-ended or sticky-ended patterns. An example of a Type I restriction endonuclease.

DNA Repair

Endonucleases play a role in DNA repair. AP endonuclease, specifically, catalyzes the incision of DNA exclusively at AP sites, and therefore prepares DNA for subsequent excision, repair synthesis and DNA ligation. For example, when depurination occurs, this lesion leaves a deoxyribose sugar with a missing base. The AP endonuclease recognizes this sugar and essentially cuts the DNA at this site and then allows for DNA repair to continue. E. coli cells contain two AP endonucleases: endonuclease IV (endoIV) and exonuclease III (exoIII). While in Eukaryotes, there is only one AP endonuclease.

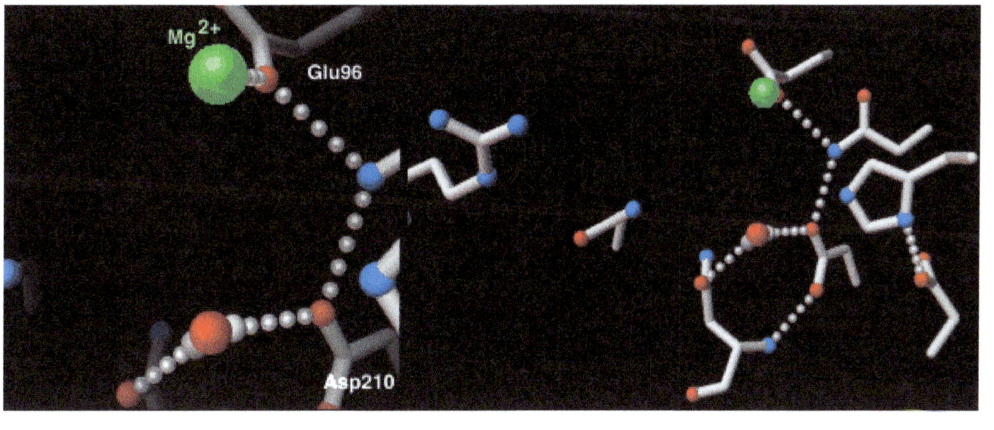

Common Endonucleases

Below are tables of common prokaryotic and eukaryotic endonucleases.

Prokaryotic Enzyme	Source	Comments
RecBCD enonuclease	E. coli	Partially ATP dependent; also an exonuclease; functions in recombination and repair
T7 endonuclease	phage T7 encoded (gene 3)	Essential for replication; preference for single stranded over double stranded DNA
T4 endonuclease IV	phage T4 encoded (denA)	Splits -TpC- sequence to yield 5'-dCMP- terminated oligonucleotides; chain length of product varies with conditions
Bal 31 endonuclease	Alteromonas espejiana	Also an exonuclease; nibbles away 3' and 5' ends of duplex DNA
EndonucleaseI (endo I)	E. coli (endA)	Periplasmic location; average chain length of product is 7; inhibited by tRNA; produces double stranded DNA break; produces nick when complexed with tRNA; endo I mutants grow normally
Micrococcal nuclease	Staphylococcus	Produces 3'-P termini; requires Ca2+; also acts on RNA; prefers single stranded DNA and AT-rich regions
Endonuclease II (endo VI, exo III)	E. coli (xth)	Cleavage next to AP site; also a 3'-->5' exonuclease; phosphomonoesterase on 3'-P termini

Eukaryotic Enzyme	Source	Comments
Neurospora endonuclease	Neurospora crassa	Also acts on RNA
S1-nuclease	Aspergillus oryzae	Also acts on RNA
P1-nuclease	Penicillium citrinum	Also acts on RNA
Mung bean nuclease I	mung bean sprouts	Also acts on RNA
Ustilago nuclease (Dnase I)	Ustilago maydis	Also acts on RNA
Dnase I	Bovine pancreas	Average chain length of product is 4; produces double strand break in presence of Mn2+
AP endonuclease	Nucleus, mitochondria	Involved in DNA Base Excision Repair pathway
Endo R	HeLa cells	Specific for GC sites

Mutations

Xeroderma pigmentosa is a rare, autosomal recessive disease caused by a defective UV-specific endonuclease. Patients with mutations are unable to repair DNA damage caused by sunlight.

Sickle Cell anemia is a disease caused by a point mutation. The sequence altered by the mutation eliminates the recognition site for the restriction endonuclease MstII that recognizes the nucleotide sequence.

tRNA splicing endonuclease mutations cause pontocerebellar hypoplasia. Pontocerebellar hypoplasias (PCH) represent a group of neurodegenerative autosomal recessive disorders that is caused by mutations in three of the four different subunits of the tRNA-splicing endonuclease complex.

Modification by Enzymes

- Polynucleotide phosphorylase was first discovered from extracts of *Azotobacter agile* by Grunberg-Manago and Ochoa.
- Polynucleotide phosphorylase (PNPase) catalyzes the synthesis of long chain polyribonucleotides (RNA) in 5' to 3' direction from nucleotide diphosphates as precursors and reversibly catalyzes phosphorolytic cleavage of polyribonucleotides in 3' to 5' direction with a release of orthophosphate in presence of inorganic phosphate.
- PNPase is a bifunctional enzyme and functions in mRNA processing and degradation inside the cell.
- Structural and physiochemical studies in enzymes showed that it is formed of subunits. The arrangements of the subunits may vary from species to species which would alter their properties.
- These enzyme can catalyze not only the synthesis of RNA from the mixtures of naturally occurring ribonucleoside diphosphates, but also that of non-naturally occurring polyribonucleotides.

Mechanism of Action

As mentioned earlier, polynucleotide phosphorylase is a bifunctional enzyme. The mechanism of action of this enzyme can be represented by following reactions:

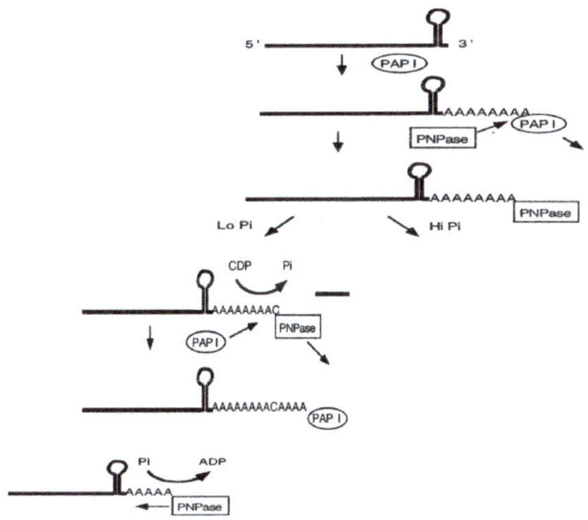

Schematic representation of the role of PNPase in poly(A) tail metabolism in *E. coli*.

In *E.coli* , polynucleotide phosphorylase regulates mRNA processing either by adding ribonucleotides to the 3' end or by cleaving bases in 3' to 5' direction. The function of PNPase depends upon inorganic phosphate (Pi) concentration inside the cell. The transcripts are polyadenylated using enzyme polyadenylate polymerase I (PAPI). After primary polyadenylylation of the transcript by PAP I, PNPase may bind to the 3′ end of the poly(A) tail. PNPase works either degradatively or biosynthetically inside the cell depending on the Pi concentration. Under high Pi concentration, it degrades the poly(A) tail releasing adenine diphosphates. If the Pi concentration is low, PAP I initiates addition of one or more nucleotides to the existing poly (A) tail and in the process generates inorganic phosphate. On dissociation of PNPase, the 3′ end again is available to PAP I for further polymerization.

Function

Different functions of Polynucleotide phosphorylase are:

- It is involved in mRNA processing and degradation in bacteria, plants, and in humans.
- It synthesizes long, highly heteropolymeric tails *in vivo* as well as accounts for all of the observed residual polyadenylation in poly(A) polymerase I deficient strains.
- PNPase function as a part of RNA degradosome in *E.coli* cell. RNA degradosome is a multicomponent enzyme complex that includes RNaseE (endoribinuclease), polynucleotide phosphorylase (3' to 5' exonuclease), RhlB helicase (a DEAD box helicase) and a glycolytic enzyme enolase. This complex catalyzes 3' to 5' exonuclease activity in presence of ATP. In eukaryotes, the exosomes are located in nucleus and cytoplasm. Degradsomes in bacteria and exosomes in eukaryotes are associated with processing, control and turnover of RNA transcripts.
- In rDNA cloning technology, it has been used to synthesize radiolabelled polyribonucleotides from nucleoside diphosphate monomers.

Deoxyribonuclease (DNase):

- A nuclease enzyme that can catalyze the hydrolytic cleavage of phosphodiester bonds in the DNA backbone are known as deoxyribonuclease (DNase).
- Based on the position of action, these enzymes are broadly classified as endodeoxyribonuclease (cleave DNA sequence internally) and exodeoxyribonuclease (cleave the terminal nucleotides).
- Unlike restriction enzymes, DNase does not have any specific recognition/restriction site and cleave DNA sequence at random locations.
- There is a wide variety of deoxyribonucleases known which have different substrate specificities, chemical mechanisms, and biological functions. They are:

Deoxyribonuclease I (DNase I):

An endonuclease which cleaves double-stranded DNA or single stranded DNA. The cleavage preferentially occurs adjacent to pyrimidine (C or T) residues. The major products are 5'-phosphorylated bi-, tri- and tetranucleotides. It requires divalent ions (Ca^{2+} and Mn^{2+} /Mg^{2+}) for its activity and creates blunt ends or 1-2 overhang sequences.

DNaseI is the most widely used enzyme in cloning experiments to remove DNA contamination from mRNA preparation (to be used for cDNA library preparation, northern hybridization, RT-PCR etc). The mode of action of DNaseI varies according to the divalent cation used.

In the presence of magnesium ions (Mg^{+2}), DNaseI hydrolyzes each strand of duplex DNA producing single stranded nicks in the DNA backbone, generating various random cleavages.

On the other hand, in the presence of manganese ions (Mn^{+2}), DNaseI cleaves both strands of a double stranded DNA at approximately the same site, producing blunt ended DNA fragments or with 1-2 base overhangs. The two major DNases found in metazoans are: deoxyribonuclease I and deoxyribonuclease II.

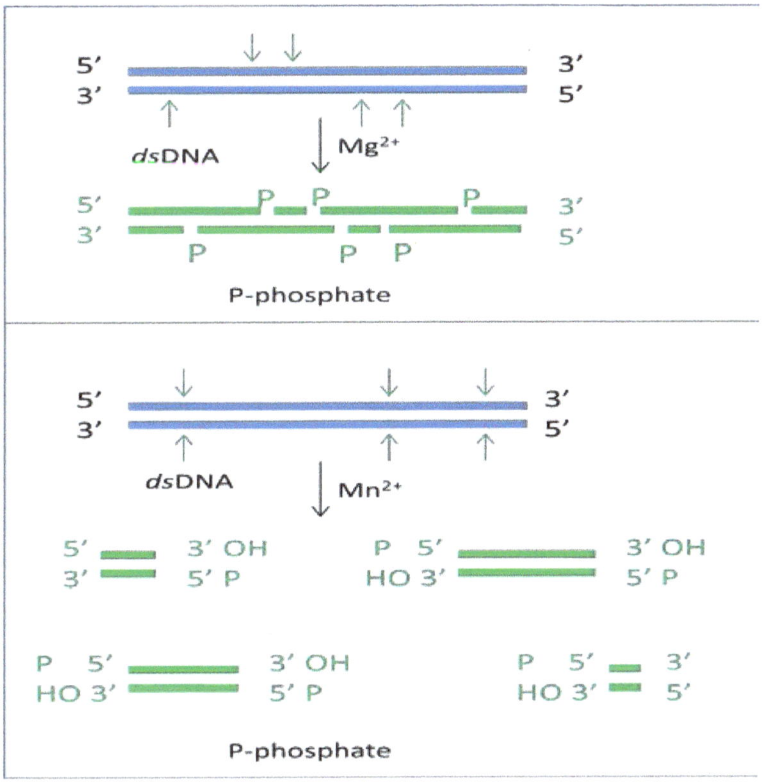

Action of DNase I in the presence of Mg^{+2} and Mn^{+2} ions. (Arrowhead denoting random site of cleavage in double stranded DNA by DNase I)

Some of the common applications of DNase I in rDNA technology have been mentioned below:

- Eliminating DNA contamination (e.g. plasmid) from preparations of RNA.
- Analyzing the DNA-protein interactions via DNA footprinting.
- Nicking DNA prior to radio-labeling by nick translation.

DeoxyribonucleaseII (DNaseII):

It is a non-specific endonuclease with optimal activity at acidic pH (4.5-5.5) and conserved from human to *C.elegans*. It does not require any divalent cation for its activity. DNaseII initially introduces multiple single stranded nicks in DNA backbone and finally generates 3' phosphate groups by hydrolyzing phosphodiester linkages.

This enzyme releases 3' phosphate groups by hydrolyzing phosphodiester linkage and creating nicks in the DNA backbone. DNaseII acts by generating multiple single stranded nicks followed by production of acid soluble nucleotides and oligonucleotides. The catalytic site of the enzyme contains three histidine residues which are essential for enzyme activity.

Some of the common applications of DNase II are as follows:

- DNA fragmentation
- Molecular weight marker
- Cell apoptosis assays etc.

Exonuclease III:

Exonuclease III is a globular enzyme which has 3′→5′ exonuclease activity in a double stranded DNA. The template DNA should be double stranded and the enzyme does not cleave single stranded DNA. The enzyme shows optimal activity with blunt ended sequences or sequences with 5' overhang.

Exonuclease III enzyme has a bound divalent cation which is essential for enzyme activity. The mechanism of the enzyme can be affected by variation in temperature, monovalent ion concentration in the reaction buffer, and structure and concentration of 3' termini. The enzyme shows optimal activity at 37°C at pH 8.0.

Various application of exonuclease III in molecular cloning experiments are:

- To generate template for DNA sequencing
- To generate substrate for DNA labeling experiments
- Directed mutagenesis
- DNA-protein interaction assays (to find blockage of exonuclease III activity by protein-DNA binding) etc.

Mung Bean Nuclease

As the name suggest, this nuclease enzyme is isolated from mung bean sprouts (*Vigna radiata*). Mung bean nuclease enzymes can degrade single stranded DNA as well RNA. Under high enzyme concentration, they can degrade double stranded DNA, RNA or even DNA/RNA hybrids.

Mung bean nuclease can cleave single stranded DNA or RNA to produce 5'-phosphoryl mono and oligonucleotides. It requires Zn^{2+} ion for its activity and shows optimal activity at 37°C. The enzyme works in low salt concentration (25mM ammonium acetate) and acidic pH (pH 5.0). Treatment with EDTA or SDS results in irreversible inactivation of the enzyme.

Mung bean nuclease is less robust than S1 nuclease and easier to handle. It has been used to create blunt end DNA by cleaving protruding ends from 5' ends. This enzyme cannot produce nicks in a double stranded DNA but at higher concentration, it can generate nicks and cleave double stranded DNA.

Phosphatase

- Phosphatase catalyses the cleavage of a phosphate (PO_4^{-2}) group from substrate by using a water molecule (hydrolytic cleavage).
- This reaction is not reversible.
- This shows totally opposite activity from enzyme like kinase and phosphorylase that add a phosphate group to their substrate. On the basis of their activity there are two types of phosphatase i.e acid phosphatase and alkaline phosphatase. In both forms the alkaline phosphatase are most common.
- Special class of phosphatase that remove a phosphate group from protein, called "Phosphoprotein phosphatase".

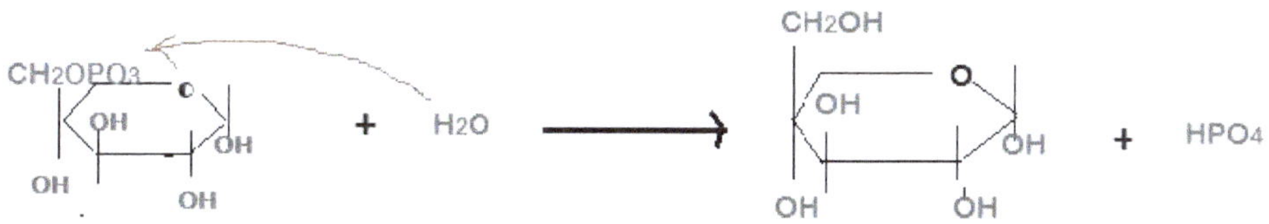

Schematic representation of hydrolytic cleavage of phosphate group (PO_4^{-2}).

Acid Phosphatase

- It shows its optimal activity at pH between 3 and 6, e.g. a lysosomal enzyme that hydrolyze organic phosphates liberating one or more phosphate groups. They are found in prostatic epithelial cells, erythrocyte, prostatic tissue, spleen, kidney etc.

Alkaline Phosphatase

- Homodimeric enzyme which catalyzes reactions like hydrolysis and transphosphophorylation of phosphate monoester.
- They show their optimal activity at pH of about 10.
- Alkaline phosphatase was the first zinc enzyme discovered having three closed spaced metal ion. Two Zn^{+2} ions and one Mg^{+2} ion, in which Zn^{+2} ions are bridges by *Asp* 51. The mechanism of action is based on reaction where a covalent serine – phosphate intermediate is formed to produce inorganic phosphate and an alcohol.
- In human body it is present in four isoforms, in which three are tissue specific isoform i.e. placental, germ cell, intestinal and one is non tissue specific isoform. The genes that encode for tissue specific isoforms are present on chromosome -2 p37-q37, while the genes for one non tissue specific are present on chromosome -1 p34- p36.1.
- During post-translational modification, alkaline phosphatase is modified by N-glycosylation. It undergoes a modification through which uptake of two Zn^{+2} ion and one Mg^{+2} ion occurs which is important in forming active site of that enzyme. Alkaline phosphatases are isolated from various sources like microorganisms, tissue of different organs, connective tissue of invertebrate and vertebrate, and human body.

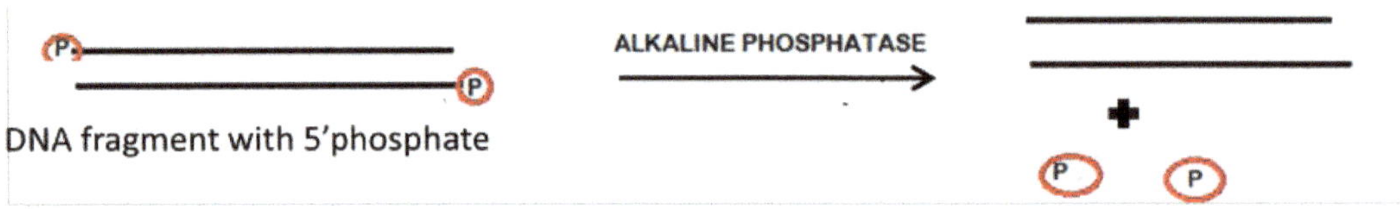

DNA fragment and separated PO_4^{-2} group

There are several AP that are used in gene manipulation:

- Bacterial alkaline phosphatase (BAP) - Bacterial alkaline phosphate is a phosphomonoester that hydrolyzes 3' and 5' phosphate from nucleic acid (DNA/ RNA). It more suitably removes phosphate group before end labeling and remove phosphate from vector prior to insert ligation. BAP generally shows optimum activity at temperature 65°C. BAP is sensitive to inorganic phosphate so in presence of inorganic phosphates activity may reduce.
- Calf intestinal alkaline phosphatase (CIP) – It is isolated from calf intestine, which catalyzes the removal of phosphate group from 5' end of DNA as well as RNA. This enzyme is highly used in gene cloning experiments, as to make a construct that could not undergo self-ligation. Hence after the treatment with CIP, without having a phosphate group at 5' ends a vector cannot self ligate and recircularise. This step improves the efficiency of vector containing desired insert.
- Shrimp alkaline phosphatase (SAP) - Shrimp alkaline phosphatase is highly specific, heat labile phosphatase enzyme isolated from arctic shrimp (*Pandalus borealis*). It removes 5' phosphate group from DNA, RNA, dNTPs and proteins. SAP has similar specificity as CIP but unlike CIP, it can be irreversibly inactivated by heat treatment at 65°C for 15mins. SAP is used for 5' dephosphorylation during cloning experiments for various application as follows:
    - Dephosphorylate 5'-phosphate group of DNA/RNA for subsequent labeling of the ends.
    - To prevent self-ligation of the linearized plasmid.
    - To prepare PCR product for sequencing.
    - To inactivate remaining dNTPs from PCR product (for downstream sequencing appication).

Two Primary uses for Alkaline Phosphatase in DNA Modification

- Removing 5' phosphate from different vector like plasmid, bacteriophage after treating with restriction enzyme. This treatment prevents self ligation because unavailability of phosphate group at end. So, this treatment greatly enhances the ligation of desired insert. During ligation of desired insert, the complementary ends of the insert and vector will come to proximity of each other (only for sticky ends but not for blunt ends). One strand of the insert having 5'-phosphate will ligate with the 3'OH of the vector and the remaining strand will have a nick. This nick will be sealed in the next step by ligase enzyme in the presence of ATP. It is used to remove 5' phosphate from fragment of DNA prior to labeling with radioactive phosphate.

Methylase

- Methyltransferase or methylase catalyzes the transfer of methyl group (-CH_3) to its substrate. The process of transfer of methyl group to its substrate is called methylation.
- Methylation is a common phenomenon in DNA and protein structure.
- Methyltransferase uses a reactive methyl group that is bound to sulfur in S-adenosyl methionine (SAM) which acts as the methyl donor.
- Methylation normally occurs on cytosine (C) residue in DNA sequence. In protein, methylation occurs on nitrogen atom either on N-terminus or on the side chain of protein.
- DNA methylation regulates gene or silence gene without changing DNA sequences, as a part of epigenetic regulation.
- In bacterial system, methylation plays a major role in preventing their genome from degradation by restriction enzymes. It is a part of restriction – modification system in bacteria.

Methyltransferase can be classified in three groups:

a) m6A-generates N6 methyladenosine,

b) m4C-generates N4 methylcytosine,

c) m5C-generates N5 methylcytosine.

m6A and m4C methyltransferase are primarily found in prokaryotes. These enzymes are responsible for methylation of DNA sequences in order to prevent the host from digesting its own genome via its restriction enzyme.

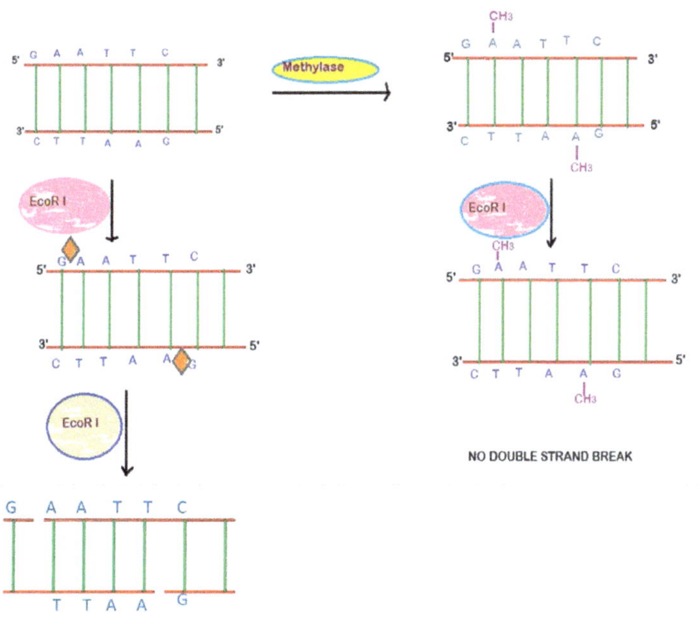

Activity of restriction and methylase enzymes

Restriction enzyme *EcoR* I cleaves within the recognition sequence if the DNA is unmethylated. On methylation by methylases, the restriction enzyme *EcoR* I is inhibited from cleaving within the restriction site.

Some common examples of methytransferases are: DNA adenyl methytransferase (DAM), histone methyltransferase, O-methyltransferase etc. DAM methylase is generally used in recombinant DNA technology which can methylate adenine (A) in the sequence 5'GATC3'. This enzyme can methylate a newly synthesized DNA strand on specific sites.

Ligases

- DNA ligase catalyses the formation of phosphodiester bond between two deoxynucleotide residues of two DNA strands.
- DNA ligase enzyme requires a free hydroxyl group at the 3´-end of one DNA chain and a phosphate group at the 5´-end of the other and requires energy in the process.
- *E.coli* and other bacterial DNA ligase utilizes NAD^+ as energy donor, whereas in T4 bacteriophage, T4 DNA ligase uses ATP as cofactor.
- The role of DNA ligase is to seal nicks in the backbone of double-stranded DNA after lagging strand formation to join the okazaki fragments.
- This joining process is essential for the normal synthesis of DNA and for repairing damaged DNA. It has been exploited by genetic engineers to join DNA chains to form recombinant DNA molecules. Usually single stranded break are repaired using the complimentary strand as the template but sometimes double stranded breaks can also be repaired with the help of DNA ligase IV.
- The most widely used DNA ligase is isolated from T4 bacteriophage. T4 DNA ligase needs ATP as a cofactor. The enzyme from *E. coli* uses cofactor NAD. Except this, the catalysis mechanism is somewhat similar for both the ligases. The role of cofactor is splitting and forming an enzyme-AMP complex which further aids in formation of phosphodiester bonds between hydroxyl and phosphate groups by exposing them.

Mechanism of Action of DNA Ligases

- ATP, or NAD^+, reacts with the ligase enzyme to form a covalent enzyme–AMP complex in which the AMP is linked to ε -amino group of a lysine residue in the active site of the enzyme through a phosphoamide bond.
  - The AMP moiety activates the phosphate group at the 5´-end of the DNA molecule to be joined. It is called as the donor.
  - The final step is a nucleophilic attack by the 3´-hydroxyl group on this activated phosphorus atom which acts as the acceptor. A phosphodiester bond is formed and AMP is released.
  - The reaction is driven by the hydrolysis of the pyrophosphate released during the formation of the enzyme–adenylate complex. Two high-energy phosphate bonds are spent

in forming a phosphodiester bond in the DNA backbone with ATP serving as energy source.

- The temperature optimum for T4 DNA ligase mediated ligation *in vitro* is 16°C. However ligation is also achieved by incubation at 4°C by incubating over night or at room temperature condition by incubating for 30 minutes.
- Adenylate and DNA-adenylate are the important intermediates of the phosphodiester bond forming pathway.

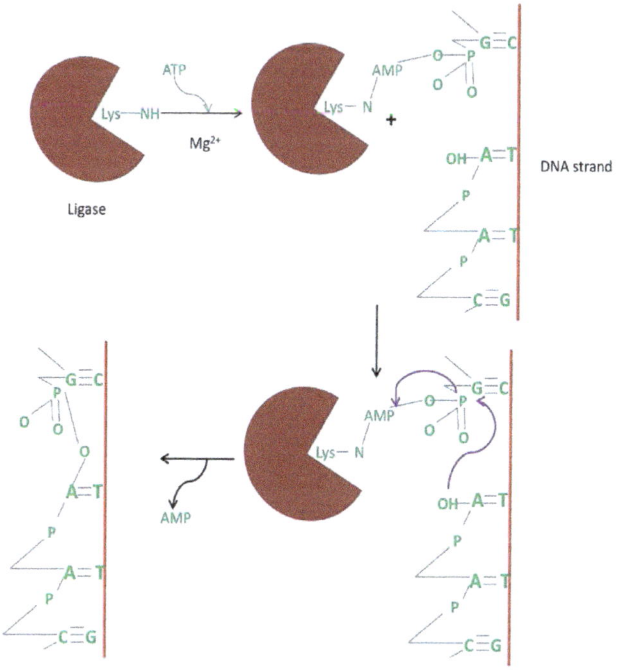

The mechanism of DNA joining by DNA ligase.

Application

- DNA ligase enzyme is used by cells to join the "okazaki fragments" during DNA replication process. In molecular cloning, ligase enzyme has been routinely used to construct a recombinant DNA. Followings are some of the examples of application of ligase enzyme in molecular cloning.Joining of adapters and linkers to blunt end DNA molecule.
- Cloning of restricted DNA to vector to construct recombinant vector.

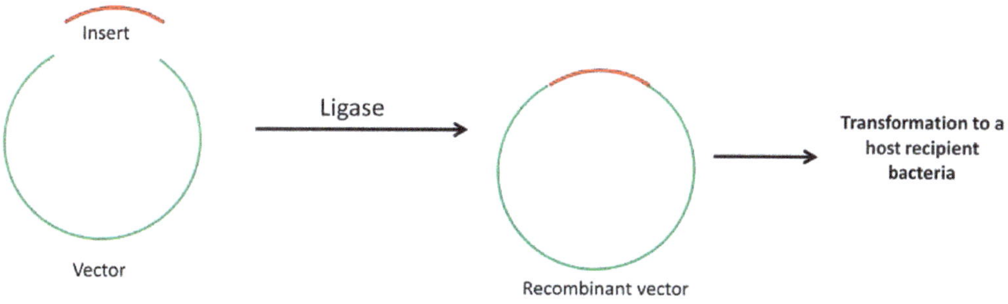

Ligation of a gene fragment into the vector and transformation of the cell.

Polynucleotide Kinase

- PNK is a homotetramer with phosphatase activity at 3' end and kinase activity at 5' end with a tunnel like active site. The active site has side chains which interact with NTP donor's beta-phosphate and 3' phosphate of acceptor with an acid which activated 5' –OH. Lys-15 and Ser-16 are important for the kinase activity of the enzyme.
- The basic residues of active site of PNK interact with the negatively charged phosphates of the DNA.
- Polynucleotide kinase (PNK) catalyzes the transfer of a phosphate group (PO_4^{-2}) from γposition of ATP to the 5' end of either DNA or RNA and nucleoside monophosphate .
- PNK can convert 3' PO_4 /5› OH ends into 3› PO_4 /5› PO_4 ends which blocks further ligation by ligase enzyme.
- PNK is used to label the ends of DNA or RNA with radioactive phosphate group.
- T4 polynucleotide kinase is the most widely used PNK in molecular cloning experiments, which was isolated from T4 bacteriophage infected *E.coli.*

PNK carries out two types of enzymatic activity:

- Forward reaction: γ-phosphate is transferred from ATP to the 5' end of a polynucleotide (DNA or RNA). 5' phosphate is not present either due to chemical synthesis or dephosphorylation. The 5' OH nucleophile is activated by abstraction of the proton. Asp35 of PNK forms the co-ordinate bond with 5' OH and attacks γ phosphorus forming an intermediate.
- Exchange reaction: target DNA or RNA having a 5' phosphate is incubated with an excess of ADP - where PNK transfers the phosphate from the nucleic acid to an ADP, forming ATP. PNK then performs a forward reaction and transfer a phosphate from ATP to the target nucleic acid. Exchange reaction is used to label with radioactive phosphate group.

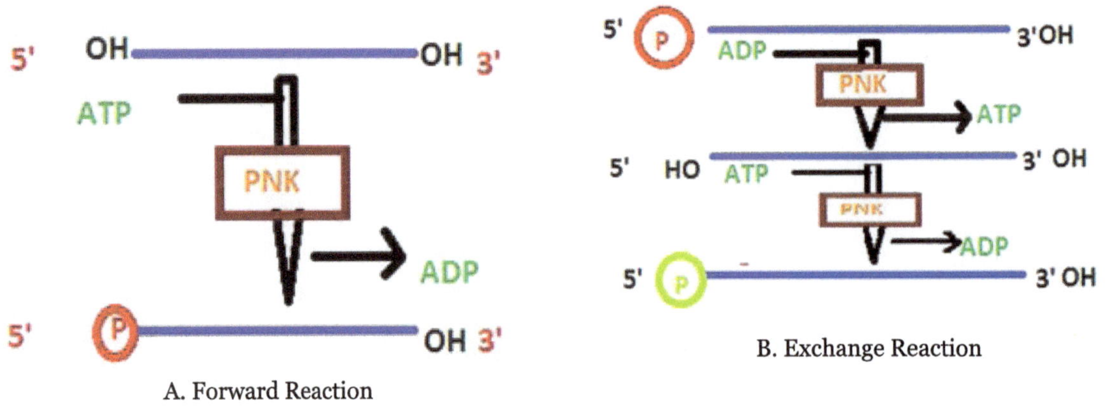

Polynucleotide kinase reaction (A) forward (B) exchange.

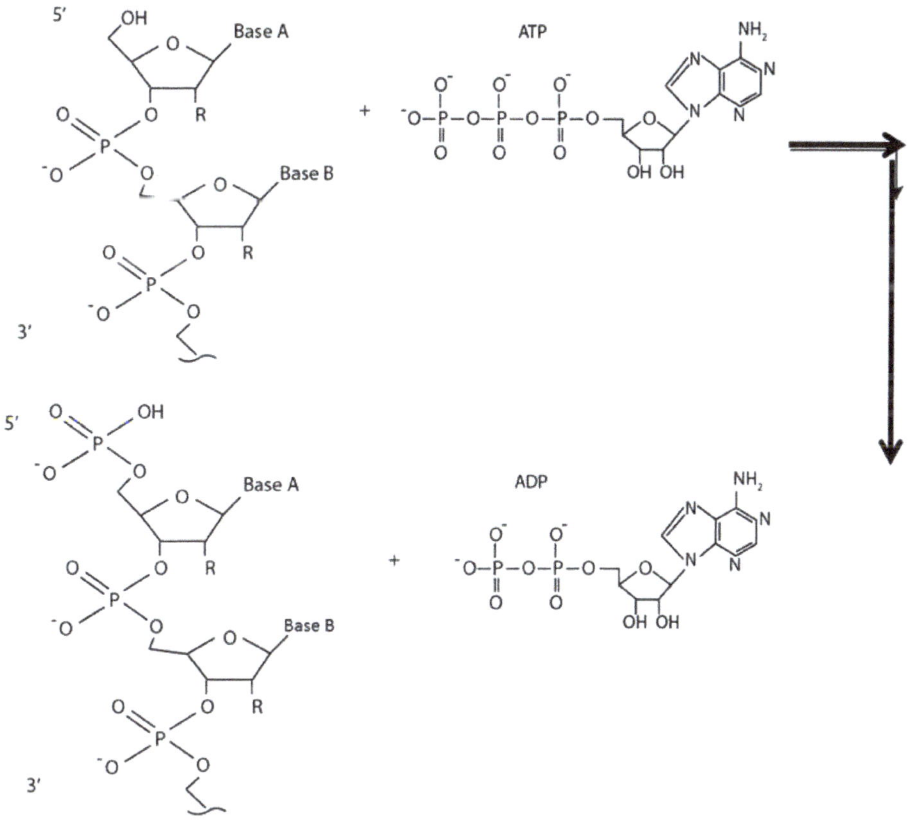

Conversion of 5' dephospho-(deoxy) ribonucleic acid to
5' phospho-(deoxy) ribonucleic acid by the action of PNK.

The efficiency of phosphorylation is less in exchange reaction compared to forward reaction. Along with the phosphorylating activity, PNK also has 3' phosphatase activity.

There are two major uses of PNK:

- The linkers and adopters are phosphorylated along with the fragments of DNA before ligation, which requires a 5' phosphate. This includes products of polymerase chain reaction, which are generated by using non-phosphorylated primers.
- PNK is also used for radio labelling oligonucleotides, generally with ^{32}P for preparing hybridization probes.

PNK is inhibited by ammonium ions, so ammonium acetate cannot be used to precipitate nucleic acids before phosphorylation. Sometimes phosphate ions or NaCl of greater than 50 mM concentration can also inhibit this enzyme.

Ribonuclease (RNase)

- Nuclease that can catalyze hydrolysis of ribonucleotides from either single stranded or double stranded RNA sequence are called ribonucleotides (RNase).
- RNase are classified into two types depending on position of cleavage, i.e. endoribonuclease (cleave internal bond) and exoribonuclease (cleave terminal bond).

- RNase is important for RNA maturation and processing.
- RNaseA and RNaseH play important role in initial defence mechanism against RNA viral infection.

Two common types of ribonucleases are discussed below:

RibonucleaseA (RNaseA):

- An endo-ribonuclease that cleaves specifically single-stranded RNA at the 3' end of pyrimidine residues.
- The RNA is degraded into 3'-phosphorylated mononucleotides C and U residues and oligonucleotides in the form of 2', 3'-cyclic monophosphate intermediates.
- Optimal temperature for RNaseA is 60°C (activity range 15-70°C) and optimal pH is 7.6.
- RNaseA has two histidine residues in its active site (His12 and His119). In the first step, His12 acts as a base; accepting proton forming a nucleophile which then attacks positively charged phosphorus atom. His119 acts as an acid in this case, donating a proton to oxygenated P-O-R' bond. The imidazole side chain acts as base in His 12 here.
- The side chain of Lys41 and Phe120 further stabilize the transition state. Nitrogen of the main chain of Phe120 donates hydrogen, thus bonding with the unbound oxygen atom.
- In the second step the acid base activities get reversed and His119 accepts proton from water causing hydroxyl attack on cyclic intermediate.
- Activity of RNaseA can be inhibited by alkylation of His12 and His119 residue essential for activity of the enzyme.

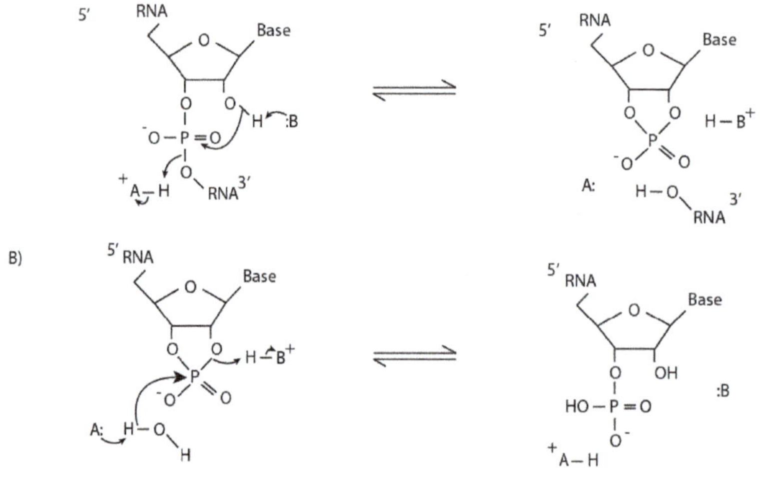

Fig: (A) Transphosphorylation reaction by RNase A
(B) Hydrolysis reaction catalyzed by RNase A

["B" → His 12
"A" → His 119]

Mechanism of action of RNase A

Application

- It is used to remove RNA contamination from DNA sample.

RibonucleaseH

- Non-specific endoribonuclease that degrades RNA by hydrolytic mechanism from DNA/RNA duplex resulting in single stranded DNA.
- Enzyme bound divalent metal ion is a cofactor here. The product formed is 5' phosphorylated ssDNA.
- During cDNA library preparation from RNA sample, RNaseH enzyme is used to cleave RNA strand of DNA-RNA duplex.

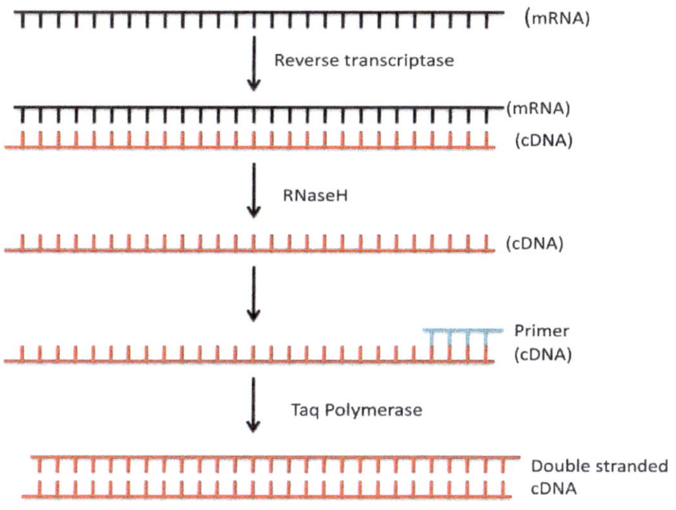

Schematic representation of cDNA preparation from mRNA

References

- Roberts RJ (Nov 1976). "Restriction endonucleases". CRC Critical Reviews in Biochemistry. 4 (2): 123–64. PMID 795607. doi:10.3109/10409237609105456
- Luria SE, Human ML (Oct 1952). "A nonhereditary, host-induced variation of bacterial viruses". Journal of Bacteriology. 64 (4): 557–69. PMC 169391. PMID 12999684
- Pingoud A, Alves J, Geiger R (1993). "Chapter 8: Restriction Enzymes". In Burrell M. Enzymes of Molecular Biology. Methods of Molecular Biology. 16. Totowa, NJ: Humana Press. pp. 107–200. ISBN 0-89603-234-5
- "The Nobel Prize in Physiology or Medicine". The Nobel Foundation. 1978. Retrieved 2008-06-07. for the discovery of restriction enzymes and their application to problems of molecular genetics
- Arber W, Linn S (1969). "DNA modification and restriction". Annual Review of Biochemistry. 38: 467–500. PMID 4897066. doi:10.1146/annurev.bi.38.070169.002343
- Lederberg S, Meselson M (May 1964). "Degradation of non-replicating bacteriophage dna in non-accepting cells". Journal of Molecular Biology. 8 (5): 623–8. PMID 14187389. doi:10.1016/S0022-2836(64)80112-1
- Primrose SB, Old RW (1994). Principles of gene manipulation: an introduction to genetic engineering. Oxford: Blackwell Scientific. ISBN 0-632-03712-1

- Naito T, Kusano K, Kobayashi I (Feb 1995). "Selfish behavior of restriction-modification systems". Science. 267 (5199): 897–9. Bibcode:1995Sci...267..897N. PMID 7846533. doi:10.1126/science.7846533
- Smith HO, Wilcox KW (Jul 1970). "A restriction enzyme from Hemophilus influenzae. I. Purification and general properties". Journal of Molecular Biology. 51 (2): 379–91. PMID 5312500. doi:10.1016/0022-2836(70)90149-X
- Massey A, Kreuzer H (2001). Recombinant DNA and Biotechnology: A Guide for Students. Washington, D.C: ASM Press. ISBN 1-55581-176-0
- Geerlof A. "Cloning using restriction enzymes". European Molecular Biology Laboratory - Hamburg. Retrieved 2008-06-07
- Kelly TJ, Smith HO (Jul 1970). "A restriction enzyme from Hemophilus influenzae. II". Journal of Molecular Biology. 51 (2): 393–409. PMID 5312501. doi:10.1016/0022-2836(70)90150-6
- Ninfa J., Alexander; Balou, David P.; Benore, Marilee (2010). Fundamental Laboratory Approaches for Biochemistry and Biotechnology. Hoboken, NJ: John Wiley & Sons. p. 341. ISBN 978-0-470-08766-4
- Horvath P, Barrangou R (Jan 2010). "CRISPR/Cas, the immune system of bacteria and archaea". Science. 327 (5962): 167–70. PMID 20056882. doi:10.1126/science.1179555
- Tovkach A, Zeevi V, Tzfira T (Jan 2011). "Expression, purification and characterization of cloning-grade zinc finger nuclease". Journal of Biotechnology. 151 (1): 1–8. PMID 21029755. doi:10.1016/j.jbiotec.2010.10.071
- Russell DW, Sambrook J (2001). Molecular cloning: a laboratory manual. Cold Spring Harbor, N.Y: Cold Spring Harbor Laboratory. ISBN 0-87969-576-5
- Smith HO, Nathans D (Dec 1973). "Letter: A suggested nomenclature for bacterial host modification and restriction systems and their enzymes". Journal of Molecular Biology. 81 (3): 419–23. PMID 4588280. doi:10.1016/0022-2836(73)90152-6
- Stephen T. Kilpatrick; Jocelyn E. Krebs; Lewin, Benjamin; Goldstein, Elliott (2011). Lewin's genes X. Boston: Jones and Bartlett. ISBN 0-7637-6632-1
- Gasior SL, Roy-Engel AM, Deininger PL (Jun 2008). "ERCC1/XPF limits L1 retrotransposition". DNA Repair. 7 (6): 983–989. PMC 2483505. PMID 18396111. doi:10.1016/j.dnarep.2008.02.006

3

Techniques of Genetic Engineering

Nucleic acids are polymers that store DNA. It is classified into types, DNA and RNA. The nucleic acids are essential in researches for which they need to be isolated. There are many methods to detect nucleic acids. A few of them are fluorometric quantification, UV spectroscopy, silver staining, ethidium bromide staining, etc. The aspects elucidated in this chapter are of vital importance, and provide a better understanding of genetic engineering.

Isolation and Purification of Nucleic Acid

Every gene manipulation procedure requires genetic materials like DNA and RNA. Nucleic acids occur naturally in association with proteins and lipoprotein organelles. The dissociation of a nucleoprotein into nucleic acid and protein moieties and their subsequent separation, are the essential steps in the isolation of all species of nucleic acids. Isolation of nucleic acids is followed by quantitation of nucleic acids generally done by either spectrophotometric or by using fluorescent dyes to determine the average concentrations and purity of DNA or RNA present in a mixture. Isolating the genetic material (DNA) from cells (bacterial, viral, plant or animal) involves three basic steps-

- Rupturing of cell membrane to release the cellular components and DNA
- Separation of the nucleic acids from other cellular components
- Purification of nucleic acids

Isolation and Purification of Genomic DNA

Genomic DNA is found in the nucleus of all living cells with the structure of double-stranded DNA remaining unchanged (helical ribbon). The isolation of genomic DNA differs in animals and plant cells. DNA isolation from plant cells is difficult due to the presence of cell wall, as compared to animal cells. The amount and purity of extracted DNA depends on the nature of the cell.

The method of isolation of genomic DNA from a bacterium comprises following steps-

1. Bacterial culture growth and harvest.
2. Cell wall rupture and cell extract preparation.
3. DNA Purification from the cell extract.
4. Concentration of DNA solution.

Growth and Harvest of Bacterial Culture

Bacterial cell culture is more convenient than any other microbe, as it requires only liquid medium (broth) containing essential nutrients at optimal concentrations, for the growth and division of bacterial cells. The bacterial cells are usually grown on a complex medium like Luria-Bertani (LB), in which the medium composition is difficult to decipher. Later, the cells are separated by centrifugation and resuspended in 1% or less of the initial culture volume.

Preparation of Cell Extract

Bacterial cell is surrounded by an additional layer called cell wall, apart from plasma membrane with some species of *E. coli* comprising multilayered cell wall. The lysis of cell wall to release the genetic material i.e. DNA can be achieved by following ways-

- Physical method by mechanical forces.
- Chemical method by metal chelating agents i.e. EDTA and surfactant i.e. SDS or enzyme (e.g. lysozyme).

Lysozyme

- Present in egg-white, salivary secretion and tears.
- Catalyzes the breakdown of cell wall i.e. the peptidoglycan layer.

EDTA (Ethylene Diamine Tetra-acetic Acid)

- A chelating agent necessary for destabilizing the integrity of cell wall.
- Inhibits the cellular enzymes that degrade DNA.

SDS (Sodium Dodecyl Sulphate)

- Helps in removal of lipid molecules and denaturation of membrane proteins. Generally, a mixture of EDTA and lysozyme is used. Cell lysis is followed by centrifugation to pellet down the cell wall fractions leaving a clear supernatant containing cell extract.

Purification of DNA

In addition to DNA, a cell extract contains significant quantities of protein and RNA which can be further purified by following methods-

Organic Extraction and Enzymatic Digestion for the Removal of Contaminants

It involves the addition of a mixture of phenol and chloroform (1:1) to the cell lysate for protein separation. The proteins aggregate as a white mass in between the aqueous phase containing DNA and RNA, and the organic layer. Treatment of lysate with pronase or protease, in addition to phenol/chloroform, ensures complete removal of proteins from the extract. The RNA can be effectively removed by using Ribonuclease, an enzyme which rapidly degrades RNA into its ribonucleotide subunits. Repeated phenol extraction is not desirable, as it damages the DNA.

Using Ion-exchange Chromatography

This involves the separation of ions and polar molecules (proteins, small nucleotides and amino acids) based on their charge. DNA carrying negative charge binds to the cationic resin or matrix which can be eluted from the column by salt gradient. Gradual increase in salt concentration detaches molecules from the resin one after another.

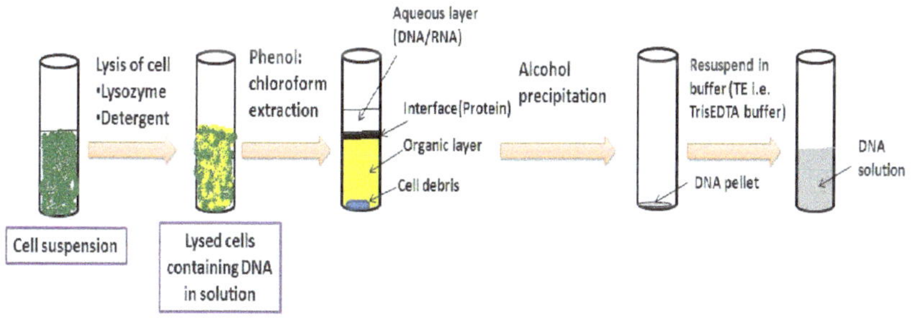

Preparation of genomic DNA

Concentration of DNA Samples

Concentration of DNA can be done using ethanol along with salts such as sodium acetate, potassium acetate etc. These salts provide metal ions like sodium ions (Na+), potassium ions (K+) which help in aggregation and hence precipitation of DNA molecules.

Advantage

It leaves short-chain and monomeric nucleic acid components in solution. Ribonucleotides produced by the ribonuclease treatment are separated from DNA.

Isolation and Purification of Plasmid DNA

1. Growth of the bacterial cell.
2. Harvesting and lysis of the bacteria.
3. Purification of the plasmid DNA.

Growth of the Bacterial Cell

It involves growth of the bacterial cells in a media containing essential nutrients.

Harvest and Lysis of Bacteria

Lysis of bacteria results in the precipitation of DNA and cellular proteins. Addition of acetate-containing neutralization buffer results in the precipitation of large and less supercoiled chromosomal DNA and proteins leaving the small bacterial DNA plasmids in solution.

Purification of Plasmid DNA

This step is same for both plasmid and genomic but former involves an additional step i.e.the separation of plasmid DNA from the large bacterial chromosomal DNA.

Methods for Separation of Plasmid DNA

Separation of plasmid DNA is based on the several features like size and conformation of plasmid DNA and bacterial DNA. Plasmids are much smaller than the bacterial main chromosomes, the largest plasmids being only 8% of the size of the *E. coli* chromosome. The separation of small molecules (i.e. plasmids) from larger ones (i.e. bacterial chromosome) is based on the fact that plasmids and the bacterial chromosomes are circular but bacterial chromosomes break into linear fragments during the preparation of the cell extract resulting in separation of pure plasmids. The methods of separation of plasmid DNA are described as below-

Separation Based on Size Difference

- It involves lysis of cells with lysozyme and EDTA (function as described above in point.) in the presence of sucrose (prevents the immediate bursting of cell).
- Cells with partially degraded cell walls are formed that retain an intact cytoplasmic membrane called as sphaeroplasts.
- Cell lysis is then induced by the addition of a non-ionic detergent (e.g. Triton X-100) or ionic detergents (e.g.SDS) causing chromosomal breakage.
- Bacterial chromosome attached to cell membrane, upon lysis gets removed with the cell debris.
- A cleared lysate consisting almost entirely of plasmid DNA is formed with very little breakage of the bacterial DNA..

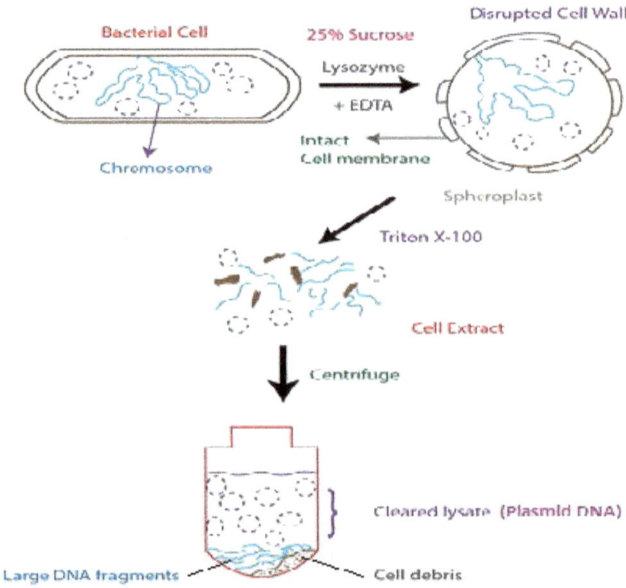

Separation of plasmid DNA on the basis of size.

Separation Based on Conformation

Plasmids are supercoiled molecules formed by partial unwinding of double helix of the plasmid DNA during the plasmid replication process by enzymes called topoisomerases. The supercoiled conformation can be maintained when both polynucleotide strands are intact, hence called covalently closed-circular (ccc) DNA. If one of the polynucleotide strands is broken, the double helix reverts to its normal relaxed state taking an alternative conformation, called open-circular (oc). Super coiling is important in plasmid preparation due to the easy separation of supercoiled molecules from non-supercoiled ones.

The commonly used methods of separation based on conformation arc as follows-

(a) Alkaline denaturation method

- This method is based on maintaining a very narrow pH range for the denaturation of non-supercoiled DNA but not the supercoiled plasmid.
- Addition of sodium hydroxideto cell extract or cleared lysate (pH12.0-12.5) results in disruption of the hydrogen bonds of non-supercoiled DNA molecules.
- As a result, the double helix unwinds and two polynucleotide chains separate.
- Further addition of acid causes the aggregation of these denatured bacterial DNA strands into a tangled mass which can be pelleted by centrifugation, leaving plasmid DNA in the supernatant.

Advantage

- Most of the RNA and protein under defined conditions (specifically cell lysis by SDS and neutralization with sodium acetate) can be removed by the centrifugation.
- No requirement of organic extraction.

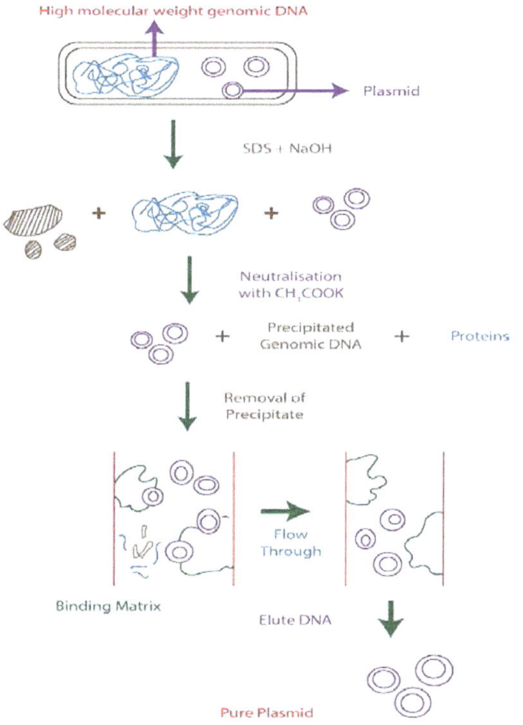

(a): Separation of plasmid DNA by Alkaline denaturation method

(b) Ethidium bromide-cesium chloride density gradient centrifugation

- Density gradient centrifugation can separate DNA, RNA and protein. It is a very efficient method for obtaining pure plasmid DNA.
- A density gradient is produced by centrifuging a solution of cesium chloride at a very high speed which pulls the CsCl ions towards the bottom. This process is referred as isopycnic centrifugation.
- The DNA migrates to the point at which it has density similar to that of CsCI i.e.1.7 g/cm^3 in the gradient.
- In contrast, protein molecules having lower buoyant densities float at the top of the tube whereas RNA gets pelleted at the bottom.

Density gradient centrifugation in the presence of ethidium bromide (EtBr) can be used to separate supercoiled DNA from non-super coiled molecules. Ethidium bromide is an intercalating dye that binds to DNA molecules causing partial unwinding of the double helix. Supercoiled DNA have very little freedom to unwind due to absence of free ends and bind to a limited amount of EtBr resulting in very less decrease in buoyant density (0.085 g/cm^3) than that of linear DNA (0.125 g/cm^3). As a result, they form a distinctb and separated from the linear bacterial DNA. The EtBr bound to DNA is then extracted by n-butanol and the CsCl is removed by dialysis.

Isolation and Purification of RNA

RNA (Ribonucleic acid) is a polymeric substance consisting of a long single-stranded chain of phosphate and ribose units with the nitrogen bases adenine, guanine, cytosine and uracil bonded to the ribose sugar present in living cells and many viruses. The steps for preparation of RNA involve homogenization, phase separation, RNA precipitation, washing and re-dissolving RNA.

The method for isolation and purification of RNA are as follows-

1. Organic extraction method
2. Filter-based, spin basket formats
3. Magnetic particle methods
4. Direct lysis method.

Organic Extraction Method

This method involves phase separation by addition and centrifugation of a mixture of a solution containing phenol, chloroform and a chaotropic agent (guanidinium thiocyanate) and aqueous sample. Guanidium thiocyanate results in the denaturation of proteins and RNases, separating rRNA from ribosomes. Addition of chloroform forms a colorless upper aqueous phase containing RNA, an interphase containing DNA and a lower phenol-chloroform phase containing protein. RNA is collected from the upper aqueous phase by alcohol (2-propanol or ethanol) precipitation followed by rehydration.

One of the advantages of this method is the stabilization of RNA and rapid denaturation of nucleases. Besides advantages, it has several drawbacks such as it is difficult to automate, needs labor and manual intensive processing, and use of chlorinated organic reagents.

Direct Lysis Methods

This method involves use of lysis buffer under specified conditions for the disruption of sample and stabilization of nucleic acids. If desired, samples can also be purified from stabilized lysates. This method eliminates the need of binding and elution from solid surfaces and thus avoids bias and recovery efficiency effects.

Advantages

- Extremely fast and easy.
- Highest ability for precise RNA representation.
- Easy to work on very small samples.
- Amenable to simple automation.

Drawbacks

- Unable to perform traditional analytical methods (e.g. spectrophotometric method).
- Dilution-based (most useful with concentrated samples).
- Potential for suboptimal performance unless developed/optimized with downstream analysis.
- Potential for residual RNase activity if lysates are not handled properly.

Quantification of Nucleic Acids

Quantification of nucleic acids is done to determine the average concentrations of DNA or RNA present in a mixture, as well as their purity. The accurate measurement is based on sensitivity, specificity and interference by contaminants. Various methods that can be employed to quantify the nucleic acid concentration are listed below,

(i) Spectrophotometric analysis

(ii) Nanodrop

(iii) Fluorescence based method

(iv) Fluorescence in situ hybridization (FISH)

Spectrophotometric Analysis

It is a simple and accurate method to assess the concentration and purity of nucleic acids based on their absorption at different wavelengths. According to Beer Lambert's Law, when light is passed through a substance of concentration c at a pathlength l (typically 1 cm), the absorbance is directly proportional to the concentration of the substance and the pathlength i.e.

$A = \epsilon cl$

Where, A= absorbance of nucleic acids at a particular wavelength

ϵ=Molar extinction coefficient (M^{-1} cm^{-1}) or specific absorption coefficient $(\mu g/ml)^{-1}$ cm^{-1} that measures how strongly a substance absorbs light at a particular wavelength.

l = Pathlength of the spectrophotometer cuvette

c= concentration of a substance

The value of $\epsilon$for ss-DNA is 0.027; ds-DNA is 0.020 and ss-RNA is 0.025.

The concentration of nucleic acids can be calculatedby the given equations-

$$\text{DNA concentration} (\mu g/ml) = \text{OD } 260 \frac{\text{x100 (dilution factor) x } 50\mu g/ml}{100}$$

For ssDNA, c (pmol/µl) = A_{260} /10 x S

c(µg/ml)= A_{260} /0.027

For dsDNA, c (pmol/µl) = A_{260} /13.2 x S

c (µg/ml)=A_{260} /0.020

For ssRNA, c (µg/ml)=A_{260} /0.025

For Oligonucleotide, absorbance is significantly affected by composition of oligonucleotide bases.c (pmol/µl) = A_{260} x 100/1.5 N_A + 0.71 N_C + 1.20 N_G+ 0.84 N_T

where S denotes the size of DNA in Kb and N represents the number of bases A, G, C or T.

The value of for the DNA bases is listed in the Table.

Bases	ϵ(at 260 nm) (M-1 cm-1)
Adenine (A)	15,200
Cytosine (C)	7,050
Guanine (G)	12,010
Thymine (T)	8,400

The purity of the nucleic acids can be determined by A260/A280 ratio.

Nanodrop Method

Unlike spectrophotometric method requiring 1-2 ml of the sample, this technique involves micro-volume (1-2µl) quantification of nucleic acid sample. The nucleic acids having the concentration range from 2-15000 ng/µl can be assessed by this method. For this, 1-2µl of sample is loaded between the two optical surfaces and the software automatically calculates the concentration and purity of the nucleic acid and displays sample quality (purity) as a spectral output. Generally the pathlength is taken as 1 cm even though in case of Hellma tray cell cuvette pathlength ranges from 0.2-1.0 mm during measurement cycle. Sensitivity measured by sample retention system of Nano-Drop is increased by many folds compared to a standard spectrophotometer. Automatic pathlength adjustment facilitates the direct quantification of concentrated samples avoiding the need for sample dilutions.

Quantification using Fluorescent Dyes

This method is simple and more sensitive than spectrophotometric method which measures the fluorescence intensity of the dyes that fluoresce upon interaction with the nucleic acids. Various fluorescent dyes such as EtBr, Hoechst 33258, picogreen. The sensitivity range of these dyes for the quantification is listed in Table. Quantification of nucleic acids separated by gel electrophoresis can be done by comparing the stained nucleic acids with stained standards of known concentration separated on the same gel. The dye:DNA complex shows greater fluorescence than the unbound dyeby intercalating between the stacked base pairs. The fluorescence intensity of the band estimates the concentration/amount of DNA. The intensity of the stain is dependent atleast in part on the base pair composition of the nucleic acid.

Nucleic acid	UV Absorbance (A_{260})	Fluorescence mathods		
		Hoechst 33258	EtBr	Picogreen
DNA	1-50 μg/mL	0.01-15 μg/mL	0.1-10 μg/mL	0.025-1000 ng/mL
RNA	1-50 μg/mL	Not applicable	1-40 μg/mL	Minimal sensitivity
Ratio of signal (DNA/RNA)	0.8	400	2.2	>100

Table. Sensitivity range of quantification of nucleic acids by UV spectrophotometry and fluorescence based methods.

Fluorescence in Situ Hybridization (FISH)

In a cytometric system, quantification of FISH signals and accurate estimation of fluorescence intensity of specific DNA sequences can be performed using an epi-fluorescence microscope with a multi-wavelength illuminator, fitted with a cooled charge couple device (CCD) camera. Quantitation by FISH was first applied as a basis for rudimentary cytogenetic assays. Detection using CCD camera enables the quantitative analysis of mRNA as well. The two factors affecting the fluorescence assay are reproducibility and signal irregularity as well as background noise which may vary from sample to sample and cell to cell. Various approaches have been developed to reduce background noise such as use of reducing agents e.g. sodium borohydride and pre-treatment by light irradiation, and image analysis methods. The former cannot be possible and effective always so the later are now employed for analysis. The output thus consist of a true signal and various noise components, the profile for each of them can be estimated and deleted by digital methods like independent component analysis. Several computational methods have been developed to overcome the differential intensity and color overlap caused by multi-color image analysis.

Further, FISH can be employed for interpretation using various approaches such as multi-color cytometry algotithms, dot-counting approaches, use of diagnostic probe sets etc.

Storage of Nucleic Acids

The purified DNA can be stored at -20°C or -70°C under slightly basic conditions (e.g., Tris- Cl, pH 8.0) as acidic conditions result in hydrolysis of DNA.

Diluted solutions of nucleic acids can be stored in aliquots and thawed once only. RNA preservation under frozen conditions is helpful.

Purified RNA can be stored at -20°C or -80°C in RNase-free solution such as-

- The RNA Storage Solution (1 mM sodium citrate, pH 6.4 ± 0.2): It is a buffer that delivers greater RNA stability than 0.1 mM EDTA or TE. The presence of sodium citrate and low pH minimizes base hydrolysis of RNA. Sodium citrate acts both as a chelating and buffering agent.
- 0.1 mM EDTA
- TE Buffer (10 mM Tris-HCl, 1 mM EDTA, pH 7.0)

This RNase-free solution is compatible with all RNA applications including *in vitro* translation, reverse transcription, nuclease protection assays and northern analysis.

cDNA Library

A cDNA library is a combination of cloned cDNA (complementary DNA) fragments inserted into a collection of host cells, which together constitute some portion of the transcriptome of the organism and stored as a "library". cDNA is produced from fully transcribed mRNA found in the nucleus and therefore contains only the expressed genes of an organism. Similarly, tissue-specific cDNA libraries can be produced. In eukaryotic cells the mature mRNA is already spliced, hence the cDNA produced lacks introns and can be readily expressed in a bacterial cell. While information in cDNA libraries is a powerful and useful tool since gene products are easily identified, the libraries lack information about enhancers, introns, and other regulatory elements found in a genomic DNA library.

cDNA Library Construction

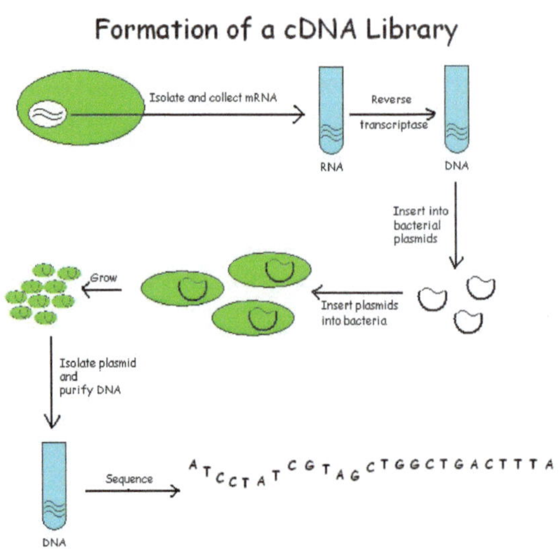

Formation of a cDNA library.

cDNA is created from a mature mRNA from a eukaryotic cell with the use of reverse transcriptase. In eukaryotes, a poly-(A) tail (consisting of a long sequence of adenine nucleotides) distinguishes mRNA from tRNA and rRNA and can therefore be used as a primer site for reverse transcription. This has the problem that not all transcripts, such as those for the histone, encode a poly-A tail.

mRNA Extraction

Firstly, the mRNA is obtained and purified from the rest of the RNAs. Several methods exist for purifying RNA such as trizol extraction and column purification. Column purification is done by using oligomeric dT nucleotide coated resins where only the mRNA having the poly-A tail will bind. The rest of the RNAs are eluted out. The mRNA is eluted by using eluting buffer and some heat to separate the mRNA strands from oligo-dT.

cDNA Construction

Once mRNA is purified, *oligo-dT* (a short sequence of deoxy-thymidine nucleotides) is tagged as a complementary primer which binds to the poly-A tail providing a free 3'-OH end that can be extended by reverse transcriptase to create the complementary DNA strand. Now, the mRNA is removed by using a RNAse enzyme leaving a single stranded cDNA (sscDNA). This sscDNA is converted into a double stranded DNA with the help of DNA polymerase. However, for DNA polymerase to synthesize a complementary strand a free 3'-OH end is needed. This is provided by the sscDNA itself by generating a *hairpin loop* at the 3' end by coiling on itself. The polymerase extends the 3'-OH end and later the loop at 3' end is opened by the scissoring action of S_1 *nuclease*. Restriction endonucleases and DNA ligase are then used to clone the sequences into bacterial plasmids.

The cloned bacteria are then selected, commonly through the use of antibiotic selection. Once selected, stocks of the bacteria are created which can later be grown and sequenced to compile the cDNA library.

cDNA Library uses

cDNA libraries are commonly used when reproducing eukaryotic genomes, as the amount of information is reduced to remove the large numbers of non-coding regions from the library. cDNA libraries are used to express eukaryotic genes in prokaryotes. Prokaryotes do not have introns in their DNA and therefore do not possess any enzymes that can cut it out during transcription process. cDNA does not have introns and therefore can be expressed in prokaryotic cells. cDNA libraries are most useful in reverse genetics where the additional genomic information is of less use. Also, it is useful for subsequently isolating the gene that codes for that mRNA.

cDNA Library vs. Genomic DNA Library

cDNA library lacks the non-coding and regulatory elements found in genomic DNA. Genomic DNA libraries provide more detailed information about the organism, but are more resource-intensive to generate and maintain.

Cloning of cDNA

cDNA molecules can be cloned by using restriction site linkers. Linkers are short, double stranded pieces of DNA (oligodeoxyribonucleotide) about 8 to 12 nucleotide pairs long that include a restriction endonuclease cleavage site e.g. BamHI. Both the cDNA and the linker have blunt ends which can be ligated together using a high concentration of T4 DNA ligase. Then sticky ends are produced in the cDNA molecule by cleaving the cDNA ends (which now have linkers with an incorporated site) with the appropriate endonuclease. A cloning vector (plasmid) is then also cleaved with the appropriate endonuclease. Following "sticky end" ligation of the insert into the vector the resulting recombinant DNA molecule is transferred into *E. coli* host cell for cloning.

Genomic Library

A genomic library is a collection of the total genomic DNA from a single organism. The DNA is stored in a population of identical vectors, each containing a different insert of DNA. In order to construct a genomic library, the organism's DNA is extracted from cells and then digested with a restriction enzyme to cut the DNA into fragments of a specific size. The fragments are then inserted into the vector using DNA ligase. Next, the vector DNA can be taken up by a host organism - commonly a population of Escherichia coli or yeast - with each cell containing only one vector molecule. Using a host cell to carry the vector allows for easy amplification and retrieval of specific clones from the library for analysis.

There are several kinds of vectors available with various insert capacities. Generally, libraries made from organisms with larger genomes require vectors featuring larger inserts, thereby fewer vector molecules are needed to make the library. Researchers can choose a vector also considering the ideal insert size to find a desired number of clones necessary for full genome coverage.

Genomic libraries are commonly used for sequencing applications. They have played an important role in the whole genome sequencing of several organisms, including the human genome and several model organisms.

History

The first DNA-based genome ever fully sequenced was achieved by two-time Nobel Prize winner, Frederick Sanger, in 1977. Sanger and his team of scientists created a library of the bacteriophage, phi X 174, for use in DNA sequencing. The importance of this success contributed to the ever-increasing demand for sequencing genomes to research gene therapy. Teams are now able to catalog polymorphisms in genomes and investigate those candidate genes contributing to maladies such as Parkinson's disease, Alzheimer's disease, multiple sclerosis, rheumatoid arthritis, and Type 1 diabetes. These are due to the advance of genome-wide association studies from the ability to create and sequence genomic libraries. Prior, linkage and candidate-gene studies were some of the only approaches.

Genomic Library Construction

Construction of a genomic library involves creating many recombinant DNA molecules. An organism's genomic DNA is extracted and then digested with a restriction enzyme. For organisms with very small genomes *(~10 kb)*, the digested fragments can be separated by gel electrophoresis. The separated fragments can then be excised and cloned into the vector separately. However, when a large genome is digested with a restriction enzyme, there are far too many fragments to excise individually. The entire set of fragments must be cloned together with the vector, and separation of clones can occur after. In either case, the fragments are ligated into a vector that has been digested with the same restriction enzyme. The vector containing the inserted fragments of genomic DNA can then be introduced into a host organism.

Below are the steps for creating a genomic library from a large genome.

1. Extract and purify DNA.
2. Digest the DNA with a restriction enzyme. This creates fragments that are similar in size, each containing one or more genes.
3. Insert the fragments of DNA into vectors that were cut with the same restriction enzyme. Use the enzyme DNA ligase to seal the DNA fragments into the vector. This creates a large pool of recombinant molecules.
4. These recombinant molecules are taken up by a host bacterium by transformation, creating a DNA library.

Below is a diagram of the above outlined steps.

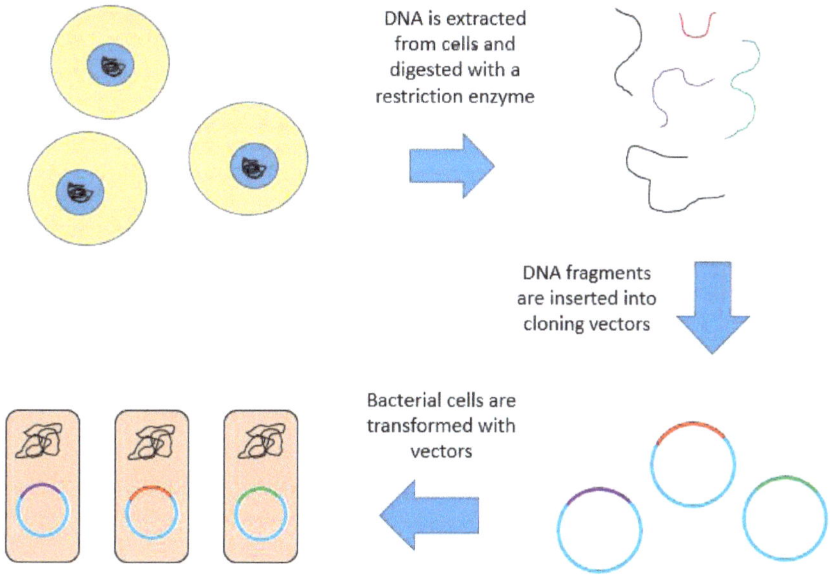

Genomic Library Construction

Determining Titer of Library

After a genomic library is constructed with a viral vector, such as lambda phage, the titer of the library can be determined. Calculating the titer allows researchers to approximate how many infectious viral particles were successfully created in the library. To do this, dilutions of the library are used to transfect cultures of E. coli of known concentrations. The cultures are then plated on agar plates and incubated overnight. The number of viral plaques are counted and can be used to calculate the total number of infectious viral particles in the library. Most viral vectors also carry a marker that allows clones containing an insert to be distinguished from those that do not have an insert. This allows researchers to also determine the percentage of infectious viral particles actually carrying a fragment of the library.

A similar method can be used to titer genomic libraries made with non-viral vectors, such as plasmids and BACs. A test ligation of the library can be used to transform E. coli. The transformation is then spread on agar plates and incubated overnight. The titer of the transformation is determined by counting the number of colonies present on the plates. These vectors generally have a selectable marker allowing the differentiation of clones containing an insert from those that do not. By doing this test, researchers can also determine the efficiency of the ligation and make adjustments as needed to ensure they get the desired number of clones for the library.

Screening Library

In order to isolate clones that contain regions of interest from a library, the library must first be screened. One method of screening is hybridization. Each transformed host cell of a library will contain only one vector with one insert of DNA. The whole library can be plated onto a filter over media. The filter and colonies are prepared for hybridization and then labeled with a probe. The target DNA- insert of interest- can be identified by detection such as autoradiography because of the hybridization with the probe as seen below.

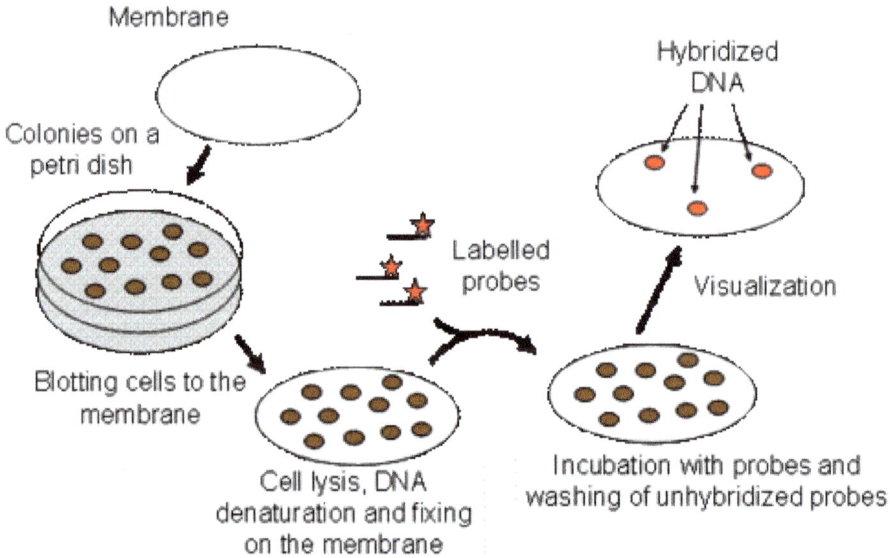

Colony Blot Hybridization

Another method of screening is with polymerase chain reaction (PCR). Some libraries are stored as pools of clones and screening by PCR is an efficient way to identify pools containing specific clones.

Types of Vectors

Genome size varies among different organisms and the cloning vector must be selected accordingly. For a large genome, a vector with a large capacity should be chosen so that a relatively small number of clones are sufficient for coverage of the entire genome. However, it is often more difficult to characterize an insert contained in a higher capacity vector.

Below is a table of several kinds of vectors commonly used for genomic libraries and the insert size that each generally holds.

Vector type	Insert size (thousands of bases)
Plasmids	up to 15
Phage lambda (λ)	up to 25
Cosmids	up to 45
Bacteriophage P1	70 to 100
P1 artificial chromosomes (PACs)	130 to 150
Bacterial artificial chromosomes (BACs)	120 to 300
Yeast artificial chromosomes (YACs)	250 to 2000

Plasmids

A plasmid is a double stranded circular DNA molecule commonly used for molecular cloning. Plasmids are generally 2 to 4 kilobase-pairs (kb) in length and are capable of carrying inserts up to 15kb. Plasmids contain an origin of replication allowing them to replicate inside a bacterium independently of the host chromosome. Plasmids commonly carry a gene for antibiotic resistance that allows for the selection of bacterial cells containing the plasmid. Many plasmids also carry a reporter gene that allows researchers to distinguish clones containing an insert from those that do not.

Phage Lambda (λ)

Phage λ is a double-stranded DNA virus that infects *E. coli*. The λ chromosome is 48.5kb long and can carry inserts up to 25kb. These inserts replace non-essential viral sequences in the λ chromosome, while the genes required for formation of viral particles and infection remain intact. The insert DNA is replicated with the viral DNA; thus, together they are packaged into viral particles. These particles are very efficient at infection and multiplication leading to a higher production of the recombinant λ chromosomes. However, due to the smaller insert size, libraries made with λ phage may require many clones for full genome coverage.

Cosmids

Cosmid vectors are plasmids that contain a small region of bacteriophage λ DNA called the cos sequence. This sequence allows the cosmid to be packaged into bacteriophage λ particles. These particles- containing a linearized cosmid- are introduced into the host cell by transduction. Once inside the host, the cosmids circularize with the aid of the host's DNA ligase and then function as plasmids. Cosmids are capable of carrying inserts up to 45kb in size.

Bacteriophage P1 Vectors

Bacteriophage P1 vectors can hold inserts 70 – 100kb in size. They begin as linear DNA molecules packaged into bacteriophage P1 particles. These particles are injected into an E. coli strain expressing Cre recombinase. The linear P1 vector becomes circularized by recombination between two loxP sites in the vector. P1 vectors generally contain a gene for antibiotic resistance and a positive selection marker to distinguish clones containing an insert from those that do not. P1 vectors also contain a P1 plasmid replicon, which ensures only one copy of the vector is present in a cell. However, there is a second P1 replicon- called the P1 lytic replicon- that is controlled by an inducible promoter. This promoter allows the amplification of more than one copy of the vector per cell prior to DNA extraction.

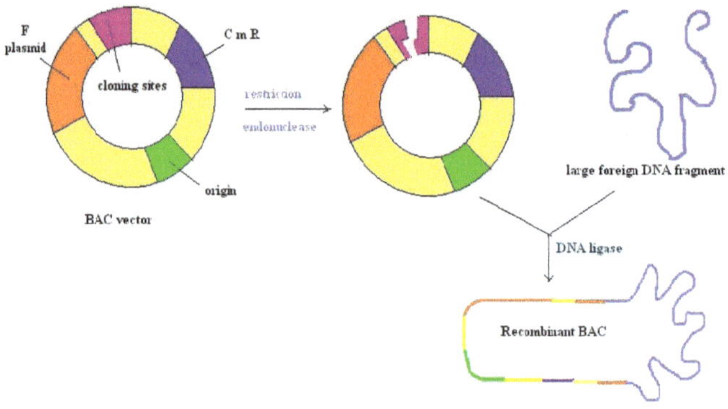

Bac vector

P1 Artificial Chromosomes

P1 artificial chromosomes (PACs) have features of both P1 vectors and Bacterial Artificial Chromosomes (BACs). Similar to P1 vectors, they contain a plasmid and a lytic replicon as described above. Unlike P1 vectors, they do not need to be packaged into bacteriophage particles for transduction. Instead they are introduced into E. coli as circular DNA molecules through electroporation just as BACs are. Also similar to BACs, these are relatively harder to prepare due to a single origin of replication.

Bacterial Artificial Chromosomes

Bacterial artificial chromosomes (BACs) are circular DNA molecules, usually about 7kb in length, that are capable of holding inserts up to 300kb in size. BAC vectors contain a replicon derived from E. coli F factor, which ensures they are maintained at one copy per cell. Once an insert is ligated into a BAC, the BAC is introduced into recombination deficient strains of E. coli by electroporation. Most BAC vectors contain a gene for antibiotic resistance and also a positive selection marker. The figure to the right depicts a BAC vector being cut with a restriction enzyme, followed by the insertion of foreign DNA that is re-annealed by a ligase. Overall, this is a very stable vector, but they may be hard to prepare due to a single origin of replication just like PACs.

Yeast Artificial Chromosomes

Yeast artificial chromosomes (YACs) are linear DNA molecules containing the necessary features of an authentic yeast chromosome, including telomeres, a centromere, and an origin of replication. Large inserts of DNA can be ligated into the middle of the YAC so that there is an "arm" of the YAC on either side of the insert. The recombinant YAC is introduced into yeast by transformation; selectable markers present in the YAC allow for the identification of successful transformants. YACs can hold inserts up to 2000kb, but most YAC libraries contain inserts 250-400kb in size. Theoretically there is no upper limit on the size of insert a YAC can hold. It is the quality in the preparation of DNA used for inserts that determines the size limit. The most challenging aspect of using YAC is the fact they are prone to rearrangement.

How to Select a Vector

Vector selection requires one to ensure the library made is representative of the entire genome. Any insert of the genome derived from a restriction enzyme should have an equal chance of being in the library compared to any other insert. Furthermore, recombinant molecules should contain large enough inserts ensuring the library size is able to be handled conveniently. This is particularly determined by the amount of clones needed to have in a library. The amount of clones to get a sampling of all the genes is determined by the size of the organism's genome as well as the average insert size. This is represented by the formula (also known as the Carbon and Clarke formula):

$$N = \frac{ln(1-P)}{ln(1-f)}$$

where,

N is the necessary number of recombinants

P is the desired probability that any fragment in the genome will occur at least once in the library created

f is the fractional proportion of the genome in a single recombinant

f can be further shown to be:

$$f = \frac{i}{g}$$

where,

i is the insert size

g is the genome size

Thus, increasing the insert size (by choice of vector) would allow for fewer clones needed to represent a genome. The proportion of the insert size versus the genome size represents the proportion of the respective genome in a single clone. Here is the equation with all parts considered:

$$N = \frac{ln(1-P)}{ln(1-\frac{i}{g})}$$

Vector Selection Example

The above formula can be used to determine the 99% confidence level that all sequences in a genome are represented by using a vector with an insert size of twenty thousand basepairs (such as the phage lambda vector). The genome size of the organism is three billion basepairs in this example.

$$N = \frac{ln(1-0.99)}{ln[1-\frac{2.0\times 10^4\, basepairs}{3.0\times 10^9\, basepairs}]}$$

$$N = \frac{-4.61}{-6.7\times 10^{-6}}$$

$$N = 688,060\,\text{clones}$$

Thus, approximately 688,060 clones are required to ensure a 99% probability that a given DNA sequence from this three billion basepair genome will be present in a library using a vector with an insert size of twenty thousand basepairs.

Applications

After a library is created, the genome of an organism can be sequenced to elucidate how genes affect an organism or to compare similar organisms at the genome-level. The aforementioned genome-wide association studies can identify candidate genes stemming from many functional traits. Genes can be isolated through genomic libraries and used on human cell lines or animal models to further research. Furthermore, creating high-fidelity clones with accurate genome representation- and no stability issues- would contribute well as intermediates for shotgun sequencing or the study of complete genes in functional analysis.

Hierarchical Sequencing

One major use of genomic libraries is hierarchichal shotgun sequencing, which is also called top-down, map-based or clone-by-clone sequencing. This strategy was developed in the 1980s for sequencing whole genomes before high throughput techniques for sequencing were available. Individual clones from genomic libraries can be sheared into smaller fragments, usually 500bp to 1000bp, which are more manageable for sequencing. Once a clone from a genomic library is sequenced, the sequence can be used to screen the library for other clones containing inserts which overlap with the sequenced clone. Any new overlapping clones can then be sequenced forming a contig. This technique, called chromosome walking, can be exploited to sequence entire chromosomes.

Whole genome shotgun sequencing is another method of genome sequencing that does not require a library of high-capacity vectors. Rather, it uses computer algorithms to assemble short sequence reads to cover the entire genome. Genomic libraries are often used in combination with whole genome shotgun sequencing for this reason. A high resolution map can be created by sequencing both ends of inserts from several clones in a genomic library. This map provides sequences of known distances apart, which can be used to help with the assembly of sequence reads acquired through shotgun sequencing. The human genome sequence, which was declared complete in 2003, was assembled using both a BAC library and shotgun sequencing.

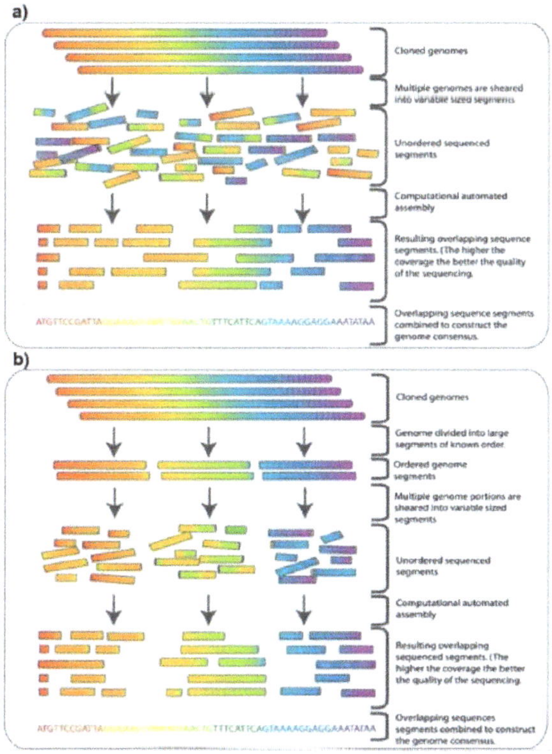

Whole genome shotgun sequencing versus Hierarchical shotgun sequencing

Genome-wide Association Studies

Genome-wide association studies are general applications to find specific gene targets and polymorphisms within the human race. In fact, the International HapMap project was created through a partnership of scientists and agencies from several countries to catalog and utilize this data. The goal of this project is to compare genetic sequences of different individuals to elucidate similarities and differences within chromosomal regions. Scientists from all of the participating nations are cataloging these attributes with data from populations of African, Asian, and European ancestry. Such genome-wide assessments may lead to further diagnostic and drug therapies while also helping future teams focus on orchestrating therapeutics with genetic features in mind. These concepts are already being exploited in genetic engineering. For example, a research team has actually constructed a PAC shuttle vector that creates a library representing two-fold coverage of the human genome. This could serve as an incredible resource to identify genes, or sets of genes, causing disease. Moreover, these studies can serve as a powerful way to investigate transcriptional regulation as it has been seen in the study of baculoviruses. Overall, advances in genome library construction and DNA sequencing has allowed for efficient discovery of different molecular targets. Assimilation of these features through such efficient methods can hasten the employment of novel drug candidates.

Library screening is the process of identification of the clones carrying the gene of interest. Screening relies on a unique property of a clone in a library. The DNA libraries consist of a collection of probably many thousand clones in the form of either plaques or colonies on a plate. Screening of libraries can be done by following approaches based on-

- Detecting a particular DNA sequence and
- Gene expression.

Methods for Screening Based on Detecting a DNA Sequence

Screening by Hybridization

- Nucleic acid hybridization is the most commonly used method of library screening first developed by Grunstein and Hogness in1975 to detect DNA sequences in transformed colonies using radioactive RNA probes.
- It relies on the fact that a single-stranded DNA molecule, used as a probe can hybridize to its complementary sequence and identify the specific sequences.
- This method is quick, can handle a very large number of clones and used in the identification of cDNA clones which are not full-length (and therefore cannot be expressed).

The commonly used methods of hybridization are:

a) Colony hybridization

b) Plaque hybridization.

(a) Colony Hybridization

Colony hybridization, also known as replica plating, allows the screening of colonies plated at high density using radioactive DNA probes. This method can be used to screen plasmid or cosmid based libraries.

(b) Plaque Hybridization

Plaque hybridization, also known as *Plaque lift*, was developed by Benton and Davis in 1977 and employs a filter lift method applied to phage plaques. This procedure is successfully applied to the isolation of recombinant phage by nucleic acid hybridization and probably is the most widely applied method of library screening. The method of screening library by plaque hybridization is described below-

- The nitrocellulose filter is applied to the upper surface of agar plates, making a direct contact between plaques and filter.
- The plaques contain phage particles, as well as a considerable amount of unpackaged recombinant DNA which bind to the filter.
- The DNA is denatured, fixed to the filter, hybridized with radioactive probes and assayed by autoradiography.

Advantages

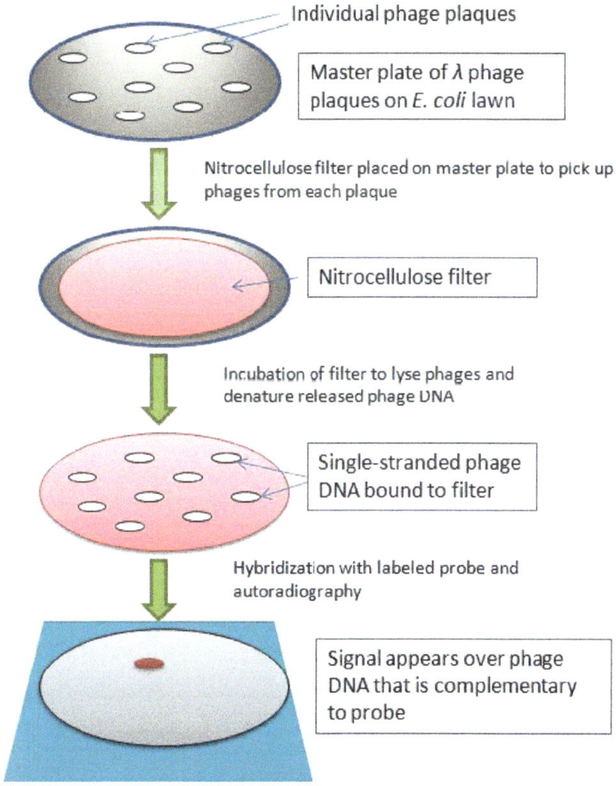

(b): Schematic process for screening libraries by Plaque hybridization.

- This method results in a 'cleaner' background and distinct signal (less background probe hybridization) for λ plaque screening due to less DNA transfer from the bacterial host to the nitrocellulose membrane while lifting plaques rather than bacterial colonies.
- Multiple screens can be performed from the same plate as plaques can be lifted several times.
- Screening can be performed at very high density by screening small plaques. High-density screening has the advantage that a large number of recombinant clones can be screened for the presence of sequences homologous to the probe in a single experiment.

Probes used for Hybridization

Cloned DNA fragments can be used as probes in hybridization reactions if a cDNA clone is available. DNA or synthetic oligonucleotide probes can be used for identification of a clone from a genomic library instead of RNA probes, for example, to study the regulatory sequences which are not part of the cDNA clone. A common method of labeling probes is the incorporation of a radioactive or other marker into the molecule. A number of alternative labeling methods are also available that involve an amplification process to detect the presence of small quantities of bound probe and avoid the use of radioactivity. These methods involve the incorporation of chemical labels such as digoxigenin or biotin into the probe which can be detected with a specific antibody or the ligand streptavidin, respectively.

Screening by PCR

PCR screening is employed for the identification of rare DNA sequences in complex mixtures of molecular clones by increasing the abundance of a particular sequence. It is possible to identify any clone by PCR only if there is available information about its sequence to design suitable primers.

DNA Sequencing Theory

DNA sequencing theory is the broad body of work that attempts to lay analytical foundations for determining the order of specific nucleotides in a sequence of DNA, otherwise known as DNA sequencing. The practical aspects revolve around designing and optimizing sequencing projects (known as "strategic genomics"), predicting project performance, troubleshooting experimental results, characterizing factors such as sequence bias and the effects of software processing algorithms, and comparing various sequencing methods to one another. In this sense, it could be considered a branch of systems engineering or operations research. The permanent archive of work is primarily mathematical, although numerical calculations are often conducted for particular problems too. DNA sequencing theory addresses *physical processes* related to sequencing DNA and should not be confused with theories of analyzing resultant DNA sequences, e.g. sequence alignment. Publications sometimes do not make a careful distinction, but the latter are primarily concerned with algorithmic issues. Sequencing theory is based on elements of mathematics, biology, and systems engineering, so it is highly interdisciplinary. The subject may be studied within the context of computational biology.

Theory and Sequencing Strategies

Sequencing as a Covering Problem

All mainstream methods of DNA sequencing rely on reading small fragments of DNA and subsequently reconstructing these data to infer the original DNA target, either via assembly or alignment to a reference. The abstraction common to these methods is that of a mathematical covering problem. For example, one can imagine a line segment representing the target and a subsequent process where smaller segments are "dropped" onto random locations of the target. The target is considered "sequenced" when adequate coverage accumulates (e.g., when no gaps remain).

The abstract properties of covering have been studied by mathematicians for over a century. However, direct application of these results has not generally been possible. Closed-form mathematical solutions, especially for probability distributions, often cannot be readily evaluated. That is, they involve inordinately large amounts of computer time for parameters characteristic of DNA sequencing. Stevens' configuration is one such example. Results obtained from the perspective of pure mathematics also do not account for factors that are actually important in sequencing, for instance detectable overlap in sequencing fragments, double-stranding, edge-effects, and target multiplicity. Consequently, development of sequencing theory has proceeded more according to the philosophy of applied mathematics. In particular, it has been problem-focused and makes expedient use of approximations, simulations, etc.

Early uses Derived From Elementary Probability Theory

The earliest result may be found directly from elementary probability theory. Suppose we model the above process taking L and G as the fragment length and target length, respectively. The probability of "covering" any given location on the target *with one particular fragment* is then L / G. (This presumes $L \ll G$, which is valid often, but not for all real-world cases.) The probability of a single fragment not covering a given location on the target is therefore $1 - L / G$, and $[1 - L / G]^N$ for N fragments. The probability of covering a given location on the target with *at least one* fragment is therefore

$$P = 1 - \left[1 - \frac{L}{G}\right]^N .$$

This equation was first used to characterize plasmid libraries, but it may appear in a modified form. For most projects $N \gg 1$, so that, to a good degree of approximation

$$\left[1 - \frac{L}{G}\right]^N \sim \exp(-NL / G),$$

where $R = NL / G$ is called the *redundancy*. Note the significance of redundancy as representing the average number of times a position is covered with fragments. Note also that in considering the covering process over all positions in the target, this probability is identical to the expected value of the random variable C, the fraction of the target coverage. The final result,

$$E\langle C\rangle = 1 - e \quad ,$$

remains in widespread use as a "back of the envelope" estimator and predicts that coverage for all projects evolves along a universal curve that is a function only of the redundancy.

Lander-Waterman Theory

In 1988, Eric Lander and Michael Waterman published an important paper examining the covering problem from the standpoint of gaps. Although they focused on the so-called mapping problem, the abstraction to sequencing is much the same. They furnished a number of useful results that were adopted as the standard theory from the earliest days of "large-scale" genome sequencing. Their model was also used in designing the Human Genome Project and continues to play an important role in DNA sequencing.

Ultimately, the main goal of a sequencing project is to close all gaps, so the "gap perspective" was a logical basis of developing a sequencing model. One of the more frequently used results from this model is the expected number of contigs, given the number of fragments sequenced. If one neglects the amount of sequence that is essentially "wasted" by having to detect overlaps, their theory yields

$$E\langle contigs\rangle = Ne^{-R}.$$

In 1995, Roach published improvements to this theory, enabling it to be applied to sequencing projects in which the goal was to completely sequence a target genome. Michael Wendl and Bob Waterston confirmed, based on Stevens' method, that both models produced similar results when the number of contigs was substantial, such as in low coverage mapping or sequencing projects. As sequencing projects ramped up in the 1990s, and projects approached completion, low coverage approximations became inadequate, and the exact model of Roach was necessary. However, as the cost of sequencing dropped, parameters of sequencing projects became easier to directly test empirically, and interest and funding for strategic genomics diminished.

The basic ideas of Lander–Waterman theory led to a number of additional results for particular variations in mapping techniques. However, technological advancements have rendered mapping theories largely obsolete except in organisms other than highly studied model organisms (e.g., yeast, flies, mice, and humans).

Parking Strategy

The parking strategy for sequencing resembles the process of parking cars along a curb. Each car is a sequenced clone, and the curb is the genomic target. Each clone sequenced is screened to ensure that subsequently sequenced clones do not overlap any previously sequenced clone. No sequencing effort is redundant in this strategy. However, much like the gaps between parked cars, unsequenced gaps less than the length of a clone accumulate between sequenced clones. There can be considerable cost to close such gaps.

Pairwise End-sequencing

In 1995, Roach *et al.* proposed and demonstrated through simulations a generalization of a set of strategies explored earlier by Edwards and Caskey. This whole-genome sequencing method became immensely popular as it was championed by Celera and used to sequence several model organisms before Celera applied it to the human genome. Today, most sequencing projects employ this strategy, often called paired end sequencing.

Post Human Genome Project advancements

The physical processes and protocols of DNA sequencing have continued to evolve, largely driven by advancements in bio-chemical methods, instrumentation, and automation techniques. There is now a wide range of problems that DNA sequencing has made in-roads into, including metagenomics and medical (cancer) sequencing. There are important factors in these scenarios that classical theory does not account for. Recent work has begun to focus on resolving the effects of some of these issues. The level of mathematics becomes commensurately more sophisticated.

Various Artifacts of Large-insert Sequencing

Biologists have developed methods to filter highly-repetitive, essentially un-sequenceable regions of genomes. These procedures are important for organisms whose genomes consist mostly of such DNA, for example corn. They yield multitudes of small islands of sequenceable DNA products. Wendl and Barbazuk proposed an extension to Lander–Waterman Theory to account for "gaps" in the target due to filtering and the so-called "edge-effect". The latter is a position-specific sampling bias, for example the terminal base position has only a $1/G$ chance of being covered, as opposed to

L/G for interior positions. For $R<1$, classical Lander–Waterman Theory still gives good predictions, but dynamics change for higher redundancies.

Modern sequencing methods usually sequence both ends of a larger fragment, which provides linking information for *de novo* assembly and improved probabilities for alignment to reference sequence. Researchers generally believe that longer lengths of data (read lengths) enhance performance for very large DNA targets, an idea consistent with predictions from distribution models. However, Wendl showed that smaller fragments provide better coverage on small, linear targets because they reduce the edge effect in linear molecules. These findings have implications for sequencing the products of DNA filtering procedures. Read-pairing and fragment size evidently have negligible influence for large, whole-genome class targets.

Individual and Population Sequencing

Sequencing is emerging as an important tool in medicine, for example in cancer research. Here, the ability to detect heterozygous mutations is important and this can only be done if the sequence of the diploid genome is obtained. In the pioneering efforts to sequence individuals, Levy *et al.* and Wheeler *et al.*, who sequenced Craig Venter and Jim Watson, respectively, outlined models for covering both alleles in a genome. Wendl and Wilson followed with a more general theory that allowed for an arbitrary number of coverings of each allele and arbitrary ploidy. These results point to the general conclusion that the amount of data needed for such projects is significantly higher than for traditional haploid projects. Generally, at least 30-fold redundancy, i.e. each nucleotide spanned by an average of 30 sequence reads, is now standard. However, requirements can be even greater, depending upon what kinds of genomic events are to be found. For example, in the so-called "discordant read pairs method", DNA insertions can be inferred if the distance between read pairs is larger than expected. Calculations show that around 50-fold redundancy is needed to avoid false-positive errors at 1% threshold.

The advent of next-generation sequencing has also made large-scale population sequencing feasible, for example the 1000 Genomes Project to characterize variation in human population groups. While common variation is easily captured, rare variation poses a design challenge: too few samples with significant sequence redundancy risks not having a variant in the sample group, but large samples with light redundancy risk not capturing a variant in the read set that is actually in the sample group. Wendl and Wilson report a simple set of optimization rules that maximize the probability of discovery for a given set of parameters. For example, for observing a rare allele at least twice (to eliminate the possibility is unique to an individual) a little less than 4-fold redundancy should be used, regardless of the sample size.

Metagenomic Sequencing

Next-generation instruments are now also enabling the sequencing of whole uncultured metagenomic communities. The sequencing scenario is more complicated here and there are various ways of framing design theories for a given project. For example, Stanhope developed a probabilistic model for the amount of sequence needed to obtain at least one contig of a given size from each novel organism of the community, while Wendl et al. reported analysis for the average contig size or the probability of completely recovering a novel organism for a given rareness within the community. Conversely, Hooper et al. propose a semi-empirical model based on the Gamma distribution.

Limitations

DNA sequencing theories often invoke the assumption that certain random variables in a model are independent and identically distributed. For example, in Lander–Waterman Theory, a sequenced fragment is presumed to have the same probability of covering each region of a genome and all fragments are assumed to be independent of one another. In actuality, sequencing projects are subject to various types of bias, including differences of how well regions can be cloned, sequencing anomalies, biases in the target sequence (which is *not* random), and software-dependent errors and biases. In general, theory will agree well with observation up to the point that enough data have been generated to expose latent biases. The kinds of biases related to the underlying target sequence are particularly difficult to model, since the sequence itself may not be known *a priori*. This presents a type of "chicken and egg" closure problem.

Maxam–Gilbert Sequencing

Maxam–Gilbert sequencing is a method of DNA sequencing developed by Allan Maxam and Walter Gilbert in 1976–1977. This method is based on nucleobase-specific partial chemical modification of DNA and subsequent cleavage of the DNA backbone at sites adjacent to the modified nucleotides.

Maxam–Gilbert sequencing was the first widely adopted method for DNA sequencing, and, along with the Sanger dideoxy method, represents the first generation of DNA sequencing methods. Maxam–Gilbert sequencing is no longer in widespread use, having been supplanted by next-generation sequencing methods.

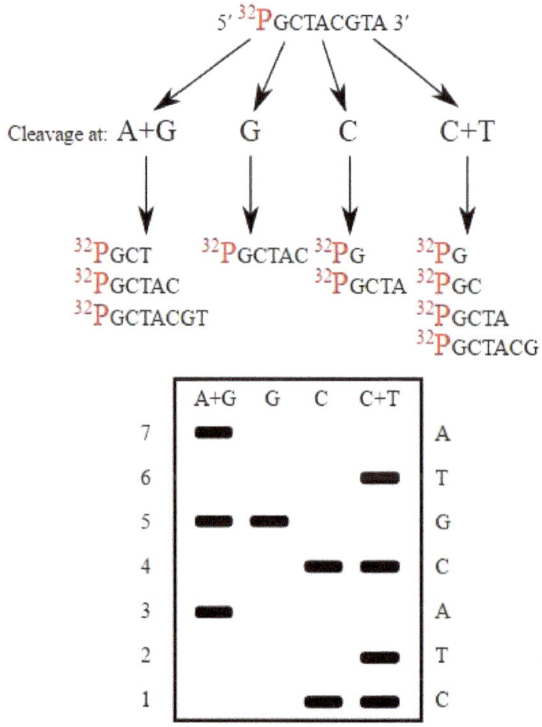

An example Maxam–Gilbert sequencing reaction. Cleaving the same tagged segment of DNA at different points yields tagged fragments of different sizes. The fragments may then be separated by gel electrophoresis.

History

Although Maxam and Gilbert published their chemical sequencing method two years after Frederick Sanger and Alan Coulson published their work on plus-minus sequencing, Maxam–Gilbert sequencing rapidly became more popular, since purified DNA could be used directly, while the initial Sanger method required that each read start be cloned for production of single-stranded DNA. However, with the improvement of the chain-termination method, Maxam–Gilbert sequencing has fallen out of favour due to its technical complexity prohibiting its use in standard molecular biology kits, extensive use of hazardous chemicals, and difficulties with scale-up.

Procedure

Maxam–Gilbert sequencing requires radioactive labeling at one 5′ end of the DNA fragment to be sequenced (typically by a kinase reaction using gamma-^{32}P ATP) and purification of the DNA. Chemical treatment generates breaks at a small proportion of one or two of the four nucleotide bases in each of four reactions (G, A+G, C, C+T). For example, the purines (A+G) are depurinated using formic acid, the guanines (and to some extent the adenines) are methylated by dimethyl sulfate, and the pyrimidines (C+T) are hydrolysed using hydrazine. The addition of salt (sodium chloride) to the hydrazine reaction inhibits the reaction of thymine for the C-only reaction. The modified DNAs may then be cleaved by hot piperidine;$(CH_2)_5NH$ at the position of the modified base. The concentration of the modifying chemicals is controlled to introduce on average one modification per DNA molecule. Thus a series of labeled fragments is generated, from the radiolabeled end to the first "cut" site in each molecule.

The fragments in the four reactions are electrophoresed side by side in denaturing acrylamide gels for size separation. To visualize the fragments, the gel is exposed to X-ray film for autoradiography, yielding a series of dark bands each showing the location of identical radiolabeled DNA molecules. From presence and absence of certain fragments the sequence may be inferred.

Related Methods

This method led to the Methylation Interference Assay, used to map DNA-binding sites for DNA-binding proteins.

An automated Maxam–Gilbert sequencing protocol was developed in 1994.

Cloning Strategies

The cloning of DNA fragment into a vector is routinely used method in recombinant DNA technology. Following are some strategies that are commonly used for cloning DNA fragment-

- TA cloning
- Gateway cloning

TA Cloning

It is a simple and more efficient cloning method without any requirement of restriction enzymes. It involves the use of terminal transferase activity of *Taq* polymerase which preferentially adds adenosine at 3' end of the PCR amplified DNA molecule. To clone PCR products having 3' overhangs, T-vector having 3'-T overhangs at both its ends can be used. This method is based on the complementary base pairing between the 3'-A overhangs of PCR product and 3'-T overhang of a vector molecule.

A single T-vector can be employed for easy and convenient cloning of any PCR amplified double-stranded DNA as well as blunt or sticky ended DNA molecules through certain minor modifications in TA cloning method. This technique, known as Universal TA cloning method, is useful in sub-cloning of DNA fragments when there are no compatible restriction sites available without compromising the cloning efficiency. Directional TA cloning can be possible by appropriate hemi-phosphorylation of the T-vectors and the inserts.

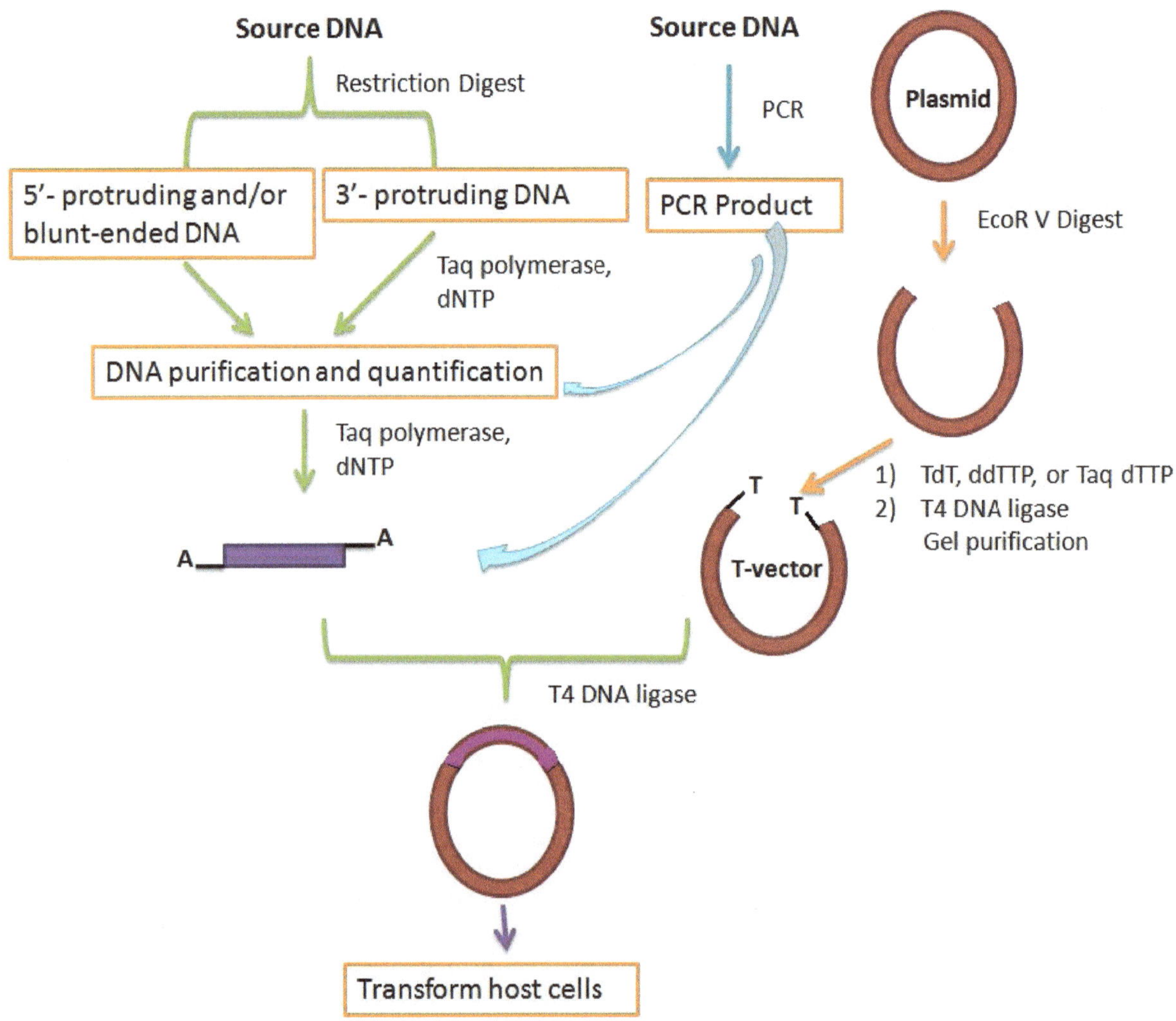

(a): Schematic representation of the cloning procedures using the universal T-vectors

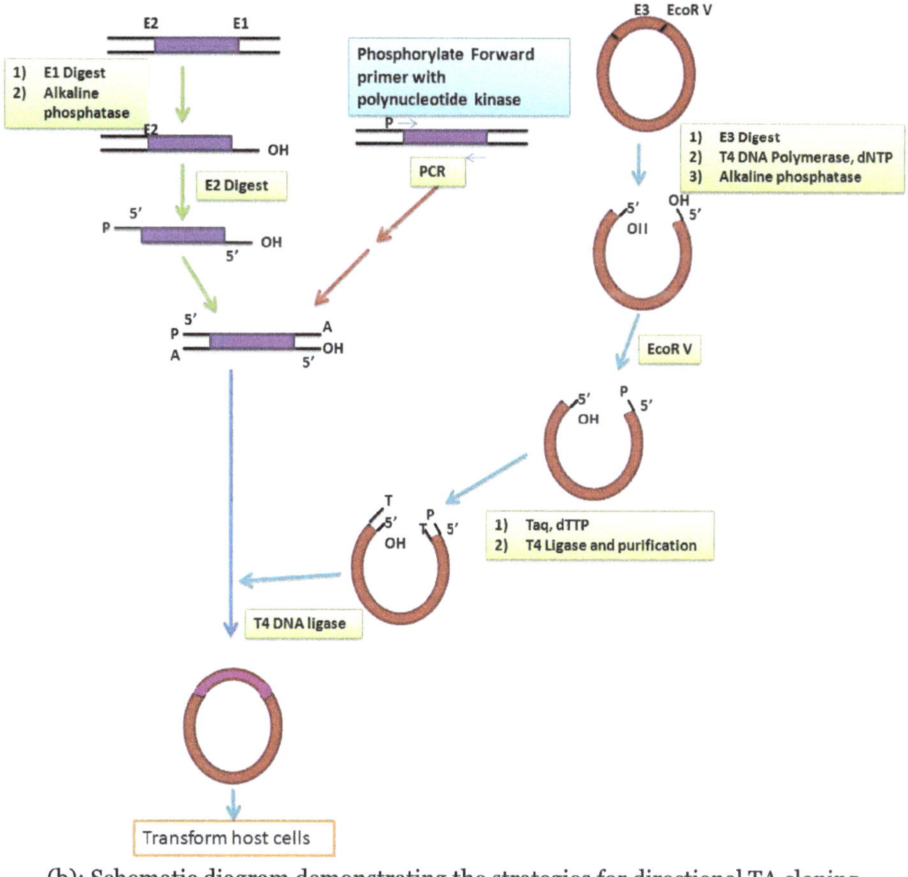

(b): Schematic diagram demonstrating the strategies for directional TA cloning
(E1, E2 and E3 - restriction enzymes, E3 - Generates blunt ends)

Gateway Cloning

The Gateway technology addresses the complications of efficiency, adaptability and compatibility of the traditional cloning procedures. It allows the cloning, combination and transfer of DNA fragments between various expression platforms in a high-throughput mode. It maintains the orientation and reading frame of the fragment/s of interest.

It is based on the site specific recombination method that mediates both integration and excision of bacteriophage λ These events occur by recombination at specific attachment site on the phage DNA (*attp*) and bacterial chromosome (*attb*). The recombination begins when the phage encoded integrase and the bacterial integration host factor binds to the *attp* sequence and now this complex couples with the *attb* in the bacterial chromosome. Staggered DNA nicks are produced at the ends of *attp* and *attb* sequence followed by strand exchange between them. This results in the formation of a tiny heteroduplex joint at the point of exchange due to the presence of a short region (7bp) of sequence homology in the two joined sequences. The holiday junction is then resolved and the prophage becomes flanked by two-hybrid *att* sites, called *attL* (for 'left') and *attR* (for 'right'). This initial reaction is referred as BP reaction. Although recombination reaction is reversible, different reaction condition prevail for each reaction ensuring that integration is not reversed by excision. Excision requires excisionase (Xis) and this reaction is known as LR reaction.

LR and BP Reactions

- The core of the Gateway is the *entry clone* which consists of a plasmid where the fragment of interest is flanked by *attL* sequences.
- Once the entry clone is obtained, the fragment of interest can be transferred to a secondary plasmid, the *destination vector* which is application-specific vector.
- This reaction is accomplished by mixing, *in vitro,* the entry clone (that usually carries *attL* sequences) with the destination vector of choice (that carries *attR* sequences) in the presence of a blend of the recombination proteins (Int, IHF and Xis), the LR Clonase™ enzyme mix.
- The product of this 'LR reaction' is the *expression clone*, which carries the DNA of interest and the donor vector, which is considered as a reaction byproduct.
- An aliquot of this reaction is used to transform competent *E. coli* .
- Two selection schemes are imposed to select the cells harboring the expression clone only.
- First, the entry clone and the destination vector contain different antibiotic resistance genes.
- Second, the destination vector contains a counter selectable marker, the *ccdB* gene, which kills the cells by inducing gyrase-mediated double-stranded DNA breakage.
- Thus, transformants selected for ampicillin resistance contain only the expression clone.

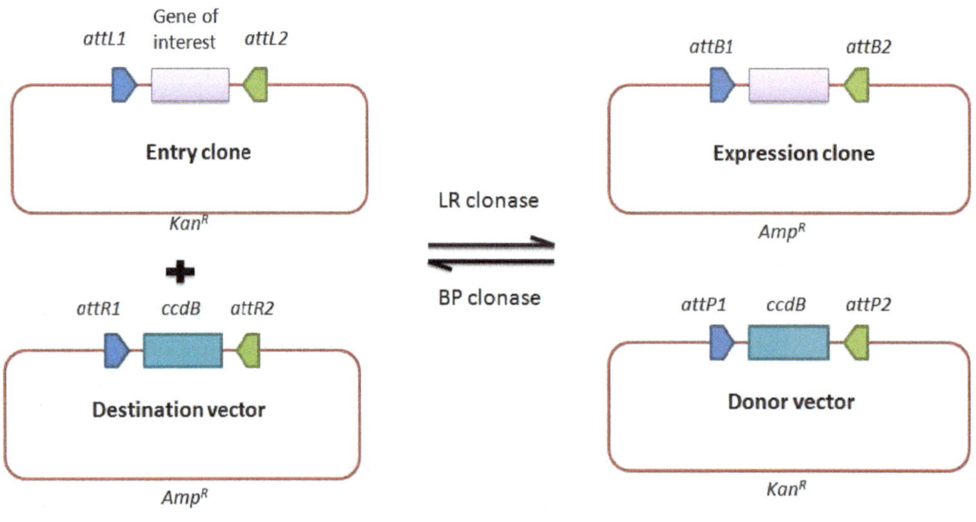

The Gateway reactions. The scheme shows the four types of plasmids and enzyme mixtures involved in Gateway cloning reactions. Red arrows represent the fragment of interest.

The Entry Clone

- The main door to enter a Gateway technology is the entry clone.

- There are 3 ways to generate this construct.
- First strategy is based on BP reaction.
- The second method is to ligate a PCR product into a TOPO- adapter entry clone. In this approach a single enzyme called vaccinia DNA topoisomerase having the ability to both cleave and rejoin DNA strands with extreme specificity at each step is used.
- Linearized vectors covalently attaches to the enzyme at 3' end to efficiently ligate the PCR product in a 5 minute reaction.
- The third method is the standard restriction cloning.

Destination Vectors

- A blunt-ended cassette comprising *attR1* -chloramphenicol resistance gene (CmR)-*ccdB* - *attR2* is inserted to any vector which results in a destination vector.
- Then the destination vector can be used in LR reaction to generate expression clone.
- Expression clones can also be generated by TOPO-cloning a PCR fragment into a TOPO-adapted expression vector.

References

- Henry RJ, Edwards M, Waters DL, et al. (November 2012). "Application of large-scale sequencing to marker discovery in plants". J. Biosci. 37 (5): 829–41. PMID 23107919. doi:10.1007/s12038-012-9253-z
- Menon R, Farina C (2011). "Shared molecular and functional frameworks among five complex human disorders: a comparative study on interactomes linked to susceptibility genes". PLoS ONE. 6 (4): e18660. PMC 3080867. PMID 21533026. doi:10.1371/journal.pone.0018660
- Losick, Richard; Watson, James D.; Tania A. Baker; Bell, Stephen; Gann, Alexander; Levine, Michael W. (2008). Molecular biology of the gene. San Francisco: Pearson/Benjamin Cummings. ISBN 0-8053-9592-X
- Pennisi E (April 2003). "Human genome. Reaching their goal early, sequencing labs celebrate". Science. 300 (5618): 409. PMID 12702850. doi:10.1126/science.300.5618.409
- Peterson, Daniel; Jeffrey Tomkins; David Frisch (2000). "Construction of Plant Bacterial Artificial Chromosome (BAC) Libraries: An Illustrated Guide". Journal of Agricultural Genomics. 5
- Russell, David W.; Sambrook, Joseph (2001). Molecular cloning: a laboratory manual. Cold Spring Harbor, N.Y: Cold Spring Harbor Laboratory. ISBN 0-87969-577-3
- Stevens WL (1939). "Solution to a Geometrical Problem in Probability". Annals of Eugenics. 9: 315–320. doi:10.1111/j.1469-1809.1939.tb02216.x
- Wendl MC (2006). "A general coverage theory for shotgun DNA sequencing". Journal of Computational Biology. 13 (6): 1177–1196. PMID 16901236. doi:10.1089/cmb.2006.13.1177
- Muse, Spencer V.; Gibson, Greg (2004). A primer of genome science. Sunderland, Mass: Sinauer Associates. ISBN 0-87893-232-1
- Clarke L, Carbon J (1976). "A colony bank containing synthetic Col-El hybrid plasmids representative of the entire E. coli genome". Cell. 9 (1): 91–99. PMID 788919. doi:10.1016/0092-8674(76)90055-6
- Levy S; et al. (2007). "The diploid genome sequence of an individual human". PLoS Biology. 5 (10): article e254. PMC 1964779. PMID 17803354. doi:10.1371/journal.pbio.0050254

- John R. McCarrey; Williams, Steven J.; Barton E. Slatko (2006). Laboratory investigations in molecular biology. Boston: Jones and Bartlett Publishers. ISBN 0-7637-3329-6
- Arratia R; et al. (1991). "Genomic mapping by anchoring random clones: a mathematical analysis". Genomics. 11 (4): 806–827. PMID 1783390. doi:10.1016/0888-7543(91)90004-X
- Stanhope SA (2010). "Occupancy modeling maximum contig size probabilities and designing metagenomics experiments". PLoS ONE. 5 (7): article e11652. Bibcode:2010PLoSO...511652S. PMC 2912229. PMID 20686599. doi:10.1371/journal.pone.0011652
- Waterman, Michael S. (1995). Introduction to Computational Biology. Boca Raton: Chapman and Hall/CRC. ISBN 0-412-99391-0
- Port E; et al. (1995). "Genomic mapping by end-characterized random clones: a mathematical analysis". Genomics. 26 (1): 84–100. PMID 7782090. doi:10.1016/0888-7543(95)80086-2
- Wendl MC; et al. (2012). "Coverage theories for metagenomic DNA sequencing based on a generalization of Stevens' theorem". Journal of Mathematical Biology. 67: 1141–1161. PMID 22965653. doi:10.1007/s00285-012-0586-x
- Solomon, H. (1978). Geometric Probability. Philadelphia: Society for Industrial and Applied Mathematics. ISBN 0-898-71025-1
- Roach JC; et al. (2000). "Parking strategies for genome sequencing". Genome Research. 10 (7): 1020–1030. PMC 310895. PMID 10899151. doi:10.1101/gr.10.7.1020

4

Nucleic Acid Detection and Hybridization

For gene manipulation, nucleic acid needs to be isolated and separated. Isolation goes through three steps. First, the cell membrane is broken to free cellular components and DNA. Then, the nucleic acid is separated from other organelles. Finally, the nucleic acid is purified. Organic extraction and enzymatic digestion, ion-exchange chromatography, DNA samples concentration are few methods of purifying nucleic acid. This chapter will provide an integrated understanding of genetic engineering.

Nucleic Acid Detection

Nucleic acid molecules like deoxyribonucleic acids (DNA), ribonucleic acids (RNA) are basic, essential and primary molecules for all molecular biology related research. Before discussing the common methods to detect nucleic acids let us we learn about the isolation of pure form of nucleic acid.

Preparation Step

First important step is to establish if a given sample contains DNA or RNA and whether it is in pure form, since many samples will contain both species as well as other contaminants such as cellular proteins. Irrespective of the method applied for detection, the optimal nucleic acid isolation protocol must provide:

a) Reproducible result

b) No degradation of the sample and

c) Safer handling.

In deciding what method of nucleic acid measurement is appropriate, three issues are critical: specificity, sensitivity, and interfering substances. Based on these features different detection techniques are to be followed.

Traditional Detection Methods

UV Spectroscopy

Majority of bio-molecules intrinsically absorb light in the ultraviolet and not in the visible range. This property of UV absorbance can be used to quickly estimate the concentration and purity of DNA and RNA (also proteins) in a analytical sample. The amount of DNA in a sample can be estimated by looking at its absorbance at a wavelength of 260nm or 280nm (in the UV region). Purines and pyrimidines have absorbance maxima slightly below and above 260 respectively. Thus the absorbance maxima of different fragments of DNA vary somewhat depending on their subunit composition. Contaminants like proteins exhibit two absorbance peaks, one between 215-230 nm (due to peptide bonds absorption) and at about 280 nm (absorption by aromatic amino acids-tyrosine, tryptophan and phenylalanine). Remember that although proteins have little absorbance at 260 nm, both proteins and nucleic acids absorb light at 280 nm. That is the reason why, if nucleic acids and proteins are mixed in the same sample, their spectra interfere (overlap) with one another.

Table: The relationship between concentration of DNA, RNA, Protein and absorptivity are as below:

Sample	Absorbance value	*Quantity (*approximate*)*
Double-stranded DNA	1at **260 nm**	**50 µg/mL**
Pure single-stranded DNA	1 at **260 nm**	**33 µg/mL**
Pure RNA	1 at **260 nm**	**40 µg/mL**
Pure protein (vary in general) Antibodies[1]	**1 at 280 nm** **1.2-1.35**	**1 mg/mL** 1 mg/mL

The purity of a solution of nucleic acid is determined by measuring the absorbance of the solution at two wavelengths, usually 260 nm and 280 nm, and calculating the ratio of A_{260}/A_{280}. Value of this ratio is 2.0, 1.8 and 0.6 for pure RNA, DNA and protein respectively. A ratio of less than 1.8 signifies that the sample is contaminated with protein or phenol and the preparation is not proper.

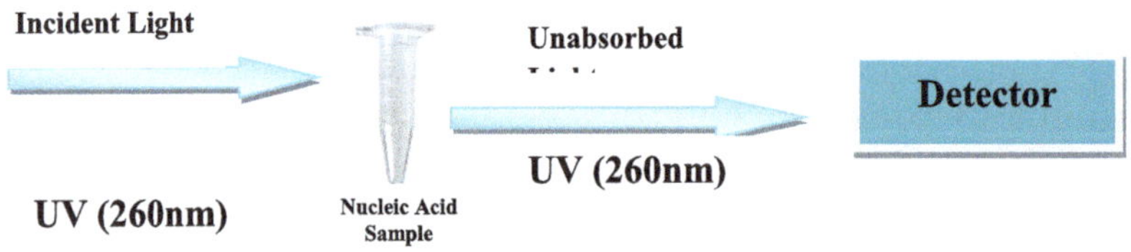

UV detection of nucleic acids

Hyperchromicity is the phenomenon of increment of absorbance when any material is exposed to UV light.The most well known application is the hyperchromicity of DNA that occurs when the DNA duplex is denatured and melting of DNA occurs. The UV absorption is increased when the duplex DNA strands are being separated into single strands, either by heat or by addition of denaturant or by increasing the pH level. On the contrary, decrease of absorbance peak is called hypochromicity.

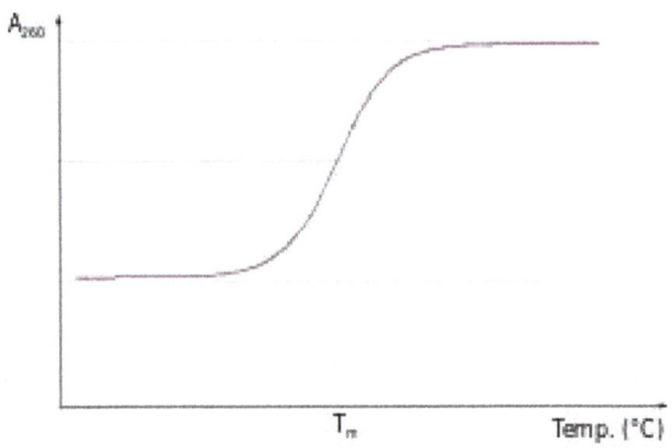

Nucleic acid melting curve showing hyperchromicity as a function of temperature

Ethidium Bromide Staining

The IUPAC name for EtBr is 2, 7-diamino-10-ethyl-9-phenylphenanthridiniumbromide. It is commonly used as a fluorescent dye for nucleic acid staining. It binds as well as intercalates with nucleic acid (mainly with major and minor groove of DNA) and gives orange fluorescence under UV radiation from 500 – 590 nm. Usually EtBr may be added in warm agarose gel before solidification. When DNA or RNA samples are run in agarose gel electrophoresis EtBr molecules will bind with nucleic acids and help in detection under UV light. The post staining can also be done for nucleic acid detection.

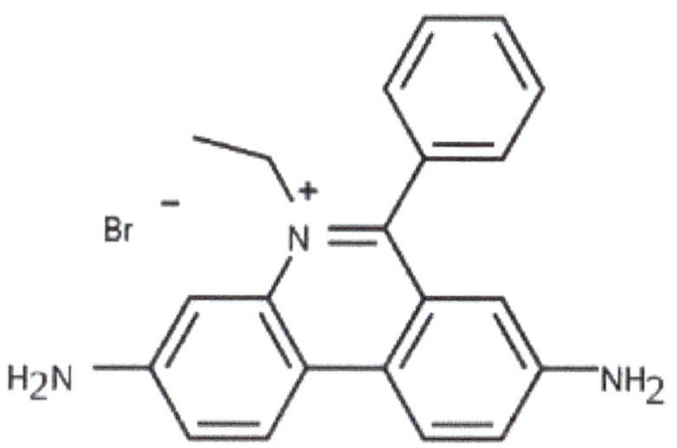

Ethidium bromide

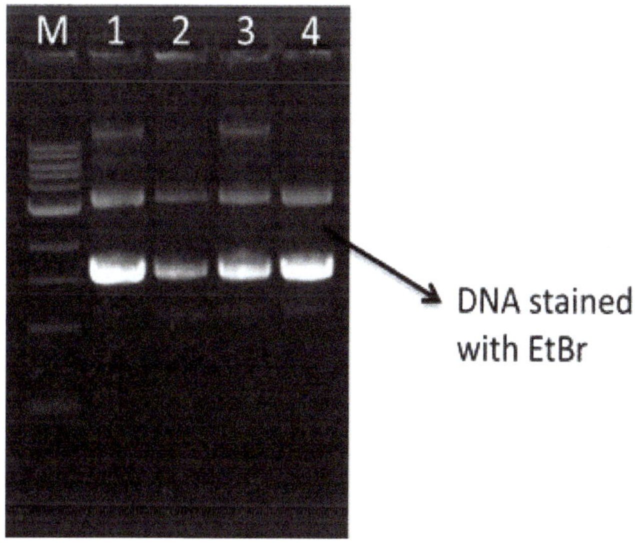

Agarose gel stained with EtBr

Ethidium bromide (EtBr) is a potent mutagen and carcinogen. Dyes to stain nucleic acids such as SYBR green, SYBR Safe etc are safer to use instead of EtBr.

Polyacrylamide gel can be used for separation of Nucleic acids and post staining of the gel with EtBr is done for detection. The sensitivity of EtBr staining is of nano-molar level.

Silver Staining

Silver staining based on reduction of silver nitrate is more sensitive than ethidium bromide for double stranded DNA, as well as detection of single stranded DNA or RNA with a good sensitivity (in picogram level). It is based on the reduction of silver cations to insoluble silver metal by nucleic acids. This chemical reaction is insensitive to the macrostructure of the DNA molecule. Reduced silver molecules deposit in the gel around the DNA bands, creating a dark black band like image (i.e. "latent image"). Then the latent image can be developed to visualize by soaking the gel in a solution of silver cations (Silver nitrate) and a reducing agent (eg. formaldehyde). The silver granules in the latent image catalyze the further reduction and deposition of silver from the solution. Bands manifest as dark brown or black regions which appear before significant background develops. Development is stopped by altering the pH of the gel to a point where silver reduction is no longer favored.

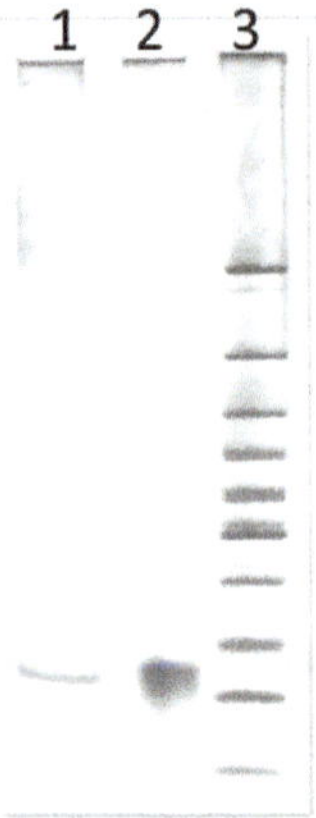

Representation of a silver staining of DNA

Lane 1: DNA of lesser concentration
Lane 2: DNA of higher concentration
Lane 3: Low molecular weight ladder

Nanodrop

Detection assays are persistently being developed that use progressively smaller amounts of nucleic acid, often precluding the use of conventional cuvette-based instruments for nucleic acid quantitation for those that can perform micro-volume quantitation. The patented NanoDrop microvolume sample retention system functions by combining fiber optic technology and natural surface tension properties to capture and retain small amounts of sample . This is a novel technology which allows us to measure nano-liter volumes (pico concentration) of the nucleic acid (DNA or RNA) sample. It is a type of spectrophotometer with a smaller sample size (as much less as 1-2 microlitre) requirement and higher sensitivity (even upto pico molar level). This is also a time saving technology widely used in basic molecular biology research.

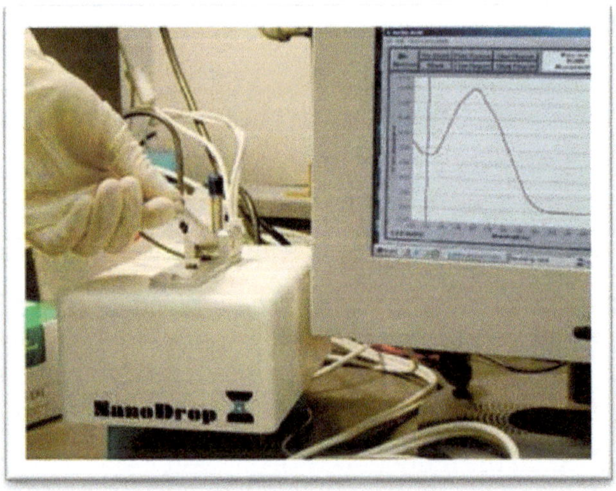

NanoDrop® ND-1000 Spectrophotometer

Fluorometric Quantification

Fluorometric method applies fluorescence dyes to detect the presence and concentration of a class of nucleic acid (DNA or RNA). This method is more sensitive and less prone to contaminants than UV spectroscopy.

An assay using Hoechst 33258 dye is specific for DNA because it is less sensitive to detect RNA.

This assay is commonly used for rapid measurement of low quantities of DNA, with a detection limit of ~ 1 ng DNA. It is useful for the measurement of both small and large amounts of DNA (verifying DNA concentrations prior to performing electrophoretic separations and Southern blots) because this assay accurately quantifies a broad range of DNA concentrations from 10 ng/ml to 15 µg/ml. The Hoechst 33258 assay can also be employed for measuring products of the polymerase chain reaction (PCR) synthesis.

Hoechst 33258 is non-intercalating reagent and binds to the minor groove of the DNA with a preference for AT sequences (Portugal and Waring, 1988). The binding to the minor groove has is dependent upon a combination of structural preferences (eg., the minor groove with a series of contiguous AT base pairs is more narrow). (Neidle (2001) ,Like other minor groove binding ligands, Hoechst 33258 is positively charged and thus form electrostatic interaction with the negative potential of stretch of AT base pairs. Upon binding to the minor groove of the double helix DNA, the fluorescence characteristics of Hoechst 33258 change dramatically, showing a large increase in emission at ~ 458 nm.

According to Daxhelet *et al*, the fluorochrome 4' ,6-diamidino-2-phenylindole (DAPI) has similar characteristics to H33258 and binds to the minor groove as well. DAPI is also appropriate for DNA or RNA quantitation, although it is not as commonly used as Hoechst 33258. DAPI is excited with a peak at 344 nm. Emission is detected at around 466 nm for DNA, similar to Hoechst 33258 but for RNA the peak shifts to ~500 nm.

Hybridization-Based Techniques

Hybridization-Based Techniques for nucleic acid detection has higher resolution (down to the actual nucleotide sequence) and utilizes "probe sequence" for DNA or RNA, and when it finds its intended target, binds to it by hybridization process. Since the "probe" is attached to a label such as a fluorescent chemical isotope, the bound sequence can thus be visualized. This is the principle of Southern Blotting method, as well as its variants, and requires the target nucleic acid sequence to undergo separation by agarose-electrophoresis, transferred to an appropriate membrane (typically nitrocellulose), and then treated with a solution containing the labeled probe (fluorescent or colorimetric).

Northern blotting involves the use of electrophoresis to separate RNA samples by size and detection with a hybridization probe complementary to part of or the entire target sequence. Since the probe specifically binds to its target, the membrane can be documented and analyzed so that the bound target sequences can be identified and studied..

FISH (Fluoroscence in-situ hybridization) for visualization of nucleic acids developed as an alternative to older methods that used radio labeled probes (Gall and Pardue, 1969). Several drawbacks of isotopic, radiolabeled hybridization stimulated the development of novel techniques like FISH. In the year 1980, RNA was first directly labeled on the 3' end with fluorophore was used as a probe for specific DNA sequences (Bauman et al., 1980). Enzymatic incorporation of fluorophore-modified bases throughout the probe has been widely used for the preparation of fluorescent probes. The use of amino-allyl modified bases (Langer et al., 1981), which could later be conjugated to any sort of hapten or fluorophores, was critical for the development of in situ technologies because it allowed production of an array of low-noise probes by simple chemistry. Methods of indirect detection result into higher magnitude signal output by the use of secondary reporter molecules that bind to the hybridization probes.

In the early 1980s, assays like nick-translated, biotinylated probes, and secondary detection by streptavidin conjugates were used for detection of DNA (Manuelidis *et al*., 1982) and mRNA (Singer and Ward, 1982) targets. Later, advanced labeling of synthetic, single-stranded DNA probes allowed the chemical preparation of hybridization probes carrying enough fluorescent molecules to allow direct detection (Kislauskis *et al*., 1993). At this current era, based on these reaction themes of indirect and direct labeling have since been introduced, a wide spectrum of nucleic acid detection schemes are available.

Whereas the initial development of FISH involved expansion of the types of probe and number of detectable targets, the outlook for future development of fluorescence techniques will include expansion of the subjects of investigation like in clinical, diagnostics, forensic applications.

Other detection methods include amplification of target region by polymerase chain reaction and various forms of chemiluminescent, fluorescent or radioactive detection method. Specific technical or sample requirements are present for each method, and the original purpose for detection the nucleic acid will determine which of the methods is most suitable.

Polymerase Chain Reaction

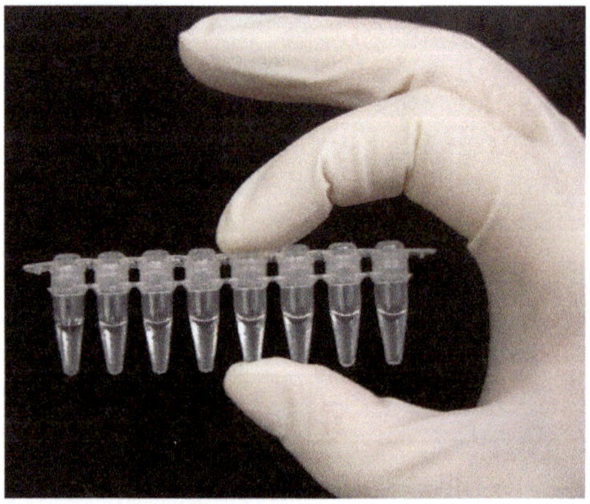

A strip of eight PCR tubes, each containing a 100 μl reaction mixture

Polymerase chain reaction (PCR) is a technique used in molecular biology to amplify a single copy or a few copies of a segment of DNA across several orders of magnitude, generating thousands to millions of copies of a particular DNA sequence. It is an easy, cheap, and reliable way to repeatedly replicate a focused segment of DNA, a concept which is applicable to numerous fields in modern biology and related sciences.

Developed in 1983 by Kary Mullis, PCR is now a common and often indispensable technique used in clinical and research laboratories for a broad variety of applications. These include DNA cloning for sequencing, construction of DNA-based phylogenies, or functional analysis of genes; diagnosis and monitoring of hereditary diseases; analysis of genetic fingerprints for DNA profiling (for example, in forensic science and parentage testing); and detection of pathogens in nucleic acid tests for the diagnosis of infectious diseases. In 1993, Mullis was awarded the Nobel Prize in Chemistry along with Michael Smith for his work on PCR.

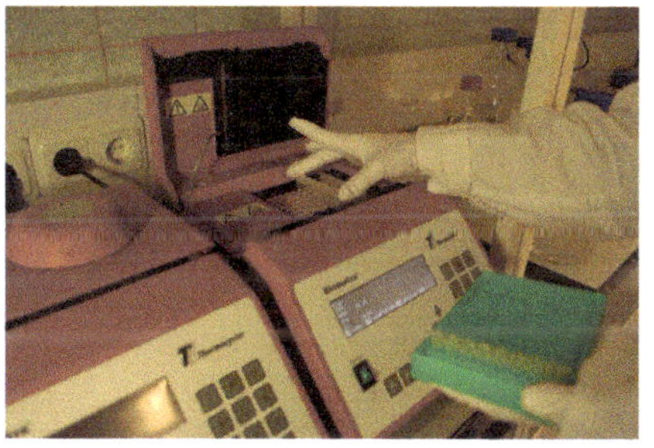

Placing a strip of eight PCR tubes into a thermal cycler

The vast majority of PCR methods rely on thermal cycling, which involves exposing the reactants to cycles of repeated heating and cooling, permitting different temperature-dependent reactions—specifically, DNA melting and enzyme-driven DNA replication—to quickly proceed many times in sequence. Primers (short DNA fragments) containing sequences complementary to the target region, along with a DNA polymerase, after which the method is named, enable selective and repeated amplification. As PCR progresses, the DNA generated is itself used as a template for replication, setting in motion a chain reaction in which the original DNA template is exponentially amplified. The simplicity of the basic principle underlying PCR means it can be extensively modified to perform a wide array of genetic manipulations. PCR is not generally considered to be a recombinant DNA method, as it does not involve cutting and pasting DNA, only amplification of existing sequences.

Almost all PCR applications employ a heat-stable DNA polymerase, such as Taq polymerase, an enzyme originally isolated from the thermophilic bacterium *Thermus aquaticus*. This DNA polymerase enzymatically assembles a new DNA strand from free nucleotides, the building blocks of DNA, by using single-stranded DNA as a template and DNA oligonucleotides (the primers mentioned above) to initiate DNA synthesis.

In the first step, the two strands of the DNA double helix are physically separated at a high temperature in a process called DNA melting. In the second step, the temperature is lowered and the two DNA strands become templates for DNA polymerase to selectively amplify the target DNA. The selectivity of PCR results from the use of primers that are complementary to the DNA region targeted for amplification under specific thermal cycling conditions.

Principles

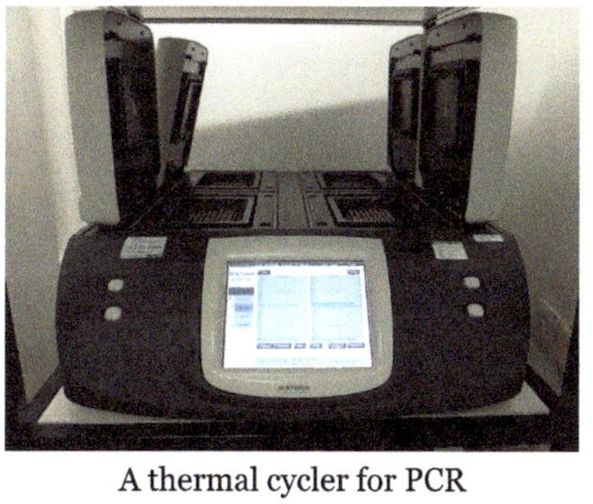

A thermal cycler for PCR

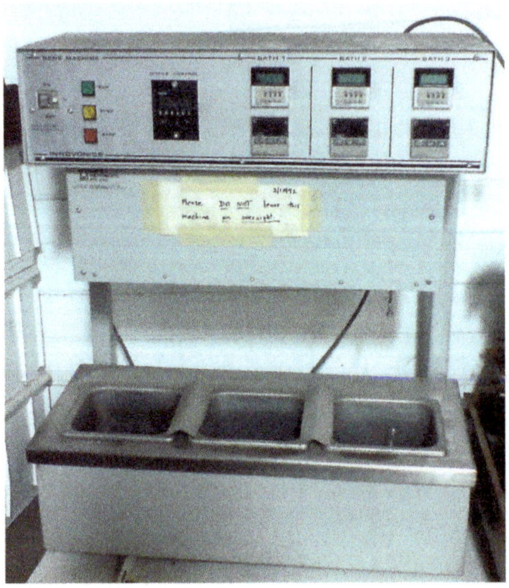

An older model three-temperature thermal cycler for PCR

PCR amplifies a specific region of a DNA strand (the DNA target). Most PCR methods amplify DNA fragments of between 0.1 and 10 kilo base pairs (kbp), although some techniques allow for amplification of fragments up to 40 kbp in size. The amount of amplified product is determined by the available substrates in the reaction, which become limiting as the reaction progresses.

A basic PCR set-up requires several components and reagents, including:

- a *DNA template* that contains the DNA target region to amplify
- a *DNA polymerase*, an enzyme that polymerizes new DNA strands; heat-resistant Taq polymerase is especially common, as it is more likely to remain intact during the high-temperature DNA denaturation process
- two DNA *primers* that are complementary to the 3' (three prime) ends of each of the sense and anti-sense strands of the DNA target (DNA polymerase can only bind to and elongate from a double-stranded region of DNA; without primers there is no double-stranded initiation site at which the polymerase can bind); specific primers that are complementary to the DNA target region are selected beforehand, and are often custom-made in a laboratory or purchased from commercial biochemical suppliers
- *deoxynucleoside triphosphates*, or dNTPs (sometimes called "deoxynucleotide triphosphates"; nucleotides containing triphosphate groups), the building blocks from which the DNA polymerase synthesizes a new DNA strand
- a *buffer solution* providing a suitable chemical environment for optimum activity and stability of the DNA polymerase
- *bivalent cations*, typically magnesium (Mg) or manganese (Mn) ions; Mg^{2+} is the most common, but Mn^{2+} can be used for PCR-mediated DNA mutagenesis, as a higher Mn^{2+} concentration increases the error rate during DNA synthesis
- *monovalent cations*, typically potassium (K) ions.

The reaction is commonly carried out in a volume of 10–200 μl in small reaction tubes (0.2–0.5 ml volumes) in a thermal cycler. The thermal cycler heats and cools the reaction tubes to achieve the temperatures required at each step of the reaction. Many modern thermal cyclers make use of the Peltier effect, which permits both heating and cooling of the block holding the PCR tubes simply by reversing the electric current. Thin-walled reaction tubes permit favorable thermal conductivity to allow for rapid thermal equilibration. Most thermal cyclers have heated lids to prevent condensation at the top of the reaction tube. Older thermal cyclers lacking a heated lid require a layer of oil on top of the reaction mixture or a ball of wax inside the tube.

Procedure

Typically, PCR consists of a series of 20–40 repeated temperature changes, called cycles, with each cycle commonly consisting of two or three discrete temperature steps. The cycling is often preceded by a single temperature step at a very high temperature (>90 °C (194 °F)), and followed by one hold at the end for final product extension or brief storage. The temperatures used and the length of time they are applied in each cycle depend on a variety of parameters, including the enzyme used for DNA synthesis, the concentration of bivalent ions and dNTPs in the reaction, and the melting temperature (*Tm*) of the primers. The individual steps common to most PCR methods are as follows:

- Initialization: This step is only required for DNA polymerases that require heat activation by hot-start PCR. It consists of heating the reaction chamber to a temperature of 94–96 °C (201–205 °F), or 98 °C (208 °F) if extremely thermostable polymerases are used, which is then held for 1–10 minutes.

- Denaturation: This step is the first regular cycling event and consists of heating the reaction chamber to 94–98 °C (201–208 °F) for 20–30 seconds. This causes DNA melting, or denaturation, of the double-stranded DNA template by breaking the hydrogen bonds between complementary bases, yielding two single-stranded DNA molecules.

- Annealing: In the next step, the reaction temperature is lowered to 50–65 °C (122–149 °F) for 20–40 seconds, allowing annealing of the primers to each of the single-stranded DNA templates. Two different primers are typically included in the reaction mixture: one for each of the two single-stranded complements containing the target region. The primers are single-stranded sequences themselves, but are much shorter than the length of the target region, complementing only very short sequences at the 3' end of each strand.

  It is critical to determine a proper temperature for the annealing step because efficiency and specificity are strongly affected by the annealing temperature. This temperature must be low enough to allow for hybridization of the primer to the strand, but high enough for the hybridization to be specific, i.e., the primer should bind *only* to a perfectly complementary part of the strand, and nowhere else. If the temperature is too low, the primer may bind imperfectly. If it is too high, the primer may not bind at all. A typical annealing temperature is about 3–5 °C below the *Tm* of the primers used. Stable hydrogen bonds between complementary bases are formed only when the primer sequence very closely matches the template sequence. During this step, the polymerase binds to the primer-template hybrid and begins DNA formation.

- Extension/elongation: The temperature at this step depends on the DNA polymerase used; the optimum activity temperature for Taq polymerase is approximately 75–80 °C (167–176 °F), though a temperature of 72 °C (162 °F) is commonly used with this enzyme. In this step, the DNA polymerase synthesizes a new DNA strand complementary to the DNA template strand by adding free dNTPs from the reaction mixture that are complementary to the template in the 5'-to-3' direction, condensing the 5'-phosphate group of the dNTPs with the 3'-hydroxy group at the end of the nascent (elongating) DNA strand. The precise time required for elongation depends both on the DNA polymerase used and on the length of the DNA target region to amplify. As a general rule-of-thumb, at their optimal temperature, most DNA polymerases polymerize a thousand bases per minute. Under optimal conditions (i.e., if there are no limitations due to limiting substrates or reagents), at each extension/elongation step, the number of DNA target sequences is doubled. With each successive cycle, the original template strands plus all newly generated strands become template strands for the next round of elongation, leading to exponential (geometric) amplification of the specific DNA target region.

  The processes of denaturation, annealing and elongation constitute a single cycle. Multiple cycles are required to amplify the DNA target to millions of copies. The formula used to calculate the number of DNA copies formed after a given number of cycles is 2^n, where n is the number of cycles. Thus, a reaction set for 30 cycles results in 2^{30}, or 1073741824, copies of the original double-stranded DNA target region.

- Final elongation: This single step is optional, but is performed at a temperature of 70–74 °C (158–165 °F) (the temperature range required for optimal activity of most polymerases used in PCR) for 5–15 minutes after the last PCR cycle to ensure that any remaining single-stranded DNA is fully elongated.

- Final hold: The final step cools the reaction chamber to 4–15 °C (39–59 °F) for an indefinite time, and may be employed for short-term storage of the PCR products.

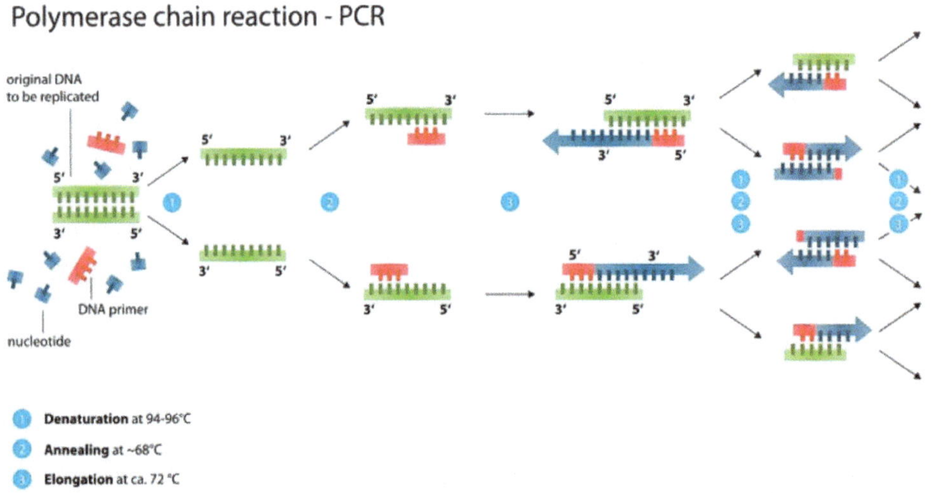

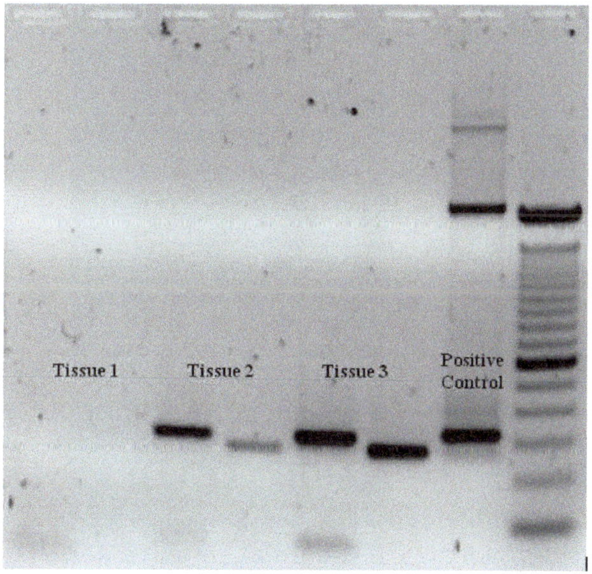

Ethidium bromide-stained PCR products after gel electrophoresis. Two sets of primers were used to amplify a target sequence from three different tissue samples. No amplification is present in sample #1; DNA bands in sample #2 and #3 indicate successful amplification of the target sequence. The gel also shows a positive control, and a DNA ladder containing DNA fragments of defined length for sizing the bands in the experimental PCRs.

To check whether the PCR successfully generated the anticipated DNA target region (also sometimes referred to as the amplimer or amplicon), agarose gel electrophoresis may be employed for size separation of the PCR products. The size(s) of PCR products is determined by comparison with a DNA ladder, a molecular weight marker which contains DNA fragments of known size run on the gel alongside the PCR products.

Stages

As with other chemical reactions, the reaction rate and efficiency of PCR are affected by limiting factors. Thus, the entire PCR process can further be divided into three stages based on reaction progress:

- *Exponential amplification*: At every cycle, the amount of product is doubled (assuming 100% reaction efficiency). The reaction is very sensitive: only minute quantities of DNA must be present.
- *Leveling off stage*: The reaction slows as the DNA polymerase loses activity and as consumption of reagents such as dNTPs and primers causes them to become limiting.
- *Plateau*: No more product accumulates due to exhaustion of reagents and enzyme.

Optimization

In practice, PCR can fail for various reasons, in part due to its sensitivity to contamination causing amplification of spurious DNA products. Because of this, a number of techniques and procedures have been developed for optimizing PCR conditions. Contamination with extraneous DNA is addressed with lab protocols and procedures that separate pre-PCR mixtures from potential DNA contaminants. This usually involves spatial separation of PCR-setup areas from areas for analysis or purification of PCR products, use of disposable plasticware, and thoroughly cleaning the work surface between reaction setups. Primer-design techniques are important in improving PCR product yield and in avoiding the formation of spurious products, and the usage of alternate buffer components or polymerase enzymes can help with amplification of long or otherwise problematic regions of DNA. Addition of reagents, such as formamide, in buffer systems may increase the specificity and yield of PCR. Computer simulations of theoretical PCR results (Electronic PCR) may be performed to assist in primer design.

Applications

Selective DNA Isolation

PCR allows isolation of DNA fragments from genomic DNA by selective amplification of a specific region of DNA. This use of PCR augments many ways, such as generating hybridization probes for Southern or northern hybridization and DNA cloning, which require larger amounts of DNA, representing a specific DNA region. PCR supplies these techniques with high amounts of pure DNA, enabling analysis of DNA samples even from very small amounts of starting material.

Other applications of PCR include DNA sequencing to determine unknown PCR-amplified sequences in which one of the amplification primers may be used in Sanger sequencing, isolation of a DNA sequence to expedite recombinant DNA technologies involving the insertion of a DNA sequence into a plasmid, phage, or cosmid (depending on size) or the genetic material of another organism. Bacterial colonies *(such as E. coli)* can be rapidly screened by PCR for correct DNA vector constructs. PCR may also be used for genetic fingerprinting; a forensic technique used to identify a person or organism by comparing experimental DNAs through different PCR-based methods.

Some PCR 'fingerprints' methods have high discriminative power and can be used to identify genetic relationships between individuals, such as parent-child or between siblings, and are used in paternity testing. This technique may also be used to determine evolutionary relationships among organisms when certain molecular clocks are used (i.e., the 16S rRNA and recA genes of microorganisms).

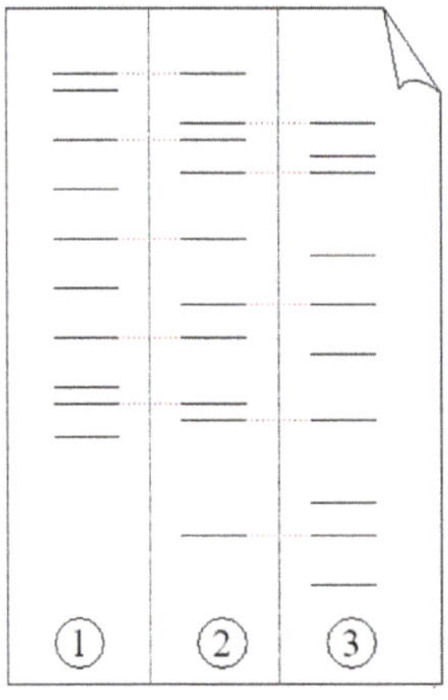

Electrophoresis of PCR-amplified DNA fragments. (1) Father. (2) Child. (3) Mother. The child has inherited some, but not all of the fingerprint of each of its parents, giving it a new, unique fingerprint.

Amplification and Quantification of DNA

Because PCR amplifies the regions of DNA that it targets, PCR can be used to analyze extremely small amounts of sample. This is often critical for forensic analysis, when only a trace amount of DNA is available as evidence. PCR may also be used in the analysis of ancient DNA that is tens of thousands of years old. These PCR-based techniques have been successfully used on animals, such as a forty-thousand-year-old mammoth, and also on human DNA, in applications ranging from the analysis of Egyptian mummies to the identification of a Russian tsar and the body of English king Richard III.

Quantitative PCR methods allow the estimation of the amount of a given sequence present in a sample—a technique often applied to quantitatively determine levels of gene expression. Quantitative PCR is an established tool for DNA quantification that measures the accumulation of DNA product after each round of PCR amplification.

Medical Applications

PCR has been applied to a large number of medical procedures:

- The first application of PCR was for genetic testing, where a sample of DNA is analyzed for the presence of genetic disease mutations. Prospective parents can be tested for being genetic carriers, or their children might be tested for actually being affected by a disease. DNA samples for prenatal testing can be obtained by amniocentesis, chorionic villus sampling, or even by the analysis of rare fetal cells circulating in the mother's bloodstream. PCR analysis is also essential to preimplantation genetic diagnosis, where individual cells of a developing embryo are tested for mutations.

- PCR can also be used as part of a sensitive test for tissue typing, vital to organ transplantation. As of 2008, there is even a proposal to replace the traditional antibody-based tests for blood type with PCR-based tests.

- Many forms of cancer involve alterations to oncogenes. By using PCR-based tests to study these mutations, therapy regimens can sometimes be individually customized to a patient. PCR permits early diagnosis of malignant diseases such as leukemia and lymphomas, which is currently the highest-developed in cancer research and is already being used routinely. PCR assays can be performed directly on genomic DNA samples to detect translocation-specific malignant cells at a sensitivity that is at least 10,000 fold higher than that of other methods.

Infectious Disease Applications

PCR allows for rapid and highly specific diagnosis of infectious diseases, including those caused by bacteria or viruses. PCR also permits identification of non-cultivatable or slow-growing microorganisms such as mycobacteria, anaerobic bacteria, or viruses from tissue culture assays and animal models. The basis for PCR diagnostic applications in microbiology is the detection of infectious agents and the discrimination of non-pathogenic from pathogenic strains by virtue of specific genes.

Characterization and detection of infectious disease organisms have been revolutionized by PCR in the following ways:

- The human immunodeficiency virus (or HIV), is a difficult target to find and eradicate. The earliest tests for infection relied on the presence of antibodies to the virus circulating in the bloodstream. However, antibodies don't appear until many weeks after infection, maternal antibodies mask the infection of a newborn, and therapeutic agents to fight the infection don't affect the antibodies. PCR tests have been developed that can detect as little as one viral genome among the DNA of over 50,000 host cells. Infections can be detected earlier, donated blood can be screened directly for the virus, newborns can be immediately tested for infection, and the effects of antiviral treatments can be quantified.

- Some disease organisms, such as that for tuberculosis, are difficult to sample from patients and slow to be grown in the laboratory. PCR-based tests have allowed detection of small numbers of disease organisms (both live or dead), in convenient samples. Detailed genetic analysis can also be used to detect antibiotic resistance, allowing immediate and effective therapy. The effects of therapy can also be immediately evaluated.

- The spread of a disease organism through populations of domestic or wild animals can be monitored by PCR testing. In many cases, the appearance of new virulent sub-types can be detected and monitored. The sub-types of an organism that were responsible for earlier epidemics can also be determined by PCR analysis.

- Viral DNA can be detected by PCR. The primers used must be specific to the targeted sequences in the DNA of a virus, and PCR can be used for diagnostic analyses or DNA sequencing of the viral genome. The high sensitivity of PCR permits virus detection soon after infection and even before the onset of disease. Such early detection may give physicians a significant lead time in treatment. The amount of virus ("viral load") in a patient can also be quantified by PCR-based DNA quantitation techniques.

Forensic Applications

The development of PCR-based genetic (or DNA) fingerprinting protocols has seen widespread application in forensics:

- In its most discriminating form, genetic fingerprinting can uniquely discriminate any one person from the entire population of the world. Minute samples of DNA can be isolated from a crime scene, and compared to that from suspects, or from a DNA database of earlier evidence or convicts. Simpler versions of these tests are often used to rapidly rule out suspects during a criminal investigation. Evidence from decades-old crimes can be tested, confirming or exonerating the people originally convicted.

- Less discriminating forms of DNA fingerprinting can help in DNA paternity testing, where an individual is matched with their close relatives. DNA from unidentified human remains can be tested, and compared with that from possible parents, siblings, or children. Similar testing can be used to confirm the biological parents of an adopted (or kidnapped) child. The actual biological father of a newborn can also be confirmed (or ruled out).

Research Applications

PCR has been applied to many areas of research in molecular genetics:

- PCR allows rapid production of short pieces of DNA, even when not more than the sequence of the two primers is known. This ability of PCR augments many methods, such as generating hybridization probes for Southern or northern blot hybridization. PCR supplies these techniques with large amounts of pure DNA, sometimes as a single strand, enabling analysis even from very small amounts of starting material.
- The task of DNA sequencing can also be assisted by PCR. Known segments of DNA can easily be produced from a patient with a genetic disease mutation. Modifications to the amplification technique can extract segments from a completely unknown genome, or can generate just a single strand of an area of interest.
- PCR has numerous applications to the more traditional process of DNA cloning. It can extract segments for insertion into a vector from a larger genome, which may be only available in small quantities. Using a single set of 'vector primers', it can also analyze or extract fragments that have already been inserted into vectors. Some alterations to the PCR protocol can generate mutations (general or site-directed) of an inserted fragment.
- Sequence-tagged sites is a process where PCR is used as an indicator that a particular segment of a genome is present in a particular clone. The Human Genome Project found this application vital to mapping the cosmid clones they were sequencing, and to coordinating the results from different laboratories.
- An exciting application of PCR is the phylogenic analysis of DNA from ancient sources, such as that found in the recovered bones of Neanderthals, or from frozen tissues of mammoths. In some cases the highly degraded DNA from these sources might be reassembled during the early stages of amplification.
- A common application of PCR is the study of patterns of gene expression. Tissues (or even individual cells) can be analyzed at different stages to see which genes have become active, or which have been switched off. This application can also use quantitative PCR to quantitate the actual levels of expression
- The ability of PCR to simultaneously amplify several loci from individual sperm has greatly enhanced the more traditional task of genetic mapping by studying chromosomal crossovers after meiosis. Rare crossover events between very close loci have been directly observed by analyzing thousands of individual sperms. Similarly, unusual deletions, insertions, translocations, or inversions can be analyzed, all without having to wait (or pay for) the long and laborious processes of fertilization, embryogenesis, etc.

Limitations

One major limitation of PCR is that prior information about the target sequence is necessary in order to generate the primers that will allow its selective amplification. This means that, typically, PCR users must know the precise sequence(s) upstream of the target region on each of the two single-stranded templates in order to ensure that the DNA polymerase properly binds to the primer-template hybrids and subsequently generates the entire target region during DNA synthesis.

Like all enzymes, DNA polymerases are also prone to error, which in turn causes mutations in the PCR fragments that are generated. Additionally, the specificity of the PCR fragments can mutate to the template DNA, due to nonspecific binding of primers.

Variations

- *Allele-specific PCR*: a diagnostic or cloning technique based on single-nucleotide variations (single-base differences in a patient). It requires prior knowledge of a DNA sequence, including differences between alleles, and uses primers whose 3' ends encompass the SNV (base pair buffer around SNV usually incorporated). PCR amplification under stringent conditions is much less efficient in the presence of a mismatch between template and primer, so successful amplification with an SNP-specific primer signals presence of the specific SNP in a sequence.
- *Assembly PCR* or *Polymerase Cycling Assembly (PCA)*: artificial synthesis of long DNA sequences by performing PCR on a pool of long oligonucleotides with short overlapping segments. The oligonucleotides alternate between sense and antisense directions, and the overlapping segments determine the order of the PCR fragments, thereby selectively producing the final long DNA product.
- Asymmetric PCR: preferentially amplifies one DNA strand in a double-stranded DNA template. It is used in sequencing and hybridization probing where amplification of only one of the two complementary strands is required. PCR is carried out as usual, but with a great excess of the primer for the strand targeted for amplification. Because of the slow (arithmetic) amplification later in the reaction after the limiting primer has been used up, extra cycles of PCR are required. A recent modification on this process, known as *L*inear-*A*fter-*T*he-*E*xponential-PCR (LATE-PCR), uses a limiting primer with a higher melting temperature (Tm) than the excess primer to maintain reaction efficiency as the limiting primer concentration decreases mid-reaction.
- Convective PCR: a pseudo-isothermal way of performing PCR. Instead of repeatedly heating and cooling the PCR mixture, the solution is subjected to a thermal gradient. The resulting thermal instability driven convective flow automatically shuffles the PCR reagents from the hot and cold regions repeatedly enabling PCR. Parameters such as thermal boundary conditions and geometry of the PCR enclosure can be optimized to yield robust and rapid PCR by harnessing the emergence of chaotic flow fields. Such convective flow PCR setup significantly reduces device power requirement and operation time.
- *Dial-out PCR*: a highly parallel method for retrieving accurate DNA molecules for gene synthesis. A complex library of DNA molecules is modified with unique flanking tags before massively parallel sequencing. Tag-directed primers then enable the retrieval of molecules with desired sequences by PCR.
- *Digital PCR (dPCR)*: used to measure the quantity of a target DNA sequence in a DNA sample. The DNA sample is highly diluted so that after running many PCRs in parallel,

some of them do not receive a single molecule of the target DNA. The target DNA concentration is calculated using the proportion of negative outcomes. Hence the name 'digital PCR'.

- *Helicase-dependent amplification*: similar to traditional PCR, but uses a constant temperature rather than cycling through denaturation and annealing/extension cycles. DNA helicase, an enzyme that unwinds DNA, is used in place of thermal denaturation.
- Hot start PCR: a technique that reduces non-specific amplification during the initial set up stages of the PCR. It may be performed manually by heating the reaction components to the denaturation temperature (e.g., 95 °C) before adding the polymerase. Specialized enzyme systems have been developed that inhibit the polymerase's activity at ambient temperature, either by the binding of an antibody or by the presence of covalently bound inhibitors that dissociate only after a high-temperature activation step. Hot-start/cold-finish PCR is achieved with new hybrid polymerases that are inactive at ambient temperature and are instantly activated at elongation temperature.
- *In silico PCR* (digital PCR, virtual PCR, electronic PCR, e-PCR) refers to computational tools used to calculate theoretical polymerase chain reaction results using a given set of primers (probes) to amplify DNA sequences from a sequenced genome or transcriptome. In silico PCR was proposed as an educational tool for molecular biology.
- Intersequence-specific PCR (ISSR): a PCR method for DNA fingerprinting that amplifies regions between simple sequence repeats to produce a unique fingerprint of amplified fragment lengths.
- Inverse PCR: is commonly used to identify the flanking sequences around genomic inserts. It involves a series of DNA digestions and self ligation, resulting in known sequences at either end of the unknown sequence.
- Ligation-mediated PCR: uses small DNA linkers ligated to the DNA of interest and multiple primers annealing to the DNA linkers; it has been used for DNA sequencing, genome walking, and DNA footprinting.
- *Methylation-specific PCR* (MSP): developed by Stephen Baylin and Jim Herman at the Johns Hopkins School of Medicine, and is used to detect methylation of CpG islands in genomic DNA. DNA is first treated with sodium bisulfite, which converts unmethylated cytosine bases to uracil, which is recognized by PCR primers as thymine. Two PCRs are then carried out on the modified DNA, using primer sets identical except at any CpG islands within the primer sequences. At these points, one primer set recognizes DNA with cytosines to amplify methylated DNA, and one set recognizes DNA with uracil or thymine to amplify unmethylated DNA. MSP using qPCR can also be performed to obtain quantitative rather than qualitative information about methylation.
- *Miniprimer PCR*: uses a thermostable polymerase (S-Tbr) that can extend from short primers ("smalligos") as short as 9 or 10 nucleotides. This method permits PCR targeting to smaller primer binding regions, and is used to amplify conserved DNA sequences, such as the 16S (or eukaryotic 18S) rRNA gene.

- Multiplex ligation-dependent probe amplification (*MLPA*): permits amplifying multiple targets with a single primer pair, thus avoiding the resolution limitations of multiplex PCR.

- *Multiplex-PCR*: consists of multiple primer sets within a single PCR mixture to produce amplicons of varying sizes that are specific to different DNA sequences. By targeting multiple genes at once, additional information may be gained from a single test-run that otherwise would require several times the reagents and more time to perform. Annealing temperatures for each of the primer sets must be optimized to work correctly within a single reaction, and amplicon sizes. That is, their base pair length should be different enough to form distinct bands when visualized by gel electrophoresis.

- *Nanoparticle-Assisted PCR (nanoPCR)*: In recent years, it has been reported that some nanoparticles (NPs) can enhance the efficiency of PCR (thus being called nanoPCR), and some even perform better than the original PCR enhancers. It was also found that quantum dots (QDs) can improve PCR specificity and efficiency. Single-walled carbon nanotubes (SWCNTs) and multi-walled carbon nanotubes (MWCNTs) are efficient in enhancing the amplification of long PCR. Carbon nanopowder (CNP) was reported be able to improve the efficiency of repeated PCR and long PCR. ZnO, TiO_2, and Ag NPs were also found to increase PCR yield. Importantly, already known data has indicated that non-metallic NPs retained acceptable amplification fidelity. Given that many NPs are capable of enhancing PCR efficiency, it is clear that there is likely to be great potential for nanoPCR technology improvements and product development.

- *Nested PCR*: increases the specificity of DNA amplification, by reducing background due to non-specific amplification of DNA. Two sets of primers are used in two successive PCRs. In the first reaction, one pair of primers is used to generate DNA products, which besides the intended target, may still consist of non-specifically amplified DNA fragments. The product(s) are then used in a second PCR with a set of primers whose binding sites are completely or partially different from and located 3' of each of the primers used in the first reaction. Nested PCR is often more successful in specifically amplifying long DNA fragments than conventional PCR, but it requires more detailed knowledge of the target sequences.

- *Overlap-extension PCR* or *Splicing by overlap extension (SOEing)* : a genetic engineering technique that is used to splice together two or more DNA fragments that contain complementary sequences. It is used to join DNA pieces containing genes, regulatory sequences, or mutations; the technique enables creation of specific and long DNA constructs. It can also introduce deletions, insertions or point mutations into a DNA sequence.

- PAN-AC: uses isothermal conditions for amplification, and may be used in living cells.

- *quantitative PCR* (qPCR): used to measure the quantity of a target sequence (commonly in real-time). It quantitatively measures starting amounts of DNA, cDNA, or RNA. quantitative PCR is commonly used to determine whether a DNA sequence is present in a sample and the number of its copies in the sample. *Quantitative PCR* has a very high degree of precision. Quantitative PCR methods use fluorescent dyes, such as Sybr Green, EvaGreen or fluorophore-containing DNA probes, such as TaqMan, to measure the amount of amplified product in real time. It is also sometimes abbreviated to RT-PCR (*real-time* PCR) but this abbreviation should be used only for reverse transcription PCR. qPCR is the appropriate contractions for quantitative PCR (real-time PCR).

- *Reverse Transcription PCR (RT-PCR)*: for amplifying DNA from RNA. Reverse transcriptase reverse transcribes RNA into cDNA, which is then amplified by PCR. RT-PCR is widely used in expression profiling, to determine the expression of a gene or to identify the sequence of an RNA transcript, including transcription start and termination sites. If the genomic DNA sequence of a gene is known, RT-PCR can be used to map the location of exons and introns in the gene. The 5' end of a gene (corresponding to the transcription start site) is typically identified by RACE-PCR (*Rapid Amplification of cDNA Ends*).

- *Solid Phase PCR*: encompasses multiple meanings, including Polony Amplification (where PCR colonies are derived in a gel matrix, for example), Bridge PCR (primers are covalently linked to a solid-support surface), conventional Solid Phase PCR (where Asymmetric PCR is applied in the presence of solid support bearing primer with sequence matching one of the aqueous primers) and Enhanced Solid Phase PCR (where conventional Solid Phase PCR can be improved by employing high Tm and nested solid support primer with optional application of a thermal 'step' to favour solid support priming).

- *Suicide PCR*: typically used in paleogenetics or other studies where avoiding false positives and ensuring the specificity of the amplified fragment is the highest priority. It was originally described in a study to verify the presence of the microbe Yersinia pestis in dental samples obtained from 14th Century graves of people supposedly killed by plague during the medieval Black Death epidemic. The method prescribes the use of any primer combination only once in a PCR (hence the term "suicide"), which should never have been used in any positive control PCR reaction, and the primers should always target a genomic region never amplified before in the lab using this or any other set of primers. This ensures that no contaminating DNA from previous PCR reactions is present in the lab, which could otherwise generate false positives.

- *Thermal asymmetric interlaced PCR (TAIL-PCR)*: for isolation of an unknown sequence flanking a known sequence. Within the known sequence, TAIL-PCR uses a nested pair of primers with differing annealing temperatures; a degenerate primer is used to amplify in the other direction from the unknown sequence.

- *Touchdown PCR (Step-down PCR)*: a variant of PCR that aims to reduce nonspecific background by gradually lowering the annealing temperature as PCR cycling progresses. The annealing temperature at the initial cycles is usually a few degrees (3–5 °C) above the T_m of the primers used, while at the later cycles, it is a few degrees (3–5 °C) below the primer T_m. The higher temperatures give greater specificity for primer binding, and the lower temperatures permit more efficient amplification from the specific products formed during the initial cycles.

- Universal Fast Walking: for genome walking and genetic fingerprinting using a more specific 'two-sided' PCR than conventional 'one-sided' approaches (using only one gene-specific primer and one general primer—which can lead to artefactual 'noise') by virtue of a mechanism involving lariat structure formation. Streamlined derivatives of UFW are LaNe RAGE (lariat-dependent nested PCR for rapid amplification of genomic DNA ends), 5'RACE LaNe and 3'RACE LaNe.

History

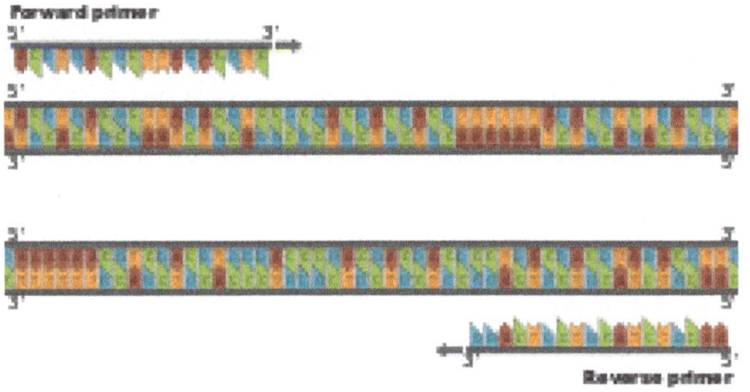

Diagrammatic representation of an example primer pair. The use of primers in an in vitro assay to allow DNA synthesis was a major innovation that allowed the development of PCR.

A 1971 paper in the Journal of Molecular Biology by Kjell Kleppe (no) and co-workers in the laboratory of H. Gobind Khorana first described a method using an enzymatic assay to replicate a short DNA template with primers *in vitro*. However, this early manifestation of the basic PCR principle did not receive much attention at the time, and the invention of the polymerase chain reaction in 1983 is generally credited to Kary Mullis.

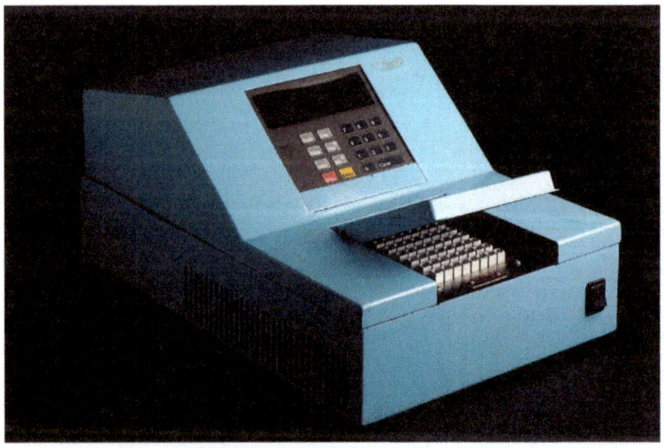

"Baby Blue", a 1986 prototype machine for doing PCR

When Mullis developed the PCR in 1983, he was working in Emeryville, California for Cetus Corporation, one of the first biotechnology companies. There, he was responsible for synthesizing short chains of DNA. Mullis has written that he conceived of PCR while cruising along the Pacific Coast Highway one night in his car. He was playing in his mind with a new way of analyzing changes (mutations) in DNA when he realized that he had instead invented a method of amplifying any DNA region through repeated cycles of duplication driven by DNA polymerase. In *Scientific American*, Mullis summarized the procedure: "Beginning with a single molecule of the genetic material

DNA, the PCR can generate 100 billion similar molecules in an afternoon. The reaction is easy to execute. It requires no more than a test tube, a few simple reagents, and a source of heat." He was awarded the Nobel Prize in Chemistry in 1993 for his invention, seven years after he and his colleagues at Cetus first put his proposal to practice. However, some controversies have remained about the intellectual and practical contributions of other scientists to Mullis' work, and whether he had been the sole inventor of the PCR principle.

At the core of the PCR method is the use of a suitable DNA polymerase able to withstand the high temperatures of >90 °C (194 °F) required for separation of the two DNA strands in the DNA double helix after each replication cycle. The DNA polymerases initially employed for in vitro experiments presaging PCR were unable to withstand these high temperatures. So the early procedures for DNA replication were very inefficient and time-consuming, and required large amounts of DNA polymerase and continuous handling throughout the process.

The discovery in 1976 of Taq polymerase — a DNA polymerase purified from the thermophilic bacterium, *Thermus aquaticus*, which naturally lives in hot (50 to 80 °C (122 to 176 °F)) environments such as hot springs — paved the way for dramatic improvements of the PCR method. The DNA polymerase isolated from *T. aquaticus* is stable at high temperatures remaining active even after DNA denaturation, thus obviating the need to add new DNA polymerase after each cycle. This allowed an automated thermocycler-based process for DNA amplification.

Patent Disputes

The PCR technique was patented by Kary Mullis and assigned to Cetus Corporation, where Mullis worked when he invented the technique in 1983. The *Taq* polymerase enzyme was also covered by patents. There have been several high-profile lawsuits related to the technique, including an unsuccessful lawsuit brought by DuPont. The pharmaceutical company Hoffmann-La Roche purchased the rights to the patents in 1992 and currently holds those that are still protected.

A related patent battle over the Taq polymerase enzyme is still ongoing in several jurisdictions around the world between Roche and Promega. The legal arguments have extended beyond the lives of the original PCR and Taq polymerase patents, which expired on March 28, 2005.

Reverse Transcription Polymerase Chain Reaction

Reverse transcription polymerase chain reaction (RT-PCR), a variant of polymerase chain reaction (PCR), is a technique commonly used in molecular biology to detect RNA expression. RT-PCR is often confused with real-time polymerase chain reaction (qPCR) by students and scientists alike, but they are separate and distinct techniques. While RT-PCR is used to qualitatively detect gene expression through creation of complementary DNA (cDNA) transcripts from RNA, qPCR is used to quantitatively measure the amplification of DNA using fluorescent dyes. qPCR is also referred to as quantitative PCR, quantitative real-time PCR, and real-time quantitative PCR.

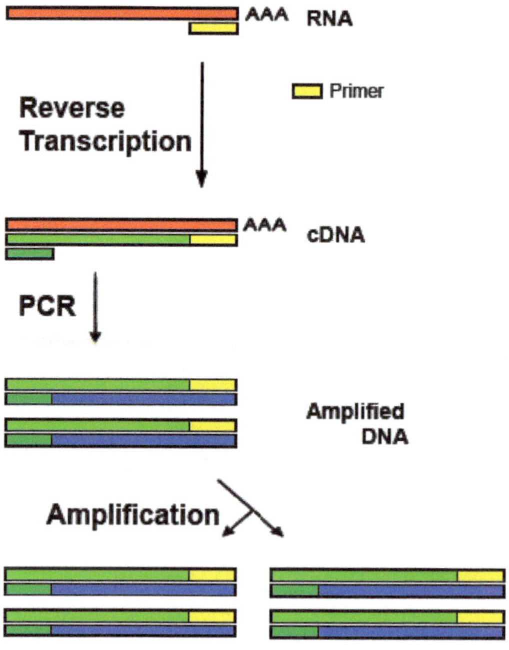

RT-PCR

Although RT-PCR and the traditional PCR both produce multiple copies of particular DNA isolates through amplification, the applications of the two techniques are fundamentally different. Traditional PCR is used to exponentially amplify target DNA sequences. RT-PCR is used to clone expressed genes by reverse transcribing the RNA of interest into its DNA complement through the use of reverse transcriptase. Subsequently, the newly synthesized cDNA is amplified using traditional PCR.

In addition to the qualitative study of gene expression, quantitative PCR can be utilized for quantification of RNA, in both relative and absolute terms, by incorporating qPCR into the technique. The combined technique, described as quantitative RT-PCR or real-time RT-PCR (sometimes even quantitative real-time RT-PCR), is often abbreviated as qRT-PCR, RT-qPCR, or RRT-PCR. Compared to other RNA quantification methods, such as northern blot, qRT-PCR is considered to be the most powerful, sensitive, and quantitative assay for the detection of RNA levels. It is frequently used in the expression analysis of single or multiple genes, and expression patterns for identifying infections and diseases.

In order to avoid confusion, the following abbreviations will be used consistently throughout this article:

Technique	Abbreviation
Polymerase chain reaction	PCR
Reverse transcription polymerase chain reaction	RT-PCR
Real-time polymerase chain reaction	qPCR
RT-PCR / qPCR combined technique	qRT-PCR

History

Since its introduction in 1977, Northern blot had been used extensively for RNA quantification despite its shortcomings: (a) time-consuming technique, (b) requires a large quantity of RNA for detection, and (c) quantitatively inaccurate in the low abundance of RNA content. However, the discovery of reverse transcriptase during the study of viral replication of genetic material led to the development of RT-PCR, which has since displaced Northern blot as the method of choicc for RNA detection and quantification.

RT-PCR has risen to become the benchmark technology for the detection and/or comparison of RNA levels for several reasons: (a) it does not require post PCR processing, (b) a wide range (>10^7-fold) of RNA abundance can be measured, and (c) it provides insight into both qualitative and quantitative data. Due to its simplicity, specificity and sensitivity, RT-PCR is used in a wide range of applications from experiments as simple as quantification of yeast cells in wine to more complex uses as diagnostic tools for detecting infectious agents such as the avian flu virus.

Principles

In RT-PCR, the RNA template is first converted into a complementary DNA (cDNA) using a reverse transcriptase. The cDNA is then used as a template for exponential amplification using PCR. QT-NASBA is currently the most sensitive method of RNA detection available. The use of RT-PCR for the detection of RNA transcript has revolutionalized the study of gene expression in the following important ways:

- Made it theoretically possible to detect the transcripts of practically any gene
- Enabled sample amplification and eliminated the need for abundant starting material that one faces when using northern blot analysis
- Provided tolerance for RNA degradation as long as the RNA spanning the primer is intact

One-step RT-PCR vs. Two-step RT-PCR

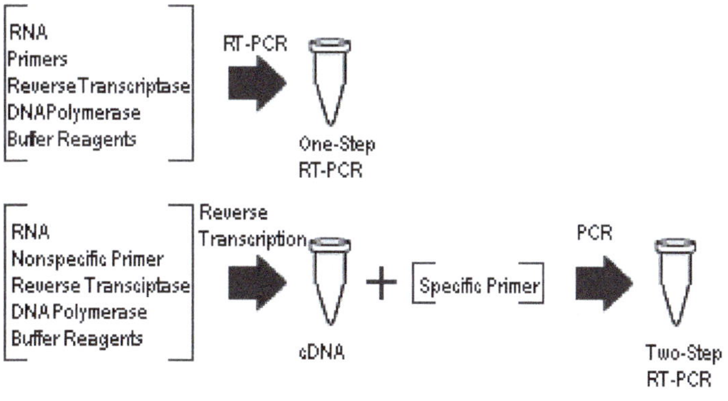

One-step vs Two-step RT-PCR

The quantification of mRNA using RT-PCR can be achieved as either a one-step or a two-step reaction. The difference between the two approaches lies in the number of tubes used when performing the procedure. In the one-step approach, the entire reaction from cDNA synthesis to PCR amplification occurs in a single tube. On the other hand, the two-step reaction requires that the reverse transcriptase reaction and PCR amplification be performed in separate tubes. The one-step approach is thought to minimize experimental variation by containing all of the enzymatic reactions in a single environment. However, the starting RNA templates are prone to degradation in the one-step approach, and the use of this approach is not recommended when repeated assays from the same sample is required. Additionally, one-step approach is reported to be less accurate compared to the two-step approach. It is also the preferred method of analysis when using DNA binding dyes such as SYBR Green since the elimination of primer-dimers can be achieved through a simple change in the melting temperature. The disadvantage of the two-step approach is susceptibility to contamination due to more frequent sample handling.

End-point RT-PCR vs. Real-time RT-PCR

Quantification of RT-PCR products can largely be divided into two categories: end-point and real-time. The use of end-point RT-PCR is preferred for measuring gene expression changes in small number of samples, but the real-time RT-PCR has become the gold standard method for validating results obtained from array analyses or gene expression changes on a global scale.

End-point RT-PCR

The measurement approaches of end-point RT-PCR requires the detection of gene expression levels by the use of fluorescent dyes like ethidium bromide, P32 labeling of PCR products using phosphorimager, or by scintillation counting. End-point RT-PCR is commonly achieved using three different methods: relative, competitive and comparative.

Relative RT-PCR: Relative quantifications of RT-PCR involves the co-amplification of an internal control simultaneously with the gene of interest. The internal control is used to normalize the samples. Once normalized, a direct comparison of relative transcript abundances across multiple samples of mRNA can be made. One precaution to note is that the internal control must be chosen so that it is not affected by the experimental treatment. The expression level should be constant across all samples and with the mRNA of interest for the results to be accurate and meaningful. Because the quantification of the results are analyzed by comparing the linear range of the target and control amplification, it is crucial to take into consideration the starting target molecules concentration and their amplification rate prior to starting the analysis. The results of the analysis are expressed as the ratios of gene signal to internal control signal, which the values can then be used for the comparison between the samples in the estimation of relative target RNA expression.

Competitive RT-PCR: Competitive RT-PCR technique is used for absolute quantification. It involves the use of a synthetic "competitor" RNA that can be distinguished from the target RNA by a small difference in size or sequence. It is important for the design of the synthetic RNA be identical in sequence but slightly shorter than the target RNA for accurate results. Once designed and synthesized, a known amount of the competitor RNA is added to experimental samples and is co-amplified with the target using RT-PCR. Then, a concentration curve of the competitor RNA is produced and it is used to compare the RT-PCR signals produced from the endogenous transcripts to determine the amount of target present in the sample.

Comparative RT-PCR: Comparative RT-PCR is similar to the competitive RT-PCR in that the target RNA competes for amplification reagents within a single reaction with an internal standard of unrelated sequence. Once the reaction is complete, the results are compared to an external standard curve to determine the target RNA concentration. In comparison to the relative and competitive quantification methods, comparative RT-PCR is considered to be the more convenient method to use since it does not require the investigator to perform a pilot experiment; in relative RT-PCR, the exponential amplification range of the mRNA must be predetermined and in competitive RT-PCR, a synthetic competitor RNA must be synthesized.

Real-time RT-PCR

The emergence of novel fluorescent DNA labeling techniques in the past few years have enabled the analysis and detection of PCR products in real-time and has consequently led to the widespread adoption of real-time RT-PCR for the analysis of gene expression. Not only is real-time RT-PCR now the method of choice for quantification of gene expression, it is also the preferred method of obtaining results from array analyses and gene expressions on a global scale. Currently, there are four different fluorescent DNA probes available for the real-time RT-PCR detection of PCR products: SYBR Green, TaqMan, Molecular Beacons, and Scorpions. All of these probes allow the detection of PCR products by generating a fluorescent signal. While the SYBR Green dye emits its fluorescent signal simply by binding to the double-stranded DNA in solution, the TaqMan probes, Molecular Beacons and Scorpions generation of fluorescence depend on Förster Resonance Energy Transfer (FRET) coupling of the dye molecule and a quencher moiety to the oligonucleotide substrates.

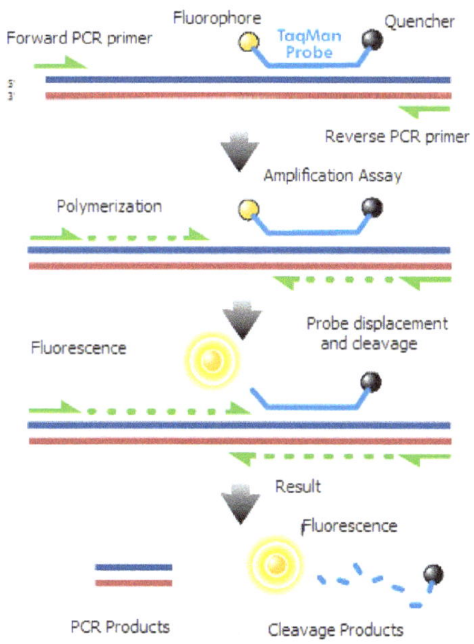

Taqman Probes

SYBR Green: When the SYBR Green binds to the double-stranded DNA of the PCR products, it will emit light upon excitation. The intensity of the fluorescence increases as the PCR products accumulate. This technique is easy to use since designing of probes is not necessary given lack of specificity of its binding. However, since the dye does not discriminate the double-stranded DNA from the PCR products and those from the primer-dimers, overestimation of the target concentration is a common problem. Where accurate quantification is an absolute necessity, further assay for the validation of results must be performed. Nevertheless, amongst the real-time RT-PCR product detection methods, SYBR Green is the most economical and easiest to use.

TaqMan Probes: TaqMan probes are oligonucleotides that have a fluorescent probe attached to the 5' end and a quencher to the 3' end. During PCR amplification, these probes will hybridize to the target sequences located in the amplicon and as polymerase replicates the template with TaqMan bound, it also cleaves the fluorescent probe due to polymerase 5'- nuclease activity. Because the close proximity between the quench molecule and the fluorescent probe normally prevents fluorescence from being detected through FRET, the decoupling results in the increase of intensity of fluorescence proportional to the number of the probe cleavage cycles. Although well-designed TaqMan probes produce accurate real-time RT-PCR results, it is expensive and time-consuming to synthesize when separate probes must be made for each mRNA target analyzed.

Molecular Beacon Probes: Similar to the TaqMan probes, Molecular Beacons also make use of FRET detection with fluorescent probes attached to the 5' end and a quencher attached to the 3' end of an oligonucleotide substrate. However, whereas the TaqMan fluorescent probes are cleaved during amplification, Molecular Beacon probes remain intact and rebind to a new target during each reaction cycle. When free in solution, the close proximity of the fluorescent probe and the quencher molecule prevents fluorescence through FRET. However, when Molecular Beacon probes hybridize to a target, the fluorescent dye and the quencher are separated resulting in the emittance of light upon excitation. As is with the TaqMan probes, Molecular Beacons are expensive to synthesize and require separate probes for each RNA target.

Scorpion Probes: The Scorpion probes, like Molecular Beacon, will not be fluorescent active in an unhybridized state, again, due to the fluorescent probe on the 5' end being quenched by the moiety on the 3' end of an oligonucleotide. With Scorpions, however, the 3' end also contains sequence that is complementary to the extension product of the primer on the 5' end. When the Scorpion extension binds to its complement on the amplicon, the Scorpion structure opens, prevents FRET, and enables the fluorescent signal to be measured.

Multiplex Probes: TaqMan probes, Molecular Beacons and Scorpions allow the concurrent measurement of PCR products in a single tube. This is possible because each of the different fluorescent dyes can be associated with a specific emission spectra. Not only does the use of multiplex probes save time and effort without compromising test utility, its application in wide areas of research such as gene deletion analysis, mutation and polymorphism analysis, quantitative analysis, and RNA detection, make it an invaluable technique for laboratories of many discipline.

Two strategies are commonly employed to quantify the results obtained by real-time RT-PCR; the standard curve method and the comparative threshold method.

Application

The exponential amplification via reverse transcription polymerase chain reaction provides for a highly sensitive technique in which a very low copy number of RNA molecules can be detected. RT-PCR is widely used in the diagnosis of genetic diseases and, semiquantitatively, in the determination of the abundance of specific different RNA molecules within a cell or tissue as a measure of gene expression.

Research Methods

RT-PCR is commonly used in research methods to measure gene expression. For example, Lin et al. used qRT-PCR to measure expression of Gal genes in yeast cells. First, Lin et al. engineered a mutation of a protein suspected to participate in the regulation of Gal genes. This mutation was hypothesized to selectively abolish Gal expression. To confirm this, gene expression levels of yeast cells containing this mutation were analyzed using qRT PCR. The researchers were able to conclusively determine that the mutation of this regulatory protein reduced Gal expression. Northern blot analysis is used to study the RNA's gene expression further.

Gene Insertion

RT-PCR can also be very useful in the insertion of eukaryotic genes into prokaryotes. Because most eukaryotic genes contain introns, which are present in the genome but not in the mature mRNA, the cDNA generated from a RT-PCR reaction is the exact (without regard to the error-prone nature of reverse transcriptases) DNA sequence that would be directly translated into protein after transcription. When these genes are expressed in prokaryotic cells for the sake of protein production or purification, the RNA produced directly from transcription need not undergo splicing as the transcript contains only exons. (Prokaryotes, such as E. coli, lack the mRNA splicing mechanism of eukaryotes).

Genetic Disease Diagnosis

RT-PCR can be used to diagnose genetic disease such as Lesch–Nyhan syndrome. This genetic disease is caused by a malfunction in the HPRT1 gene, which clinically leads to the fatal uric acid urinary stone and symptoms similar to gout. Analyzing a pregnant mother and a fetus for mRNA expression levels of HPRT1 will reveal if the mother is a carrier and if the fetus will likely to develop Lesch–Nyhan syndrome.

Cancer Detection

Scientists are working on ways to use RT-PCR in cancer detection to help improve prognosis, and monitor response to therapy. Circulating tumor cells produce unique mRNA transcripts depending on the type of cancer. The goal is to determine which mRNA transcripts serve as the best biomarkers for a particular cancer cell type and then analyze its expression levels with RT-PCR.

RT-PCR is commonly used in studying the genomes of viruses whose genomes are composed of RNA, such as Influenzavirus A and retroviruses like HIV.

Challenges

Despite its major advantages, RT-PCR is not without drawbacks. The exponential growth of the reverse transcribed complementary DNA (cDNA) during the multiple cycles of PCR produces inaccurate end point quantification due to the difficulty in maintaining linearity. In order to provide accurate detection and quantification of RNA content in a sample, qRT-PCR was developed using fluorescence-based modification to monitor the amplification products during each cycle of PCR. The extreme sensitivity of the technique can be a double edged sword since even the slightest DNA contamination can lead to undesirable results. A simple method for elimination of false positive results is to include anchors, or tags, to the 5' region of a gene specific primer. Additionally, planning and design of quantification studies can be technically challenging due to the existence of numerous sources of variation including template concentration and amplification efficiency.

Protocol

RT-PCR can be carried out by the one-step RT-PCR protocol or the two-step RT-PCR protocol.

One-step RT-PCR

One-step RT-PCR take mRNA targets (up to 6 kb) and subjects them to reverse transcription and then PCR amplification in a single test tube. Use only intact, high quality RNA for the best results. Be sure to use a sequence-specific primer.

1. Select a one-step RT-PCR kit, which should include a mix with reverse transcriptase and the PCR system such as Taq DNA Polymerase and a proofreading polymerase.
2. Obtain all necessary materials, equipment and instruments (kits should include a detailed list of necessary items).
3. Prepare a reaction mix, which will include dNTPs, primers, template RNA, necessary enzymes and a buffer solution.
4. Add the mix to a PCR tube for each reaction. Then add the template RNA.
5. Place PCR tubes in the thermal cycler to begin cycling. The first cycle is reverse transcription to synthesize cDNA. The second cycle is initial denaturation. During this cycle reverse transcriptase is inactivated. The next 40 to 50 cycles are the amplification program, which consists of three steps: (1) denaturation, (2) annealing, (3) elongation.
6. The RT-PCR products can then be analyzed with gel electrophoresis.

Two-step RT-PCR

Two-step RT-PCR, as the name implies, occurs in two steps. First the reverse transcription and then the PCR. This method is more sensitive than the one-step method. Kits are also useful for two-step RT-PCR. Just as for one-step, use only intact, high quality RNA for the best results. The primer for two-step does not have to be sequence specific.

Step One

1. Combine template RNA, primer, dNTP mix, and nuclease-free water in a PCR tube.
2. Add RNase inhibitor and reverse transcriptase to the PCR tube.
3. Place PCR tube in thermal cycler for one cycle that includes annealing, extending and then inactivating reverse transcriptase.
4. Proceed directly to PCR or store on ice until PCR can be performed.

Step Two

1. Add a master mix (containing buffer, dNTP mix, $MgCl_2$, Taq polymerase and nuclease-free water) to each PCR tube.
2. Add appropriate primer.
3. Place PCR tubes in thermal cycler for 30 cycles of the amplification program, which includes three steps: (1) denaturation, (2) annealing, (3) elongation.
4. The RT-PCR products can then be analyzed with gel electrophoresis.

Publication Guidelines

Quantitative RT-PCR assay is considered to be the gold standard for measuring the number of copies of specific cDNA targets in a sample but it is poorly standardized. As a result, while there are numerous publications utilizing the technique, many provide inadequate experimental detail and use unsuitable data analysis to draw inappropriate conclusions. Due to the inherent variability in the quality of any quantitative PCR data, reviewers not only have a difficult time evaluating these manuscripts, the studies also become impossible to replicate. Recognizing the need for the standardization of the reporting of experimental conditions, the Minimum Information for Publication of Quantitative Real-Time PCR Experiments (MIQE, pronounced mykee) guidelines have been published by the international consortium of academic scientists. The MIQE guidelines describe the minimum information necessary for evaluating quantitative PCR experiments that should be required for publication for encouraging better experimental practice and ensuring the relevance, accuracy, correct interpretation, and repeatability of quantitative PCR data.

Besides reporting guidelines, the MIQE stresses the need to standardize the nomenclature associated with quantitative PCR to avoid confusion; for example, the abbreviation qPCR should be used for quantitative real-time PCR and RT-qPCR should be used for reverse transcription-qPCR, and genes used for normalisation should be referred to as reference genes instead of housekeeping genes. It is also proposes that commercially derived terms like TaqMan® probes should not be used but instead referred to as hydrolysis probes. Additionally, it is proposed that quantification cycle (Cq) be used to describe the PCR cycle used for quantification instead of threshold cycle (Ct), crossing point (Cp), and takeoff point (TOP), which refer to the same value but were coined by different manufacturers of real-time instruments.

The guideline consists of the following elements: 1) experimental design, 2) sample, 3) nucleic acid extraction, 4) reverse transcription, 5) qPCR target information, 6) oligonucleotides, 7) protocol, 8) validation, and 9) data analysis. Specific items within each element carry a label of either E (essential) or D (desirable). Those labelled E are considered critical and indispensable while those labelled D are considered peripheral yet important for best-practices.

Nucleic Acid Hybridization

Nucleic acid hybridization is a basic technique in molecular biology which takes advantage of the ability of individual single-stranded nucleic acid molecules to form double-stranded molecules.

According to Watson-Crick base pairing, adenine binds with thymine and guanine binds with cytosine by hydrogen bonding.

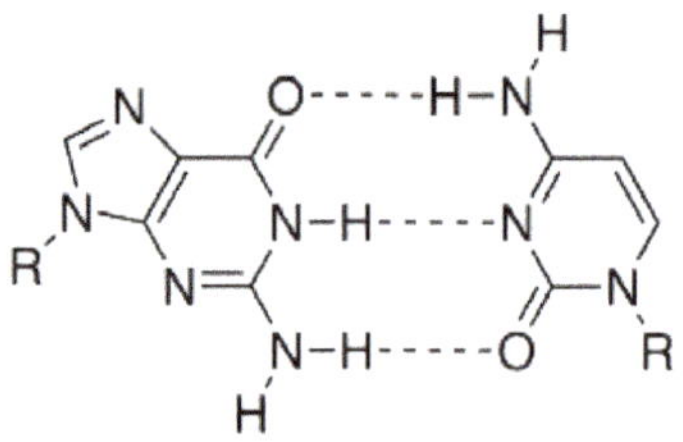

Guanine Cytosine

A GC base pair demonstrating three intermolecular hydrogen bonds

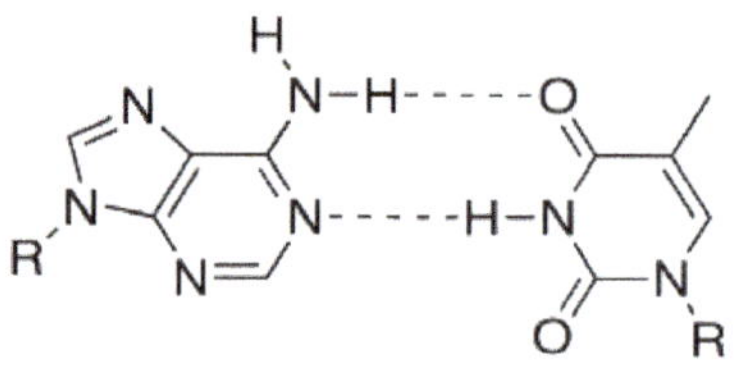

Adenine Thymine

An AT base pair demonstrating two intermolecular hydrogen bonds

Southern Hybridization

The basic principle behind the southern hybridization is the nucleic acid hybridization. Southern hybridization commonly known as southern blot is a technique employed for detection of a specific DNA sequence in DNA samples that are complementary to a given RNA or DNA sequence. It was the first blotting technique to be devised, named after its pioneer E.M Southern, a British biologist. Southern blotting involves separation of restricted DNA fragments by electrophoresis and then transferred to a nitrocellulose or a nylon membrane, followed by detection of the fragment using probe hybridization.

Separated by electrophoresis is transferred from gel to a membrane which in turn is used as a substrate for hybridization analysis employing labeled DNA or RNA probes specific to target fragments in the blotted DNA. Southern hybridization helps to detect specific fragment against a background of many other restriction fragments. Southern blotting is a technique which is used to confirm the identity of a cloned fragment or for recognition of a sub-fragment of interest from within the cloned DNA, or a genomic DNA. Southern blotting is a prerequisite to techniques such as restriction fragment length polymorphism (RFLP) analysis.

Procedure

1. The high-molecular-weight DNA strands are fractioned using restriction enzymes.
2. The DNA fragments are separated based on size by agarose gel electrophoresis.
3. The gel with the restricted fragments is then laid on a filter paper wick which serves as a connection between the membrane and the high salt buffer.
4. The nitrocellulose membrane is placed on top of the gel and a tower of filter papers is used to cover it and these are kept in place with a weight. The capillary action drives the buffer soaking through the filter paper wick, through the gel and the membrane and into the paper towels. Along with the buffer passing through the gel the DNA fragments are also carried with it into the membrane and they bind to the membrane. This causes an effective transfer of fragments (up to 15 kb in length taking around 18hours or overnight).

For DNA fragments larger than 15 kb, before blotting an acid such as diluted HCl is used to treat the gel that depurinates the DNA fragments causing breakage of DNA into smaller pieces, resulting in more efficient transfer from the gel to membrane.

Now a days blotting is also done by applying electric field. This electro blotting technique depends upon current and transfer buffer solution to nucleic acids onto a membrane. Following electrophoresis, a standard tank or semi-dry blotting transfer system is set up. A stack is put together in the following order from cathode to anode: sponge, three sheets of filter paper soaked in transfer buffer gel, PVDF or nitrocellulose membrane, three sheets of filter paper soaked in transfer buffer and then again sponge. Importantly the membrane should be located between the gel and the positively-charged anode, as the current and sample will be moving in that direction. Once the stack is prepared, it is placed in the transfer system, and suitable current is applied for a specific period of time according to the materials being used.

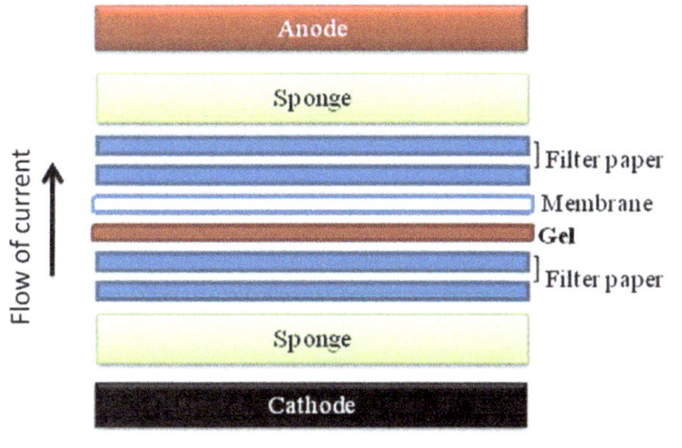

Set up for Electro-blotting system

5. For using alkaline transfer methods, the DNA gel is placed into an alkaline solution (like that of sodium hydroxide) causing denaturation of the double-stranded DNA. Denaturation in an alkaline environment enhances the binding between the negatively charged DNA and the positively charged membrane, causing separation to single DNA strands for further hybridization to the probe, alongside destroying any residual RNA that may persist in DNA. The membrane is washed with buffer to remove unbound DNA fragments.
6. The membrane which contains the transferred fragments is heated in presence or absence of vacuum at 80°C for 2 hours or exposed to ultraviolet radiation (nylon membrane) for permanent attachment of the transferred DNA to the membrane.
7. The obtained membrane is then hybridized with a probe (a DNA fragment with a specific sequence whose presence in the target DNA is to be determined).
8. Labeling of the probe DNA is done for easy detection, usually radioactivity is incorporated or the molecule is tagged with a fluorescent or chromogenic dye. The hybridization probe may be made of RNA, instead of DNA in some cases where the target is RNA specific.
9. Washing of the excess probe from the membrane is done by using saline sodium citrate(SSC buffer) after the hybridization step and the hybridization pattern is studied on an X-ray film by autoradiography (for a radioactive or fluorescent probe), or via color development on membrane if a chromogenic detection method is employed.

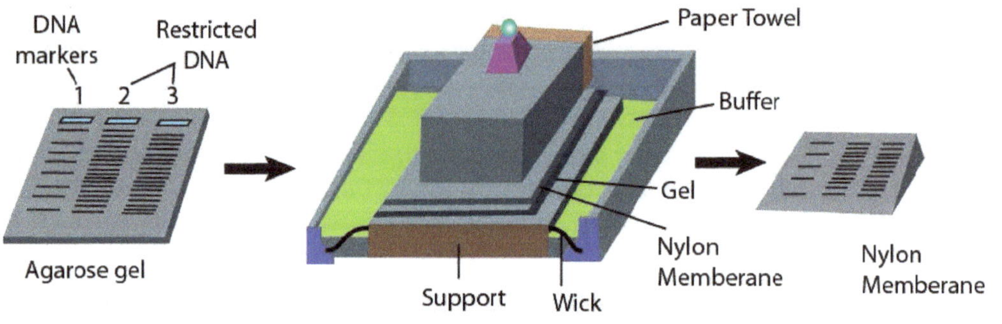

Steps of Southern Hybridization

Analysis of Southern Blot

Hybridization of the probe to a specific DNA fragment on the membrane indicates the presence of a complementary fragment in the DNA sequence. Southern hybridization performed by digestion of genomic DNA using a restriction enzyme digestion, helps in determining the number of sequences (or gene copies) in the genome. For a probe hybridizing to a single DNA segment that has not been cut by the restriction enzyme, a single band is observed and on the other hand multiple bands will likely be observed when the hybridization occurs between the probe and several highly similar target sequence (Due to sequence duplication). Alterations in the hybridization conditions like enhancing the hybridization temperature or decreasing salt concentration, helps in altering specificity and hybridization of the probe to sequences that are less than 100% similar.

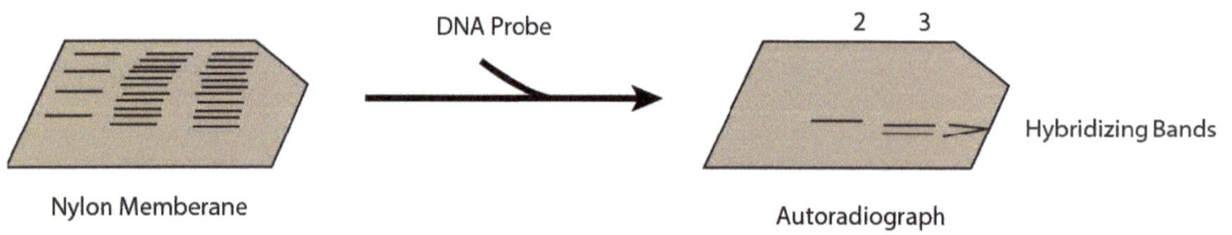

Southern hybridization analysis

Applications

Southern blotting has been exploited for various applications which include:

a. Clone identification: One of the most common applications of Southern blotting is identification and cloning of a specific gene of interest. Southern blotting is carried out for identification of one or more restriction fragments that contain the gene of interest in genomic DNA.. After cloning and tentative identification of the desired recombinant by employing colony or plaque hybridization, southern blotting is further is used to confirm the clone identification and possibly to locate a shorter restriction fragment, containing the sequence of interest.

b. Restriction fragment length polymorphism Analysis: Another major application of Southern hybridization is restriction fragment length polymorphism (RFLP) mapping, which is crucial in construction of genome maps.

Northern Hybridization

Northern blotting was developed by James Alwine, George Stark and David Kemp (1977). Northern blotting drives its name because of its similarity to the first blotting technique, which is Southern blotting, named after the biologist Edwin Southern. The major difference is that RNA being analyzed rather than DNA in the northern blot.

Expression of a particular gene can be detected by estimating the corresponding mRNA by Northern blotting. Northern blotting is a technique where RNA fragments are separated by electrophoresis and immobilized on a paper sheet.Identification of a specific RNA is then done by hybridization using a labeled nucleic acid probe. It helps to study gene expression by detection of RNA (or isolated mRNA) in a sample.

In Northern blotting, probes formed of nucleic acids with a sequence which is complementary to the sequence or to a part of the RNA of interest. The probe can be DNA, RNA or chemically synthesized oligonucleotides of minimum 25 complementary bases to the target sequence.

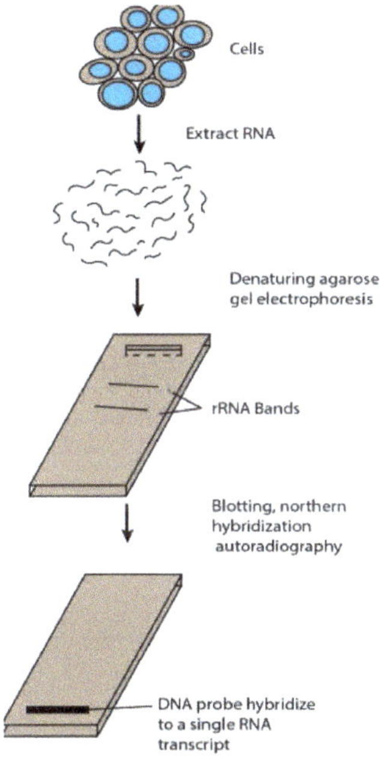

Steps of Northern Hybridization

Procedure

The northern blotting involves the following steps:

1. Total RNA is extracted from a homogenized tissue sample or cells. Further eukaryotic mRNA can then be isolated by using of oligo (dT) cellulose chromatography to isolate only those RNAs by making use of a poly A tail.

2. The isolated RNA is then separated by gel electrophoresis.

3. The RNA samples separated on the basis of size are transferred to a nylon membrane employing a capillary or vacuum based system for blotting.

Setup for Northern blotting

4. Similar to Southern blotting, the membrane filter is revealed to a labeled DNA probe that is complementary to the gene of interest and binds.

5. The labeled filter is then subjected to autoradiography for detection.

The net amount of a specific RNA in a sample can be estimated by using Northern blot. This technique is widely used for comparing the amounts of a particular mRNA in cells under different conditions. The separation of RNA samples is often done on agarose gels containing formaldehyde as a denaturing agent as it limits the RNA to form secondary structure.

Analysis of Northern Blot

RNA extract is electrophoresed in an agarose gel, using a denaturing electrophoresis buffer (containing formaldehyde) to ensure that the RNAs do not form inter- or intra-molecular base pairs, as base pairing would affect the rate at which the molecules migrate through the gel. After electrophoresis, the gel is blotted onto a nylon or nitrocellulose membrane, and hybridized with a labeled probe. If the probe is a cloned gene, the band that appears in the autoradiograph is the transcript of that gene. The size of the transcript can be determined from its position within the gel, and if RNA from different tissues is run in different lanes of the gel, then the possibility of differentially expressed gene can be examined.

Applications

Northern blotting helps in studying gene expression pattern of various tissues, organs, developmental stages, pathogen infection, and also over the course of treatment. It has been employed to study overexpression of oncogenes and down-regulation of tumor-suppressor genes in cancerous cells on comparison with healthy tissue, and also for gene expression of immune-rejection of transplanted organ.

The examination of the patterns of gene expressions obtained under given conditions can help determine the function of that gene.

Northern blotting is also used for the analysis of alternate spliced products of same gene or repetitive sequence motif by investigating the various sized RNA of the gene. This is done when only probe type with variation in one location is used to bind to the target RNA molecule.

Variations in size of a gene product may also help to identify deletions or errors in transcript processing, by altering the probe target that can be used along the known sequence and make it possible to determine the missing region of the RNA.

Colony Hybridization

It is a rapid method of isolating a colony containing a plasmid harboring a particular sequence or a gene from a mixed population. The colonies to be screened are first replica-plated on to a nitrocellulose filter disc that has been placed on the surface of an agar plate prior to inoculation. Master plate is retained for reference set of colonies. The filter bearing the colonies is removed and treated with alkali so that the bacterial colonies are lysed and the DNA they contain is denatured. The filter is then treated with proteinase K to remove protein and leave denatured DNA bound to the nitrocellulose. The DNA is fixed firmly by baking the filter at 80°C. A labeled probe is hybridized to this DNA which is monitored by autoradiography. A colony whose DNA print gives a positive auto radiographic result on X-ray film can then be picked from the reference plate.

Colony hybridization can be used to screen plasmid or cosmid based libraries.

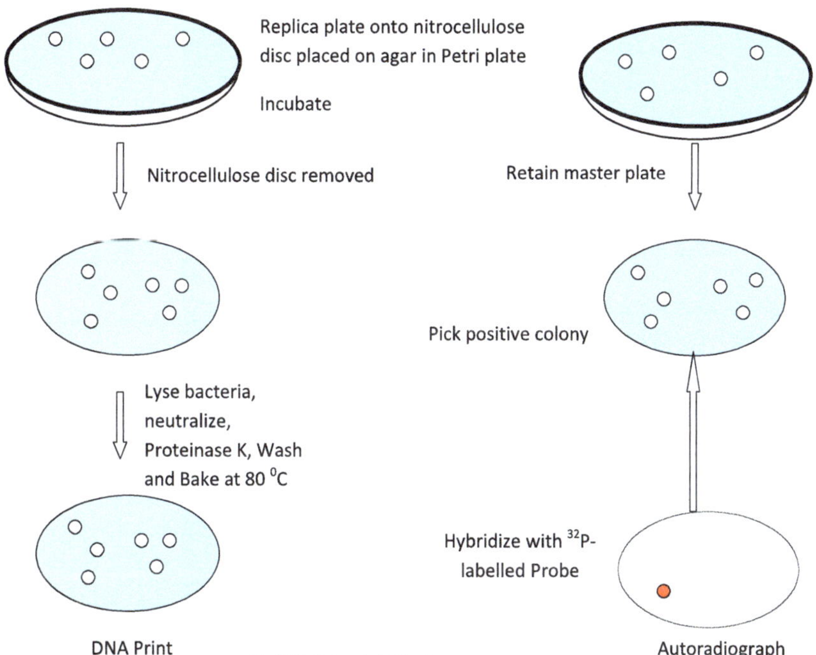

Fig 3-4.4: Colony hybridization

Colony hybridization

In Situ Hybridization (ISH)

It is a technique that employs a labeled complementary nucleotide strand (i.e. probe) for localizing specific DNA or RNA sequence targets within fixed tissues and cells (i.e *in situ*). Probes used for hybridization can be double-stranded DNA probes, single-stranded DNA probes, RNA probes, synthetic oligonucleotides. There are two ways available to detect DNA or RNA targets.

Chromogenic (CISH) in situ hybridization and Fluorescence in situ hybridization (FISH).

Chromogenic in Situ Hybridization (CISH)

It uses the labeling reactions involving alkaline phosphatase or peroxidase reactions to visualize the sample using bright-field microscopy. It is primarily used in molecular pathology diagnostics. CISH can also be employed for samples like fixed cells or tissues, blood or bone marrow smears and metaphase chromosome spreads.

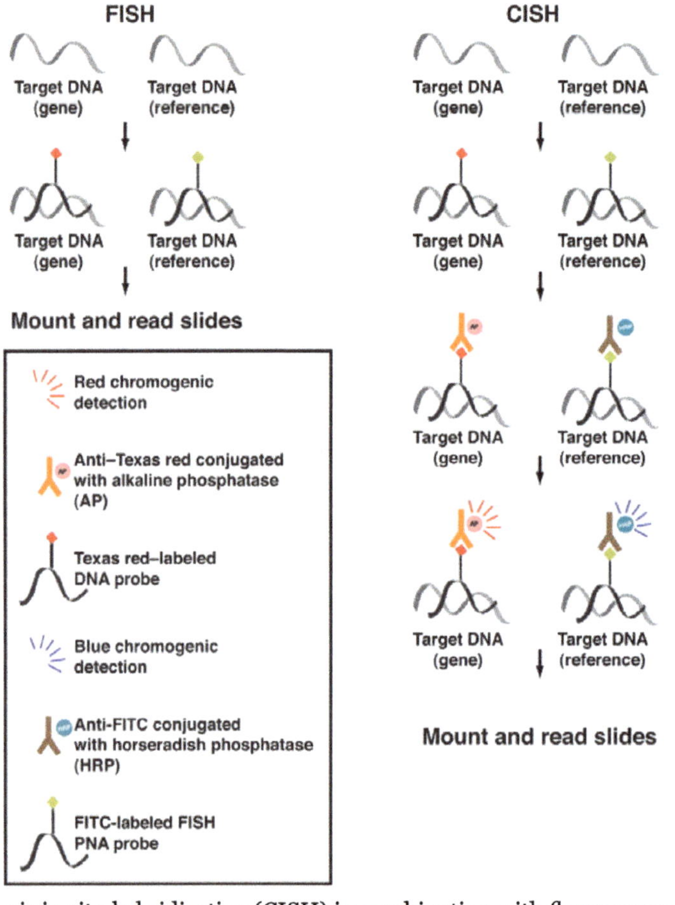

Use of dual-color Chromogenic in situ hybridization (CISH) in combination with fluorescence in situ hybridization (FISH) probes. FITC, fluorescein isothiocyanate; PNA, peptide nucleic acid.

Fluorescence in Situ Hybridization (FISH)

FISH is a cytogenetic technique that uses fluorescent probes that bind to complementary targets and sample is visualized using epi-fluorescence or confocal microscopy. Using differently labeled probes, we can visualize several targets in a single sample. It is used for spatial-temporal patterns of gene expression and resolving genetic elements in chromosomal preparations.

Cells or tissues to be analyzed are fixed and permeabilized with Proteinase K to allow target accessibility. Probe is constructed and tagged using non-radioactive labels like biotin, digoxigenin or fluorescent dye (FISH). Probe must be large enough to hybridize specifically with its target. Probe is applied to fixed sample and incubated to several hours to allow hybridization. Washing is done to remove non-specific or unbound probe removal. Results are then visualized using either bright-field or confocal microscopy.

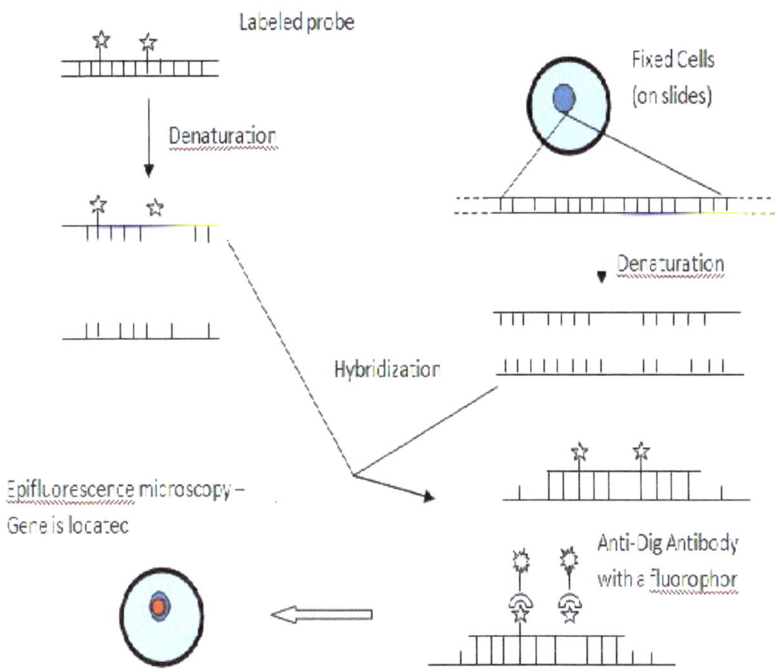

Fluorescent in-situ hybridization

Dot Blot and Slot Blot Hybridization

These two techniques represents the simplification of Southern and Western blots saving the time involved in procedures of chromatography, electrophoresis, restriction digestion and blotting of DNA or proteins from the gel to membrane. Here nucleic acid mixture is directly applied (blotted) on to the nylon or nitrocellulose membrane where hybridization between probe and target takes place, denatured to single-stranded form and baked at 80°C to bind DNA target to membrane. In dot-blot, target is blotted as circular blots whereas in slot-blots, it is in the form of rectangular blots. Due to this, slot-blot offers greater precision in observing different hybridization signals. After blotting, membrane is allowed to dry and non-specific sites are blocked by soaking in blocking buffer containing BSA. It is then followed by hybridization of labeled probe for detection of specific sequences or gene.

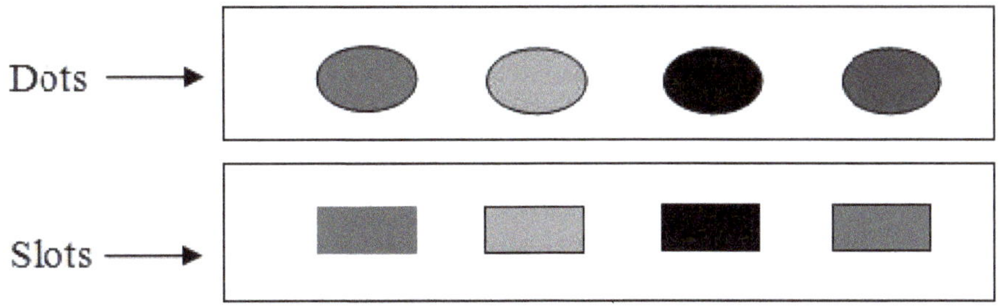

Dots and slots in dot and slot blot hybridization

These procedures can only detect presence and absence of particular sequence or gene. It cannot distinguish between two molecules of different sizes as they appear as single dot on membrane. It also has application in detecting alleles that differ in single nucleotide with the help of allele-specific oligonucleotides.

Probe and Target Sequences

Probes are short section of DNA or RNA with an additional tagged or labeled chemical entity that are used to bind to its complimentary strand and thereby allows detection of candidate nucleic acid molecule. This chemically synthesized entity can be a fluorescent molecule or it can be an attachment to a colored bead, or quantum dots (Cd-Se Qdots, Zn-SQdots), photochromic compounds, isotopic labeling, non-isotopic labeling etc. It allows us to visualize when a probe attaches to DNA, RNA or other target nucleic acids.

Homoduplex and Heteroduplex Between Probe and Target

For hybridization, the interacting single-stranded nucleic acid molecules should have a sufficiently high degree of base complementarity. It involves interaction of single strands of two sources of nucleic acids:

1. Probe consists of either chemically synthesizednucleic acids or modified oligonucleotides with known sequence that helps to identify similar or identical complementary sequences.
2. Target consists of template nucleic acid molecules with which probes will hybridize and form complex and heterogeneous mixture.

Probe or target can be in double-stranded form which must be separated before hybridization by heating or by alkaline treatment. Complementary base pairs are allowed to re-associate when single stranded probe is mixed with single stranded target. Various types of re-association can occur at this stage. Complementary probes and complementary targets can hybridize to each other forming homoduplexes. Both Homo and hetero-duplexes are useful for hybridization assay to detect a particular sequence in target.

The whole methodology revolves around the principle of hybridizing probe with a known sequence to a target which may not have related sequences to anneal with. Conditions of hybridization must be stringent enough to eliminate mismatched heteroduplexes. Low concentration of salts and high temperature increases the stringent conditions. If the probe used is small enough then hybridization reactions can be chosen such that heteroduplex is unstable even when there is single mismatch.

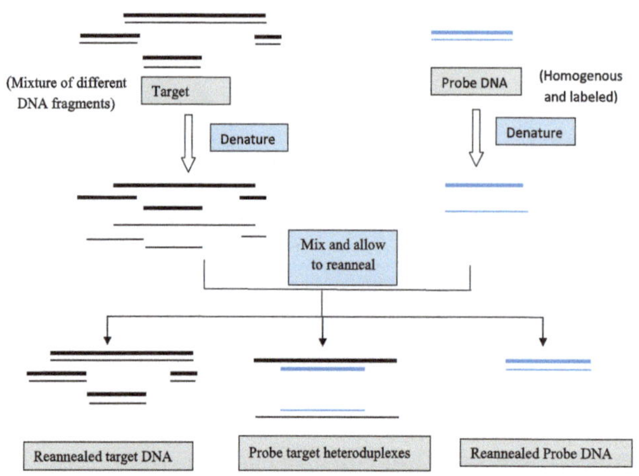

Method of Nucleic acid hybridization

Probe and target are first denatured and annealed forming the probe-probe homoduplexes and target-target homoduplexes, but the desired reaction is probe-target heteroduplex formation.

Gene Probes

Gene probes are generally longer than 500 bases and consist all or most of a target gene.

They can be generated in two ways. Cloned probes are normally used when a specific clone is available or when the DNA sequence is unknown and to be cloned first in order to be mapped and sequenced. Polymerase chain reaction is a powerful tool for making gene probes because it is possible to amplify and label simultaneously. Having the whole sequence of a gene, which can easily be obtained from Databases (Genbank, EMBL, DDBJ) primers can be designed to amplify the whole gene orgene fragments. It is more convenient when the gene of interest is PCR amplified because there is no need for restriction enzyme digestion, electrophoresis and elution of DNA fragments from vectors. If PCR amplification gives nonspecific bands, it is recommended to gel purify the specific band that will be used as a probe.

Gene probes generally provide greater specificity than oligonucleotides because of their longer sequence and because more detectable groups per probe molecule can be incorporated into them than into oligonucleotide probes.

Hybridization Probes

Probe is labeled in *standard hybridization assay* whereas target is labeled in *reverse hybridization assays* .Ease of preparation and reliability of labeled probes make the *standard hybridization* assays as prefcrable choice over reverse hybridization methods. Nucleic acid probes can be single-stranded or double-stranded molecules, but the w orking probe must be single-stranded. The probes,whichmay constitute either short sequences of DNA , RNA or synthetic oligonucleotides, are used in hybridization based techniques such as Southern, Northern blotting, colony and plaque hybridization, in situ hybridization and sequencing by hybridization.

Origin and characteristics of nucleic acid hybridization probes:

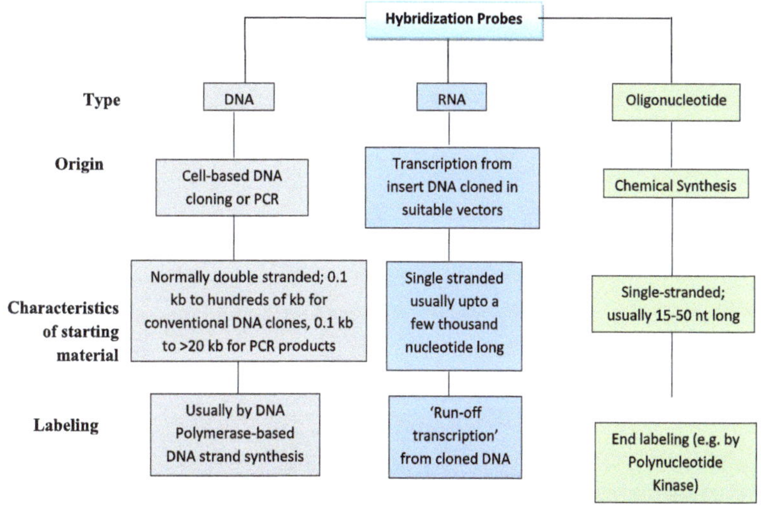

Features of Hybridization Probes

DNA Probes

Conventional DNA probes are isolated by cell-based DNA cloning or by PCR. In cell-based DNA cloning, the starting DNA can range from 0.1 kb to a range of ~10kb in length and is often double-stranded. PCR-derived DNA probes are usually 10 kb long and are usually double-stranded. These DNA probes are labeled using labeled dNTPs during in vitro synthesis.

RNA Probes

Detection of specific nucleic acid sequences can be accomplished by hybridization with a labeled RNA probe. RNA probes are sequences of a variable length that are used to detect the presence of complementary nucleotide sequences in a sample. RNA probes are first labeled with radioactive nucleotides or with modified nucleotides that can be detected by fluorescence or chemiluminescence. They can be used for Northern blotting, RNase protection assays, Southern blotting, downstream of polymerase chain reaction (PCR), and in situ hybridization analysis.

Synthetic Probes

These probes are chemically synthesized which can generate short single stranded stretches 15–50 bases in length. They are usually highly specific as they are designed particularly for target DNA. Probes with degenerate sequences can also be synthesized using protein sequences. Degeneracy is introduced when parallel synthesis of oligonucleotides is done with similarity at some sites and difference at others. These probes are usually labeled using 32 P isotope or other labeled group at the 5′ end.

Labeling of Probes

Probes can be labeled at specific location within the oligonucleotides or internally at multiple sites. Some probes are of defined length and some are heterogeneous populations of labeled molecules.

In vivo labeling: DNA and RNA can be directly labeled inside tissue culture cells by adding labeled deoxynucleotides in culture plate in vivo. This method is restricted only to prepare labeled viral DNA from virus-infected cells and to study RNA processing events.

In vitro labeling: It is a more versatile method involving *in vitro* labeling of purified RNA, DNA or oligonucleotide using DNA polymerasesfor incorporation of labeled nucleotides.

Labeling During Synthesis

In vitro labeling of DNA can be done by various methods as follows:

a) Nick-translation

b) Random primed labeling

c) PCR-mediated labeling- Labeling of RNA is generally accomplished by an *in vitro* transcription system.These procedures require DNA or RNA polymerases to add labeled nucleotides to synthesize in vitro probes. It requires at least one labeled nucleotides among four nucleotides.

DNA Nick Translation

It involves insertion of random single-strand breaks called nicks in one of the strands of double stranded target DNA which exposes 3′- OH termini and 5′-PO_4 termini.The nicks are introduced by endonuclease like pancreatic deoxyribonuclease I (DNase I).

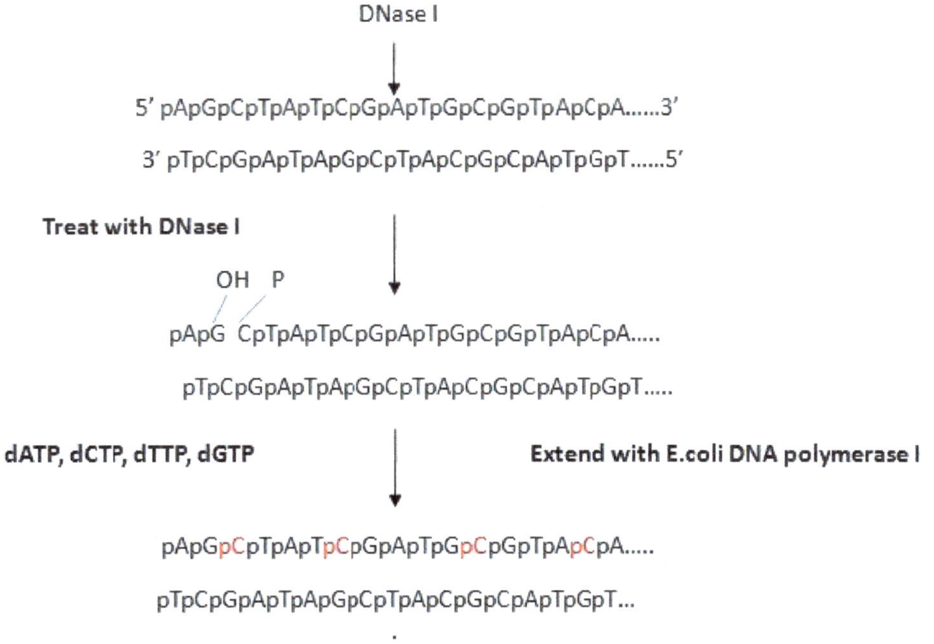

DNA Nick translation by DNase I

Addition of DNase I and multi-subunit enzyme *E. coli* DNA polymerase I is used for nick translation which contributes both activities like:

(i) $5' \rightarrow 3'$ exonuclease that attacks the exposed 5′ termini of a nick and sequentially removes the nucleotides in $5' \rightarrow 3'$ direction;

(ii) DNA polymerase adds nucleotides to the free 3′-OH group, in $5' \rightarrow 3'$ direction replacing the nucleotides removed by exonuclease and causing lateral displacement (translation) of the nick.

This method requires 100-fold less radioactive precursor than *in-vivo* labeling method. The amount of radiolabel incorporated depends on number of nicks created by DNase I. Too much or too little nicking must be avoided. At low temperatures (about 15°C), the disadvantage is that only one complete regeneration of existing nucleotide sequence takes place and reaction does not proceed further.

Random Primed DNA Labeling

This method known as oligo-labeling is based upon hybridization of a mixture of all possible hexanucleotides. The template DNA is initially denatured and then cooled slowly so as to allow random hexanucleotides to bind at complementary sequences at which extension takes place through PCR.

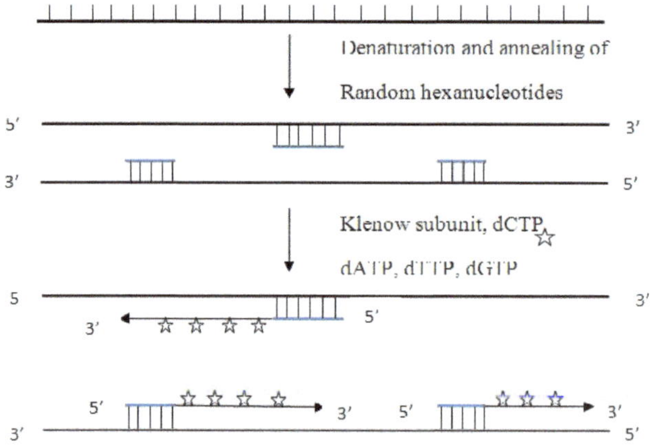

Random primed DNA labeling

The synthesis of new complementary DNA strands is primed by bound hexanucleotides and catalyzed by Klenow subunit of DNA polymerase I (contains the polymerase activity in the absence of associated 5'→3' exonuclease activities, sometimes 3'→5' exonuclease activity also removed for the reaction).This method produces labeled DNAs of high specific activity. Since binding of primer to template DNA is random and primer represents all possible sequence combinations,uniform labeling of DNA occurs.

The length of random primers is crucial. Primers shorter than 6 bases are very poor primers, whereas those longer than 7 have higher tendency to self–anneal and cause primer duplex. Thus, probe should be either 6 or 7 nucleotides in length.

Random priming is inherently simpler than nick translation because the requirements for two nuclease activities are eliminated. The probes generated by random priming method are more homogenous in size and behave more reproducibly in hybridization reactions. Average size of probe cannot be controlled with great accuracy in nick translation method.

Template DNA used preferably should not be closed, circular ds DNA because they are inefficient templates. Shorter templates generate probes of low specific activity that may not hybridize under stringent conditions.

PCR Mediated DNA Labeling

This method has several advantages over other methods.

- Defined segments of target DNA can be amplified and labeled independently of restriction sites.
- Amount of template DNA is required very small and
- No need to isolate fragments of DNA or to sub-clone into vectors containing bacteriophage promoters.

The standard PCR reaction can be modified to incorporate labeled nucleotides. The methods commonly used are:

- Standard PCR-based DNA labeling- The probe generation reaction is modified to incorporate one or more labeled nucleotide precursors at a concentration same as oligonucleotide conc.(K_m) or slightly above K_m and others at concentrations exceeding K_m, which become incorporated into the PCR product throughout its length.

- Primer-mediated 5′ end labeling. This method uses a 5' end-labeled primer, which is incorporated during PCR reaction. It is utilized in DNA sequencing, PCR based mutagenesis etc.

Radiolabeled probes can be generated for both strands using equal concentrations of primers or biased heavily in favor of one strand of DNA using higher concentration of one primer.

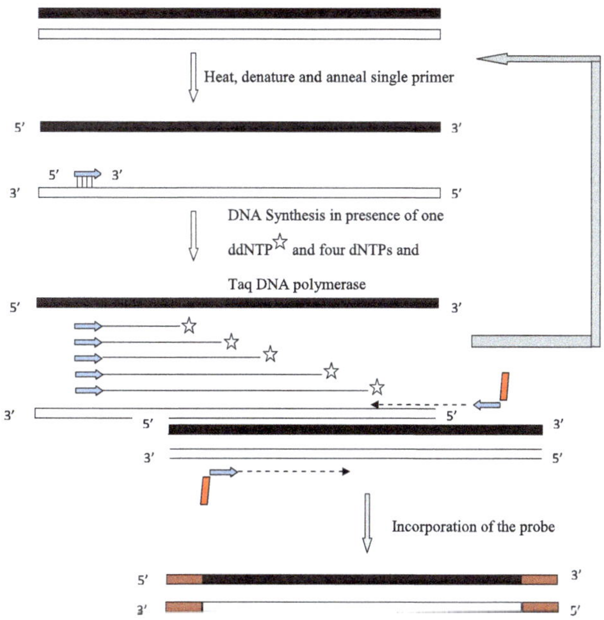

Standard PCR-based DNA labeling

Labeling of ssDNA, ssRNA and ds DNA with Protruding 5' Terminus or Blunt Ended or Recessed 5' Termini

Oligonucleotides are usually end-labeled using polynucleotide kinase. The label is in the form of a ^{32}P isotope at the γ-phosphate position of ATP. The polynucleotide kinase catalyzes an exchange reaction with the 5′-terminal phosphates at high concentration of dADP and low concentration of dATP. Alkaline phosphatase such as calf intestinal phosphatase (CIP) or shrimp alkaline phosphatase (SAP) can be used to remove 5' phosphate group and polynucleotide kinase (PNK) is used to add labeled phosphate using labeled dATP.

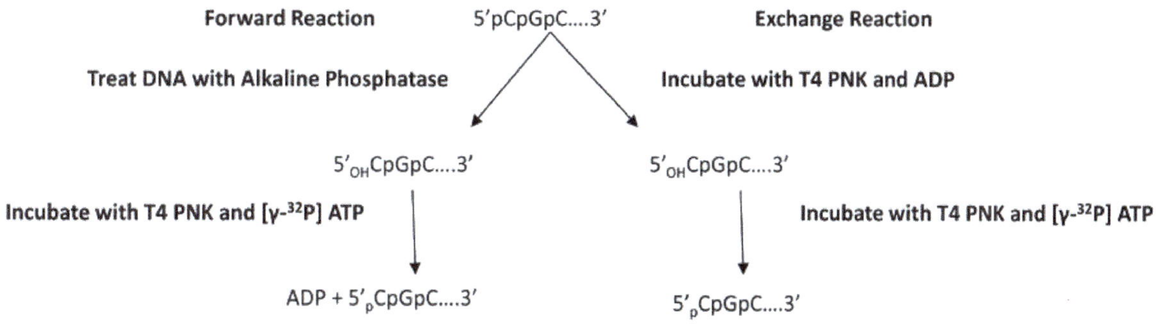

The same methodology is used for labeling double stranded DNA. Fragments which carry label at one end can only be generated using restriction enzyme which cleaves at internal sites, thus producing two fragments which can be separated by gel electrophoresis and purified. Fill-in end labeling is a popular approach to label large DNA fragments using Klenow subunit of E. coli DNA polymerase.

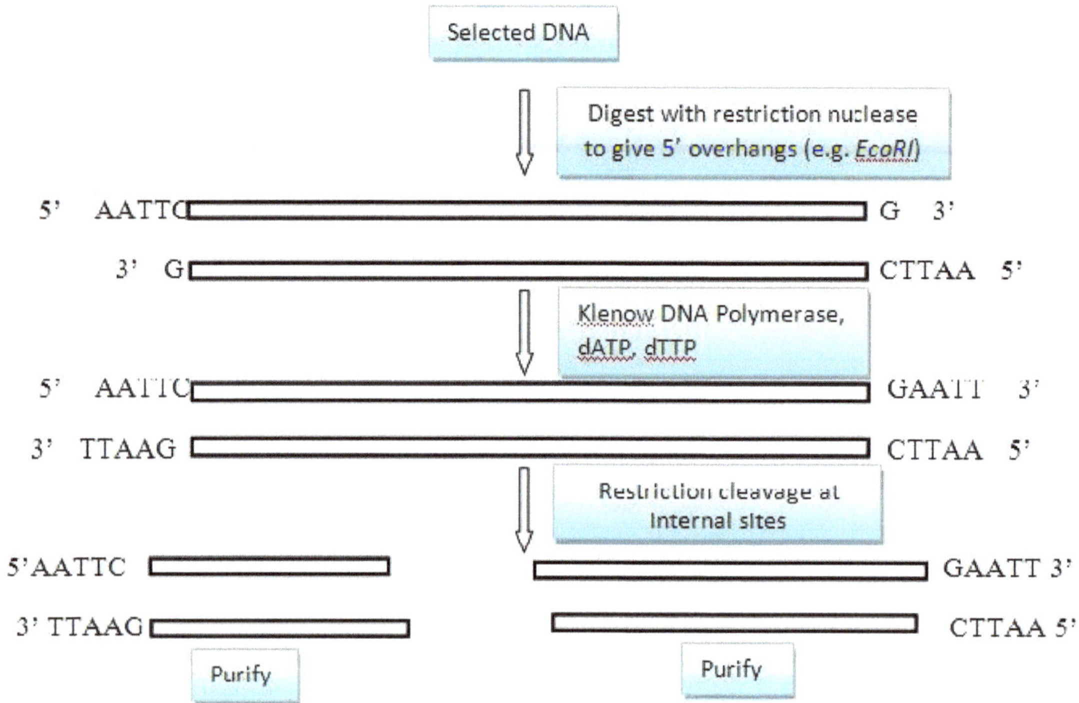

Labeling of double-stranded DNA

Labeling 3' Termini of dsDNA with Recessed 3' Termini or Blunt-ended:

Klenow fragment of *E.coli* DNA polymerase I is used to incorporate ^{32}P dNTPs into a recessed 3' terminus generated by restriction digestion. The choice of ^{32}P dNTP for labeling reaction depends on sequence of protruding 5' termini of DNA. End fragments created by restriction endonucleases can be labeled with ^{32}P dATP.

$$\begin{array}{llcll} 5'\text{-}G_{OH}\,3' \;+ & 5'_{P}\,AATTC\text{-}3' & & 5'\text{-}GAA_{OH}\text{-}3' \;+ & 5'_{P}AATTC\text{-}3' \\ 3'\text{-}CTTAA_{p}5' \;+ & 3'_{OH}G\text{-}5' & \longrightarrow & 3'\text{-}CTTAA_{p}5' \;+ & 3'_{OH}\,AAG\text{-}5' \end{array}$$

The terminal nucleotide of blunt end DNA fragment can be replaced by weak 3'→5' exonuclease and strong polymerase activity of klenow fragment.

Labeling 3' Termini of dsDNA with Protruding 3' Termini

Bacteriophage T4 DNA polymerase have strong 3'→5' exonuclease activity than of klenow fragment. It is used to digest 3'-protruding termini and then continues at slower pace to remove 3' nucleotides from ds DNA. At higher concentration of dNTPs, radiolabelled dNTP is added by polymerase activity.

Alternatively, protruding 3'termini can be labeled by calf thymus terminal transferase to catalyze the transfer of ^{32}P dideoxy ATP. Because it does not carry 3'-OH group, no additional nucleotides can be added to protruding 3'end.

Labeling RNA Probes

RNA probes yield stronger signals in hybridization reactions than DNA probes of equal specific activity which may be due to innately higher stability of hybrid involving RNA. They are the probes of choice when analyzing transcripts of mammalian genes.

RNA probes are most easily achieved by in vitro transcription in which insert DNA is cloned adjacent to multiple cloning sites in suitable plasmid expression vector. This expression vector has a phage promoter sequence which is recognized by corresponding phage RNA polymerase. In the given plasmid vector pSP64, bacteriophage SP6 promoter sequence is recognized by SP6 RNA polymerase which starts transcription from a particular initiation site after which insert has been cloned. Highly specific labeled RNA probes can be generated using radiolabeled dNTPs.

Riboprobes are mostly created by run-off transcription in which RNA insert is cloned in specialized plasmid vectors. The 5' terminus of the transcript is fixed by bacteriophage promoter, but 3' terminus is defined by downstream site of cleavage by restriction enzyme. Restriction enzymes that generate blunt or 5'-protuding termini produce best linear templates. Enzymes generating protruding 3' termini, should be avoided because transcription of such templates results in synthesis of RNA molecules that are aberrantly initiated at the termini of templates.

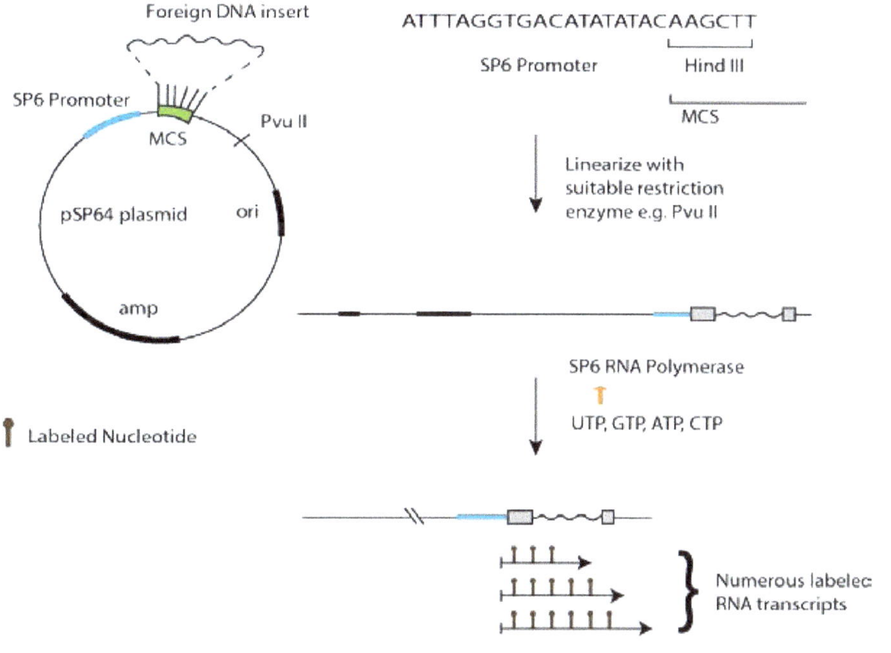

Labeling RNA probes

Commonly used systems for generation of riboprobes are Bacteriophage T3 and T7 promoter/ RNA polymerase systems. Labeled riboprobes, both sense and antisense c an be generated from any gene cloned in such vectors (in either of the two orientations). These are widely used in tissue *in situ* hybridization.

The DNA fragment to be transcribed can also be amplified in PCR with primers having 5' ends encode synthetic promoters for bacteriophage RNA polymerases. Following purification, the PCR products can be used as double stranded templates for in-vitro transcription.

Preparation of Labeled Nucleotides

Nucleotides can be labeled by isotopic and non-isotopic methods.

Isotopic Labeling

Isotopes generally used for labeling nucleotides are ^{32}P, ^{33}P, ^{35}S or ^{3}H. They can be detected directly in solution or on X-ray film using autoradiography.

Properties of radioisotopes used for labeling DNA and RNA probes:

Radioisotope	Half-life	Energy of emission
^{3}H	12.4 years	0.019 MeV
^{32}P	14.3 years	1.710 MeV
^{33}P	25.5 years	0.248 MeV
^{35}S	87.4 years	0.167 MeV

The strength of autoradiography signal depends on intensity of radiation emitted by radioisotope and duration of exposure. ^{32}P emits high energy β-particles which offer high detection sensitivity. Thus, it is widely used in Southern blot hybridization, dot-blot hybridization, colony hybridization. But it is relatively unstable and when fine resolution is required to interpret results, the image is unambiguous due to its high energy β-particle emission. Due to this, ^{35}S-labeled, ^{33}P-labeled (moderate half-lives) and ^{3}H-labeled nucleotides are used which emit less energetic β- radiation. They are used in DNA sequencing and in-situ hybridization. ^{3}H requires long exposure time due to low energy β-particle emission. Due to this, ^{35}S-labeled, ^{33}P-labeled (moderate half-lives) and ^{3}H-labeled nucleotides are used which emit less energetic β-radiation. They are used in DNA sequencing and in-situ hybridization. ^{3}H requires long exposure time due to low energy β-particle emission.

Non-isotopic Labeling

Non-isotopic labeling systems involve the use of nonradioactive probes. These methods are developed recently as compared to radioisotope labeling methods, but are finding wide variety of applications in different ways. Two types of non-radioactive labeling are conducted: direct and indirect.

Directnon-isotopic labeling, where a nucleotide containing label such as Fluorescein, Texas Red, Rhodamine that will be detected when incorporated with the help of spacer molecule. These modified nucleotides having tag, fluoresce when excited by light of certain wavelength.

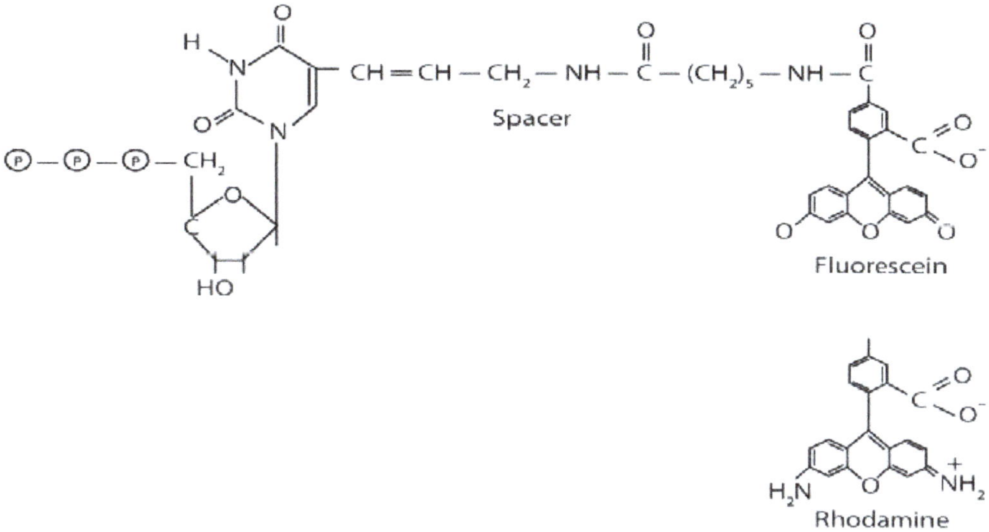

An example of fluorescein conjugated dUTP . The fluorescein group is linked to the 5′ carbon atom of the uridine by a spacer group. Similarly, Rhodamine can also be used in place of fluorescein.

Indirect non-isotopic labeling involves chemical linkage of reporter molecule to a nucleotide. When this modified nucleotide is incorporated into DNA, then it is specifically bound to a protein or other ligand which has high affinity against the reporter group. Long spacer is introduced between nucleotide and reporter so as to reduce steric hindrances for binding of affinity molecule.

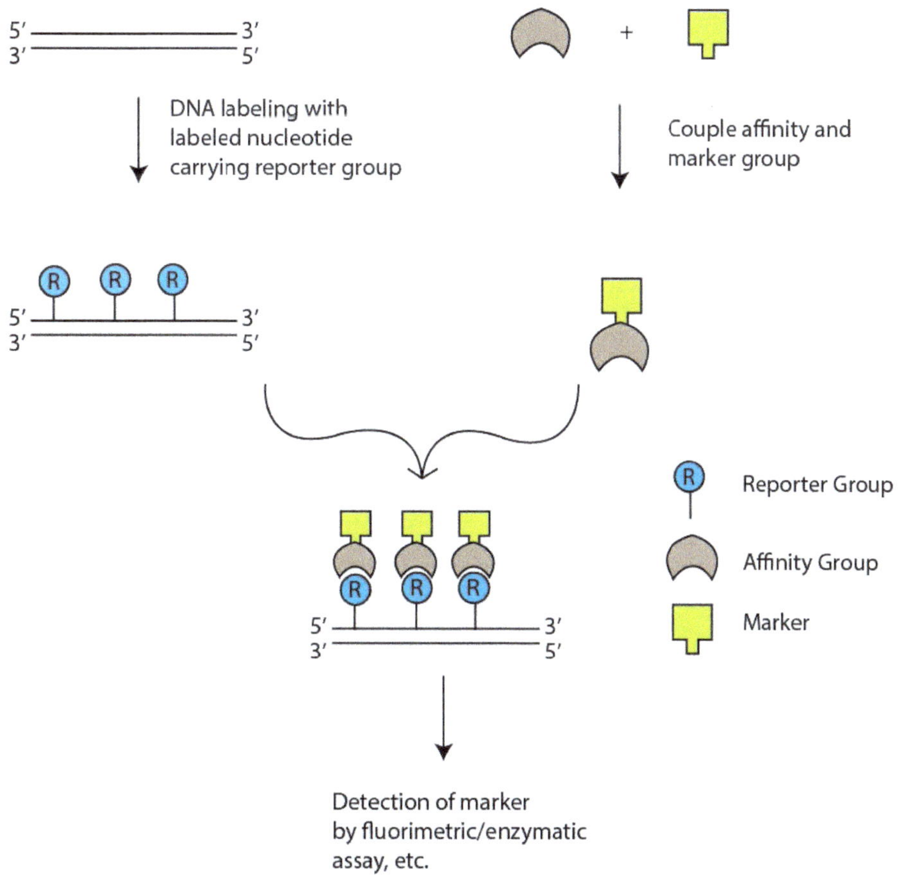

Indirect non-isotopic labeling

Two widely used non-isotopic labeling methods are:

Biotin-Streptavidin Method: This method uses two ligands which has high affinity towards each other: Biotin works as the reporter and the bacterial protein streptavidin is used as the affinity molecule. Biotinylated nucleotides like bio-11-dUTP are used as labeling agents with a spacer of 4-16 C atoms long between biotin and dNTP. However, Biotin is a ubiquitous constituent of mammalian tissues and tends to stick easily to certain type of nylon membranes which leads to high levels of background during in situ, northern and southern hybridization. To overcome this background problem, digoxigenin is used.

Digoxigenin, a plant steroid obtained from *Digitalis* plant and is used as a reporter and an affinity molecule. Digoxigenin is thus an all-purpose immuno-tag, and in particular a standard immunohistochemical marker for in situ hybridization.

Enzymatic methods are mostly used to label DNA probes with biotinylated nucleotides. Photochemical labeling of biotin to nucleic acids can also be used. The label is linked to nitrophenylazido group that is converted by UV irradiation to highly reactive nitrene that form stable covalent linkages to DNA or RNA.

End labeled DNA can be used as:

1. Molecular-weight standards in Southern blotting
2. Probes in gel-retardation experiments
3. Tracers for small quantities of DNAs on gels
4. Probes for screening bacterial colonies or plaques
5. Substrates for Maxam-Gilbert sequencing
6. Probes for RNA mapping with S1 Nuclease or Mung bean nuclease
7. Primers in primer-extension reactions.

Detection of non-radioactively labeled probes after hybridization:

Affinity molecules (streptavidin or digoxigenin-specific antibody) are conjugated with a variety of marker groups or molecules. They include various fluorophores or enzymes such as alkaline phosphatase and peroxidase which can permit detection via colorimetric assays, chemical luminescence assays or fluorescent assay.

In colorimetric assays, alkaline phosphatase catalyzes removal of phosphate group from BCIP (5-bromo-4-chloro-3-indolyl phosphate), generating a product that dimerizes to di-bromo-dichloro indigo, which reduces NBT (Nitrobluetetrazolium) to insoluble purple dye, diformazan that becomes visible at sites where probe has hybridized.

Fluorescent assays make use of HNPP (2-hydroxy-3-naphthoic acid 2'-phenylanilide phosphate). After de-phosphorylation by alkaline phosphatase, HNPP generates fluorescent precipitate on membranes that can be excited by irradiation at 290nm. The response signal emitted at 509nm are captured by CCD cameras.

Chemiluminescence is the fastest and most sensitive assay using HRP (Horseradish peroxidase) – luminal detection system. HRP catalyzes the oxidation of luminal in the presence of H_2O_2, generating reactive peroxide that emits light at 425nm during decomposition to its ground state.

Applications: This technique is used in physical mapping, karyotyping and phylogenetic analysis, gene expression profiling in developmental biology, pathogen profiling, and abnormal gene expression in pathology and for morphology and population structure of microorganisms.

Mutagenesis

Mutagenesis is a process by which the genetic information of an organism is changed, resulting in a mutation. It may occur spontaneously in nature, or as a result of exposure to mutagens. It can also be achieved experimentally using laboratory procedures. In nature mutagenesis can lead to cancer and various heritable diseases, but it is also a driving force of evolution. Mutagenesis as a science was developed based on work done by Hermann Muller, Charlotte Auerbach and J. M. Robson in the first half of the 20th century.

Background

DNA may be modified, either naturally or artificially, by a number of physical, chemical and biological agents, resulting in mutations. In 1927, Hermann Muller first demonstrated that mutation with observable changes in the chromosomes can be caused by irradiating fruit flies with X-ray, and lent support to the idea of mutation as the cause of cancer. His contemporary Lewis Stadler also showed the mutational effect of X-ray on barley in 1928, and ultraviolet (UV) radiation on maize in 1936. In 1940s, Charlotte Auerbach and J. M. Robson, found that mustard gas can also cause mutations in fruit flies.

While changes to the chromosome caused by X-ray and mustard gas were readily observable to the early researchers, other changes to the DNA induced by other mutagens were not so easily observable, and the mechanism may be complex and takes longer to unravel. For example, soot was suggested to be a cause of cancer as early as 1775, and coal tar was demonstrated to cause cancer in 1915. The chemicals involved in both were later shown to be polycyclic aromatic hydrocarbons (PAH). PAHs by themselves are not carcinogenic, and it was proposed in 1950 that the carcinogenic forms of PAHs are the oxides produced as metabolites from cellular processes. The metabolic process was identified in 1960s as catalysis by cytochrome P450 which produces reactive species that can interact with the DNA to form adducts,; the mechanism by which the PAH adducts give rise to mutation, however, is still under investigation.

Mammalian nuclear DNA may sustain more than 60,000 damage episodes per cell per day, as listed with references in DNA damage (naturally occurring). If left uncorrected, these adducts, after misreplication past the damaged sites, can give rise to mutations. In nature, the mutations that arise may be beneficial or deleterious — this is the driving force of evolution. An organism may acquire new traits through genetic mutation, but mutation may also result in impaired function of the genes, and in severe cases, cause the death of the organism. In the laboratory, however, mutagenesis is a useful technique for generating mutations that allows the functions of genes and gene products to be examined in detail, producing proteins with improved characteristics or novel functions, as well as mutant strains with useful properties. Initially, the ability of radiation and chemical mutagens to cause mutation was exploited to generate random mutations, but later techniques were developed to introduce specific mutations.

The Distinction Between a Mutation and DNA Damage

DNA damage is an abnormal alteration in the structure of DNA that cannot, itself, be replicated when DNA replicates. In contrast, a mutation is a change in the nucleic acid sequence that can be replicated; hence, a mutation can be inherited from one generation to the next. Damage can occur from chemical addition (adduct), or structural disruption to a base of DNA (creating an abnormal nucleotide or nucleotide fragment), or a break in one or both DNA strands. When DNA containing damage is replicated, an incorrect base may be inserted in the new complementary strand as it is being synthesized. The incorrect insertion in the new strand will occur opposite the damaged site in the template strand, and this incorrect insertion can become a mutation (i.e. a changed base pair) in the next round of replication. Furthermore, double-strand breaks in DNA may be repaired by an inaccurate repair process, non-homologous end joining, which produces mutations. Mutations can ordinarily be avoided if accurate DNA repair systems recognize DNA damage and repair it prior to completion of the next round of replication. At least 169 enzymes are either directly employed in DNA repair or influence DNA repair processes. Of these, 83 are directly employed in the 5 types of DNA repair processes indicated in the chart shown in the article DNA repair.

Mechanisms

Mutagenesis may occur endogenously, for example, through spontaneous hydrolysis, or through normal cellular processes that can generate reactive oxygen species and DNA adducts, or through error in replication and repair. Mutagenesis may also arise as a result of the presence of environmental mutagens that induce changes to the DNA. The mechanism by which mutation arises varies according to the causative agent, the mutagen, involved. Most mutagens act either directly, or indirectly via mutagenic metabolites, on the DNA producing lesions. Some, however, may affect the replication or chromosomal partition mechanism, and other cellular processes.

Many chemical mutagens require biological activation to become mutagenic. An important group of enzymes involved in the generation of mutagenic metabolites is cytochrome P450. Other enzymes that may also produce mutagenic metabolites include glutathione S-transferase and microsomal epoxide hydrolase. Mutagens that are not mutagenic by themselves but require biological activation are called promutagens.

Many mutations arise as a result of problems caused by DNA lesions during replication, resulting in errors in replication. In bacteria, extensive damage to DNA due to mutagens results in single-stranded DNA gaps during replication. This induces the SOS response, an emergency repair process that is also error-prone, thereby generating mutations. In mammalian cells, stalling of replication at damaged sites induces a number of rescue mechanisms that help bypass DNA lesions, but which also may result in errors. The Y family of DNA polymerases specializes in DNA lesion bypass in a process termed translesion synthesis (TLS) whereby these lesion-bypass polymerases replace the stalled high-fidelity replicative DNA polymerase, transit the lesion and extend the DNA until the lesion has been passed so that normal replication can resume. These processes may be error-prone or error-free.

Spontaneous Hydrolysis

DNA is not entirely stable in aqueous solution. Under physiological conditions the glycosidic bond may be hydrolyzed spontaneously and 10,000 purine sites in DNA are estimated to be depurinated each day in a cell. Numerous DNA repair pathways exist for DNA; however, if the apurinic site is not repaired, misincorporation of nucleotides may occur during replication. Adenine is preferentially incorporated by DNA polymerases in an apurinic site.

Cytidine may also become deaminated to uridine at one five-hundredth of the rate of depurination and can result in G to A transition. Eukaryotic cells also contain 5-methylcytosine, thought to be involved in the control of gene transcription, which can become deaminated into thymine.

Modification of Bases

Bases may be modified endogenously by normal cellular molecules. For example, DNA may be methylated by S-adenosylmethionine, and glycosylated by reducing sugars.

Many compounds, such as PAHs, aromatic amines, aflatoxin and pyrrolizidine alkaloids, may form reactive oxygen species catalyzed by cytochrome P450. These metabolites form adducts with the DNA, which can cause errors in replication, and the bulky aromatic adducts may form stable intercalation between bases and block replication. The adducts may also induce conformational changes in the DNA. Some adducts may also result in the depurination of the DNA; it is, however, uncertain how significant such depurination as caused by the adducts is in generating mutation.

Alkylation and arylation of bases can cause errors in replication. Some alkylating agents such as N-Nitrosamines may require the catalytic reaction of cytochrome-P450 for the formation of a reactive alkyl cation. N^7 and O^6 of guanine and the N^3 and N^7 of adenine are most susceptible to attack. N^7-guanine adducts form the bulk of DNA adducts, but they appear to be non-mutagenic. Alkylation at O^6 of guanine, however, is harmful because excision repair of O^6-adduct of guanine may be poor in some tissues such as the brain. The O^6 methylation of guanine can result in G to A transition, while O^4-methylthymine can be mispaired with guanine. The type of the mutation generated, however, may be dependent on the size and type of the adduct as well as the DNA sequence.

Ionizing radiation and reactive oxygen species often oxidize guanine to produce 8-oxoguanine.

DNA Damage and Spontaneous Mutation

As noted above, the number of DNA damage episodes occurring in a mammalian cell per day is high (more than 60,000 per day). Frequent occurrence of DNA damage is likely a problem for all DNA- containing organisms, and the need to cope with DNA damage and minimize their deleterious effects is likely a fundamental problem for life.

Most spontaneous mutations likely arise from error-prone trans-lesion synthesis past a DNA damage site in the template strand during DNA replication. This process can overcome potentially lethal blockages, but at the cost of introducing inaccuracies in daughter DNA. The causal relationship of DNA damage to spontaneous mutation is illustrated by aerobically growing *E. coli* bacteria, in which 89% of spontaneously occurring base substitution mutations are caused by reactive oxygen species (ROS)-induced DNA damage. In yeast, more than 60% of spontaneous single-base pair substitutions and deletions are likely caused by trans-lesion synthesis.

An additional significant source of mutations in eukaryotes is the inaccurate DNA repair process non-homologous end joining, that is often employed in repair of double strand breaks.

In general, it appears that the main underlying cause of spontaneous mutation is error prone trans-lesion synthesis during DNA replication and that the error-prone non-homologous end joining repair pathway may also be an important contributor in eukaryotes.

Crosslinking

Some alkylating agents may produce crosslinking of DNA. Some natural occurring chemicals may also promote crosslinking, such as psoralens after activation by UV radiation, and nitrous acid. Interstrand cross-linking is more damaging as it blocks replication and transcription and can cause chromosomal breakages and rearrangements. Some crosslinkers such as cyclophosphamide, mitomycin C and cisplatin are used as anticancer chemotherapeutic because of their high degree of toxicity to proliferating cells.

Dimerization

UV radiation promotes the formation of a cyclobutyl ring between adjacent thymines, resulting in the formation of pyrimidine dimers. In human skin cells, thousands of dimers may be formed in a day due to normal exposure to sunlight. DNA polymerase η may help bypass these lesions in an error-free manner; however, individuals with defective DNA repair function, such as sufferers of Xeroderma pigmentosum, are sensitive to sunlight and may be prone to skin cancer.

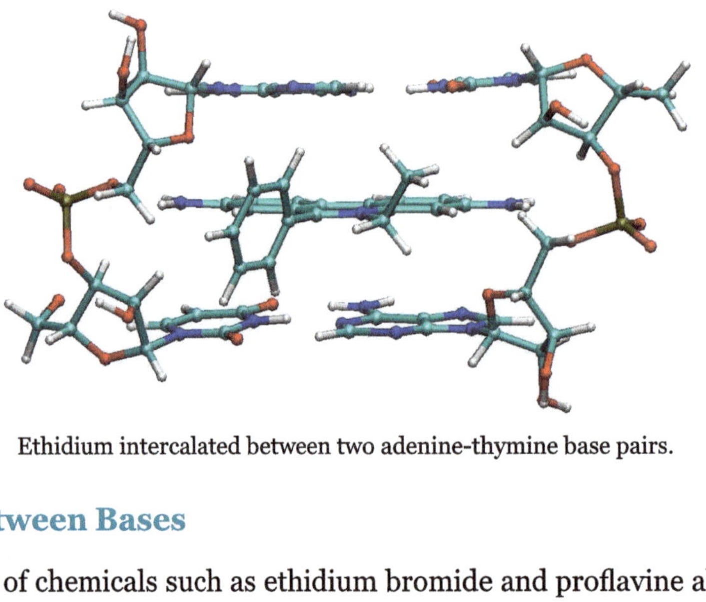

Ethidium intercalated between two adenine-thymine base pairs.

Intercalation between Bases

The planar structure of chemicals such as ethidium bromide and proflavine allows them to insert between bases in DNA. This insert causes the DNA's backbone to stretch and makes slippage in DNA during replication more likely to occur since the bonding between the strands is made less stable by the stretching. Forward slippage will result in deletion mutation, while reverse slippage will result in an insertion mutation. Also, the intercalation into DNA of anthracyclines such as daunorubicin and doxorubicin interferes with the functioning of the enzyme topoisomerase II, blocking replication as well as causing mitotic homologous recombination.

Backbone Damage

Ionizing radiation may produce highly reactive free radicals that can break the bonds in the DNA. Double-stranded breakages are especially damaging and hard to repair, producing translocation and deletion of part of a chromosome. Alkylating agents like mustard gas may also cause breakages in the DNA backbone. Oxidative stress may also generate highly reactive oxygen species that can damage DNA. Incorrect repair of other damage induced by the highly reactive species can also lead to mutations.

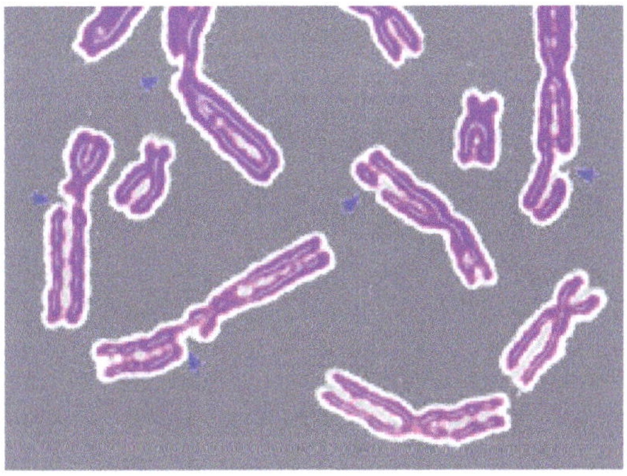

Arrows indicates chromosomal breakages due to DNA damage

Insertional Mutagenesis

Transposons and viruses may insert DNA sequences into coding regions or functional elements of a gene and result in inactivation of the gene.

Effects on Replication and DNA Repair

While most mutagens produce effects that ultimately result in errors in replication, for example creating adducts that interfere with replication, some mutagens may directly affect the replication process or reduce its fidelity. Base analog such as 5-bromouracil may substitute for thymine in replication. Metals such as cadmium, chromium, and nickel can increase mutagenesis in a number of ways in addition to direct DNA damage, for example reducing the ability to repair errors, as well as producing epigenetic changes.

As a Laboratory Technique

Mutagenesis in the laboratory is an important technique whereby DNA mutations are deliberately engineered to produce mutant genes, proteins, or strains of organism. Various constituents of a gene, such as its control elements and its gene product, may be mutated so that the functioning of a gene or protein can be examined in detail. The mutation may also produce mutant proteins with interesting properties, or enhanced or novel functions that may be of commercial use. Mutant strains may also be produced that have practical application or allow the molecular basis of particular cell function to be investigated.

Early methods of mutagenesis produced entirely random mutations; however, later methods of mutagenesis may produce site-specific mutation.

Types

- Directed mutagenesis
- Site-directed mutagenesis
- PCR mutagenesis
- Insertional mutagenesis
- Signature tagged mutagenesis
- Transposon mutagenesis
- Sequence Saturation Mutagenesis

Kunkel's Method

In the early 1980's, Thomas Kunkel developed a technique which reduced the need of selection of the mutants. Candidate DNA fragment to be mutated is inserted into a phagemid such as M13mp18/19 and then transformed into an *E. coli* strain deficient in dUTPase (*dut*) and uracil deglycosidase (*ung*). These enzymes are part of a DNA repair pathway that protects the bacterial chromosome from mutations by the spontaneous deamination of dCTP to dUTP. The dUTPase deficiency prevents the breakdown of dUTP, resulting in increased level of dUTP in the cell. The uracil deglycosidase deficiency prevents the removal of uracil from newly-synthesized DNA. As the double-mutant *E. coli* replicates the phage DNA, its enzymatic machinery may therefore mis-incorporate dUTP instead of dTTP, resulting in single stranded DNA which contains ssUDNA. The ssUDNA is extracted from the bacteriophage released into the medium, and used as a template for mutagenesis. An oligonucleotide containing the desired mutation is used for primer extension. The heteroduplex DNA containing one parental non-mutated strand of dUTP and a mutated strand containing dTTP is then transformed into an *E. coli* strain carrying the wild type *dut* and *ung* genes. Here, the uracil-containing parental DNA strand is degraded, so that nearly all resulting DNA consists of the mutated strand.

Cassette Mutagenesis

In cassette mutagenesis, a fragment of DNA containing the mutation in the gene of interest is synthesized. Then it has been inserted into a plasmid after the cleavage by a restriction enzyme at a particular site in the plasmid and subsequent ligation. Usually the restriction enzymes that cleave the plasmid and candidate oligonucleotide are same to allow generation of sticky ends of the plasmid and insert to ligate to one another. This method can generate mutants at close to 100% efficiency, but in practice limited by the availability of suitable restriction sites.

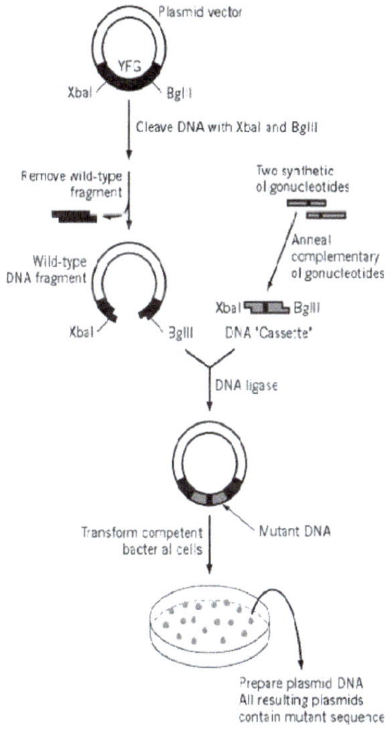

Schematic flow diagram of cassette mutagenesis

A plasmid containing a target gene YFG (black segment) is cleaved with restriction enzymes XbaI and BglII, each of which has unique restriction site in the entire plasmid. The reaction mixture is separated by agarose gel electrophoresis. Two ss oligonucleotides are synthesized by automated DNA synthesis. The sequences of the oligonucleotides are complementary to each other and differ from wild-type sequence at only a single position (black stripe) containing the desired changes. The oligonucleotides are mixed that helps in hybridization of the two strands by complementation. The ends of the duplex fragment are single-stranded, sticky ends that join with XbaI and BglII sites. The DNA cassette is mixed with the isolated fragment and two molecules are covalently joined by action of T4 DNA ligase. The ligated DNA is transformed into E. coli, and drug-resistant colonies are selected. Plasmid DNA is prepared from individual bacterial colonies. Since the two linear fragments themselves cannot transform E. coli, all colonies contain plasmids with the mutant sequence.

PCR Based site Directed Mutagenesis

The limitation of restriction sites in cassette mutagenesis may be overcome using polymerase chain reaction with specific oligonucleotide "primers", such that a larger fragment may be generated covering two convenient restriction sites. The exponential amplification in PCR produces a fragment having the desired mutation in sufficient quantity to be isolated from original, unmutated plasmid by gel electrophoresis. There are many variations of the same technique. The simplest method results the mutation site towards one of the ends of the fragment whereby one of two oligonucleotides used for generating the fragment contains the mutation. This involves a single step PCR, but still has the problem of requiring a suitable restriction site near the mutation site unless a very long primer is used. Other variations therefore employ three or four oligionucleotides, two may be non-mutagenic oligonucleotides that cover two convenient restriction sites and generate a fragment that can be digested and ligated. Mutagenic oligonucleotides may be complementary to a location within that fragment well away from any convenient restriction site. These methods require multiple steps of PCR so that the final fragment to be ligated can contain the desired mutation.

In-vivo Site Directed Mutagenesis

In vivo genome manipulation through site-directed mutagenesis can be done by various methods described as below.

a) *In vivo* site-directed mutagenesis with synthetic oligonucleotides:

This system combines the versatility of synthetic oligonucleotides for targeting with the practicality of a general selection system. It provides for an enormously wide variety of genome modifications via homologous recombination. Exceptional high frequencies of mutations can be obtained when a site-specific double-strand break (DSB) is induced within the locus targeted by the synthetic oligonucleotides. (Storici F *et al, 2006*)

b) Trans-placement "pop-in pop-out"

Most gene targeting experiments have been used to disrupt endogenous loci, resulting in targeted null alleles (strategy is often termed as 'gene knockout'). Two types of vector have been developed for this purpose: insertion vectors and replacement (or transplacement) vectors.

Insertion vectors are linearized within the homology region, resulting in insertion of entire vector into the target locus. This type of vector disrupts the target gene but leads to a duplication of the sequences adjacent to the selectable marker.

Replacement vectors are designed so that the homology region is collinear with the target. The vector is linearized outside the homology region prior to transfection, thus crossover events in which endogenous DNA is replaced by the insert DNA. With this type of vector, only sequences within the homology region (not the vector backbone) are inserted. Thus for gene knockout, the homology region itself must be interrupted. Insertion and replacement vectors are equally efficient, but replacement vectors have been used in the majority of knockout experiments.

c) Direct gene deletion and site-specific mutagenesis with PCR and one recyclable marker using long homologous regions.

Several methods have been developed till now to obtain a successful mismatch mutation (transition/ transversion) in the targeted site. Some of them are discussed below:

Spontaneous Hydrolysis

DNA is not entirely stable in aqueous solution. Under certain physiological conditions glycosidic bond may be hydrolyzed impulsively and thousands of purine sites in DNA are estimated to be depurinated daily in a cell. Several DNA repair pathways exist for the DNA, however, if apurinic site is irrepairable, mis-incorporation of nucleotide may take place during replication. Adenine is incorporated by DNA polymerases in an apurinic site. Cytidine may also deaminated to uridine at lower rate of depurination and can form G to A transition. Eukaryotic cells also contain

5'-methylcytosine, may be involved in the control of gene transcription, which can become deaminated into thymine.

Modification of Bases

Bases may be modified endogenously by normal cellular molecules. For example DNA may be methylated by S-adenosylmethionine (SAM) and glycosylated by reducing sugars.

Many compounds, such as PAHs, aromatic amines, aflatoxin and pyrrolizidine alkaloids, may form reactive oxygen species (ROS) catalyzed by cytochrome P450. These metabolites form adducts with the DNA, which can cause errors in replication, and thus bulky aromatic adducts may form stable intercalation between bases and block replication. The adducts may also induce conformational changes in the DNA. Some adducts also result in the depurination of the DNA, however it may vary how significant such depurination is in generating mutation.

Alkylation and arylation of bases can cause errors in replication. Some alkylating agents such as N-Nitrosamines may require the catalytic reaction of cytochrome-P450 for the formation of a reactive alkyl cation. N7 and O6 of guanine and the N3 and N7 of adenine are most susceptible to attack. N7 of guanine adducts form the bulk of DNA adducts, but they appear to be non-mutagenic. Alkylation at O6 of guanine however is harmful because excision repair of O6-adduct of guanine may be poor in some tissues such as the brain. The O6 methylation of guanine can result in G to A transition, while O4-methylthymine can be mis-paired with guanine. The type of the mutation generated however may be dependent on the size and type of the adduct as well as the DNA sequence. Ionizing radiations and reactive oxygen species often oxidize guanine to produce 8-oxoguanine.

Cross Linking of DNA

Some alkylating agents produce cross linking of DNA. Some natural occurring chemicals such as psoralens after activation by UV radiation, and nitrous acid may also promote cross linking. Inter-strand cross-linking is more lethal as it blocks replication and transcription and can cause chromosomal breakages and rearrangements. Some cross linkers such as cyclophosphamide, mitomycin C and cisplatin are used as anticancer agent because of their high degree of toxicity to proliferating cells.

Dimerization

UV radiation promotes the formation of a cyclobutyl ring between adjacent thymines, resulting in the formation of pyrimidine dimers. In human skin cells, thousands of dimers may be formed in a day due to normal exposure to sunlight. Human DNA polymerase kappa help bypass these lesions in an error-free manner however, individuals with defective DNA repair function, such as sufferers of *Xeroderma pigmentosum* (autosomal recessive genetic disorder), are sensitive to sunlight and may be prone to skin cancer.

Intercalation Between Bases

The planar structure of chemicals such as ethidium bromide and proflavine allows them to insert between stacked bases in ds DNA. This insert causes the DNA's backbone to stretch and makes slippage during replication more likely to occur since the bonding between the strands is made less stable by the stretching. Forward slippage results in deletion mutation, while reverse slippage results in an insertion mutation. The intercalation of anthracyclines such as daunorubicin and doxorubicin also interferes with the functioning of the enzyme topoisomerase II, blocking replication as well as causing mitotic homologous recombination.

Backbone Damage

Ionizing radiations may produce highly reactive free radicals that can break bonds in the DNA. Double-stranded breakages are particularly damaging and tough to repair, forms translocation and deletion of part of a chromosome. Alkylating agents like mustard gas cause breakages in the DNA backbone. Oxidative stress produces highly reactive oxygen species (ROS) which cause damage to DNA. Incorrect repair of other damages induced by the highly reactive species also lead to mutations.

T-DNA Mediated Mutagenesis

Transfer-DNA (T-DNA) insertion is a highly effective mutagen for genome-wide mutagenesis (Krysan *et al.* 1999). It has been widely exploited to produce insertion mutants in *Arabidopsis thaliana* for functional characterization of the every gene in the genome. Even if mRNA is transcribed, T-DNA sequence may contain stop codons, resulting in early translation termination (Krysan *et al.* 1999). Although it is not an ideal technique, T-DNA insertion mutagenesis has been a powerful technique to link genes to phenotypes.

For plant system, homologous recombination machinery is not very effective. Moreover, active transposons do not exist in every species including *Arabidopsis*. Under such limitations, T-DNA mediated insertional mutagenesis provides a suitable tool for insertional mutagenesis. T-DNA is the transferred DNA from tumour inducing plasmid from *Agrobacterium sp.* T-DNA is widely used for successful transformation in plants. During T-DNA mediated transformation, the mutated locus also gets tagged which helps further isolation of the mutated gene.

Transposon Mediated Mutagenesis

Transposons are DNA elements which have ability to move from one region to another region in the genome. The shifting of transposon element from a heterochromatin region to a transcriptionally active region results in insertional mutagenesis. It can be exploited as a genetic tool for analysis of gene and protein function. The study of transposons is well described in the case of Drosophila (in which P elements are most commonly used) and in Thale cress (*Arabidopsis thaliana*) and bacteria such as *Escherichia coli*.

Transposon based mutagenesis is a biological process that helps in genes to be transferred to a host organism's chromosome, modifying the function of an extant gene on the chromosome and causing mutation. Transposon mutagenesis was first studied by *Barbara McClintock* in the mid-20th century with corn for which she was awarded Nobel prize also.

In case of bacteria, transposition mutagenesis is usually done by the help of a plasmid from which a transposon is extracted and inserted into the host chromosome. This usually requires a set of enzymes including transposase to be translated.

Currently transposons can be used in genetic research and recombinant genetic engineering for insertional mutagenesis. Insertional mutagenesis is when transposons act like vectors to help remove and integrate genetic sequences. Due to their simple and intricate design and inherent ability to move DNA sequences, transposons are highly compatible for transducing genetic material, enable them as an ideal genetic tool .

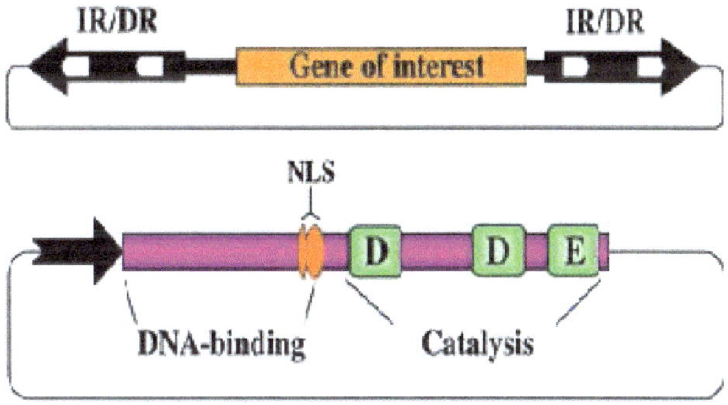

[A]

Figure [A] describes the components and structure of a two-component gene transfer system based on Sleeping Beauty. A gene of interest (orange box) to be mobilized is cloned between the terminal inverted repeats (IR/DR, black arrows) that contain binding sites for the transposase (white arrows). The transposase gene (purple box) is physically separated from the IR/DRs, and is expressed in cells from a suitable promoter (black arrow). The transposase consists of an N-terminal DNA-binding domain, a nuclear localization signal (NLS) and a catalytic domain characterized by the DDE signature.

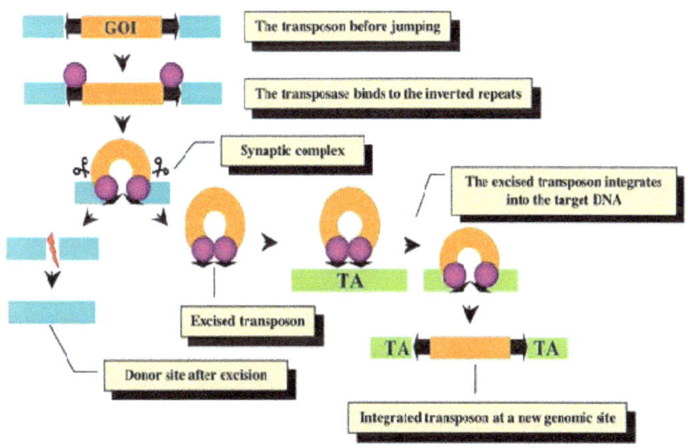

[B] The Sleeping Beauty transposon system.

Fig. [B] illustrates the mechanism of Sleeping Beauty transposition system. The transposable element carrying a gene of interest (GOI, orange box) is maintained and delivered as part of a DNA vector (blue DNA). The transposase (purple circle) binds to its sites within the transposon inverted repeats (black arrows). Excision takes place in a synaptic complex. Excision separates the transposon from the donor DNA, and the double-strand DNA breaks that are generated during this process are repaired by host factors. The excised element integrates into a TA site in the target DNA (green DNA) that will be duplicated and will be flanking the newly integrated transposon.

References

- "Chemical Synthesis, Sequencing, and Amplification of DNA (class notes on MBB/BIO 343)". Arizona State University. Archived from the original on 1997-10-09. Retrieved 2007-10-29
- Carr AC, Moore SD (2012). Lucia, Alejandro, ed. "Robust quantification of polymerase chain reactions using global fitting". PLoS ONE. 7 (5): e37640. PMC 3365123. PMID 22701526. doi:10.1371/journal.pone.0037640
- Bartlett, J. M. S.; Stirling, D. (2003). "A Short History of the Polymerase Chain Reaction". PCR Protocols. Methods in Molecular Biology. 226 (2nd ed.). pp. 3–6. ISBN 1-59259-384-4. doi:10.1385/1-59259-384-4:3
- Priye, Aashish; Hassan, Yassin; Ugaz, Victor (2013). "Microscale chaotic advection enables robust convective DNA replication.". Analytical Chemistry. 85 (21): 10536–10541. doi:10.1021/ac402611s
- Bustin SA, Benes V, Nolan T, Pfaffl MW (June 2005). "Quantitative real-time RT-PCR--a perspective". J. Mol. Endocrinol. 34 (3): 597–601. PMID 15956331. doi:10.1677/jme.1.01755. Retrieved 2012-11-06
- Mueller PR, Wold B (1988). "In vivo footprinting of a muscle specific enhancer by ligation mediated PCR". Science. 246 (4931): 780–786. PMID 2814500. doi:10.1126/science.2814500
- Varkonyi-Gasic E, Hellens RP (2010). "qRT-PCR of Small RNAs". Methods Mol. Biol. 631: 109–22. PMID 20204872. doi:10.1007/978-1-60761-646-7_10. Retrieved 2012-11-06
- Garibyan, Lilit; Avashia, Nidhi (2013-03-01). "Polymerase Chain Reaction". Journal of Investigative Dermatology. 133 (3): 1–4. ISSN 0022-202X. PMC 4102308. PMID 23399825. doi:10.1038/jid.2013.1
- P., Ballou, David; Marilee., Benore, (2010). Fundamental laboratory approaches for biochemistry and biotechnology. John Wiley. ISBN 9780470087664. OCLC 420027217
- Khan Z, Poetter K, Park DJ (2008). "Enhanced solid phase PCR: mechanisms to increase priming by solid support primers". Analytical Biochemistry. 375 (2): 391–393. PMID 18267099. doi:10.1016/j.ab.2008.01.021
- Garibyan L, Avashia N (2013). "Polymerase Chain Reaction". Journal of Investigative Dermatology. 133: 1–4. PMC 4102308. PMID 23399825. doi:10.1038/jid.2013.1
- Tomar, Rukam (2010). Molecular Markers and Plant Biotechnology. Pitman Pura, New Delhi: New India Publishing Agency. p. 188. ISBN 978-93-80235-25-7
- Park DJ (2005). "A new 5' terminal murine GAPDH exon identified using 5'RACE LaNe". Molecular Biotechnology. 29 (1): 39–46. PMID 15668518. doi:10.1385/MB:29:1:39
- Brown, J. R.; Thornton, J. L. (1957). "Percivall Pott (1714-1788) and Chimney Sweepers' Cancer of the Scrotum". British journal of industrial medicine. 14 (1): 68–70. PMC 1037746. PMID 13396156. doi:10.1136/oem.14.1.68
- Salis AD (2009). "Applications in Clinical Microbiology". Real-Time PCR: Current Technology and Applications. Caister Academic Press. ISBN 978-1-904455-39-4
- Joyce C (2002). "Quantitative RT-PCR. A review of current methodologies". Methods Mol. Biol. 193: 83–92. PMID 12325527. doi:10.1385/1-59259-283-X:083
- Ninfa, Alexander; Ballou, David P.; Benore, Marilee (2009). Fundamental Laboratory Approaches for Biochemistry and Biotechnology. United States: Wiley. pp. 408–410. ISBN 9780470087664
- Wong ML, Medrano JF (July 2005). "Real-time PCR for mRNA quantitation". BioTechniques. 39 (1): 75–85. PMID 16060372. doi:10.2144/05391rv01
- Bernstein, H; Bernstein, C (2013). "Evolutionary Origin and Adaptive Function of Meiosis". In Bernstein, Carol. Meiosis. InTech. ISBN 978-953-51-1197-9
- Bustin SA (April 2010). "Why the need for qPCR publication guidelines?--The case for MIQE". Methods. 50 (4): 217–26. PMID 20025972. doi:10.1016/j.ymeth.2009.12.006
- Muller, H. J. (1927). "Artificial Transmutation of the Gene" (PDF). Science. 66 (1699): 84–87. PMID 17802387. doi:10.1126/science.66.1699.84

5

Various Methods of Gene Transformation

Genes can be transferred using biological, chemical and mechanical methods. Biological methods include bactofection and transduction; chemical methods include calcium phosphate, cationic lipid, etc. A few mechanical methods are microinjection, sonoporation, bead transfection, electroporation, etc. In order to completely understand genetic engineering, it is necessary to understand the processes related to it. The following chapter elucidates the varied processes and methods associated with this area of study.

Gene Transfer

The main gene transfer methods using biological means are as follows:

- Bacterial gene delivery i.e. bactofection.
- Delivery using a viral vector i.e. transduction

Bactofection

It is a method of direct gene transfer using bacteria into the target tissue, organ or organism. Various bacterial strains that can be used as vectors in gene therapy are listed in Table. The genes located on the plasmids of the transformed bacterial strains are delivered and expressed into the cells. The gene delivery may be intracellular or extracellular. It has a potential to express various plasmid-encoded heterologous proteins (antigens, toxins, hormones, enzymes etc.) in different cell types. Strains that are invasive and having better cell to cell spread are more efficient.

Vector	Target gene	Disease	Model
L. monocytogenes	IL-12	*L. major*-infection	*Mus musculus*
L. monocytogenes	CFTR	Cystic fibrosis	CHO-K1 cells
S. typhimurium	VEGFR-2 (FLK-1)	Various carcinomas	*Mus musculus*
S. choleraesuis	Thrombospondin-1	Melanoma	*Mus musculus*
S. typhimurium	IFNY	Immunodeficiency	*Mus musculus*
S. typhimurium	CD40L	B-cell lymphoma	*Mus musculus*

Table: Bactofection in various disease models.

The efficiency of bactofection mediated gene transfer can be increased using integrin receptors. Integrin receptors are the transmembrane surface receptors present on the mammalian cell surface.

Another method, lipofectamine-mediated bactofection has also been employed for enhancing the gene transfer efficiency in *E. coli* strains, particularly in the transfer of large intact DNA for gene expression. This method is also effective on various widely used bacterial vectors such as *L. monocytogenes* and *S. typhimurium*.

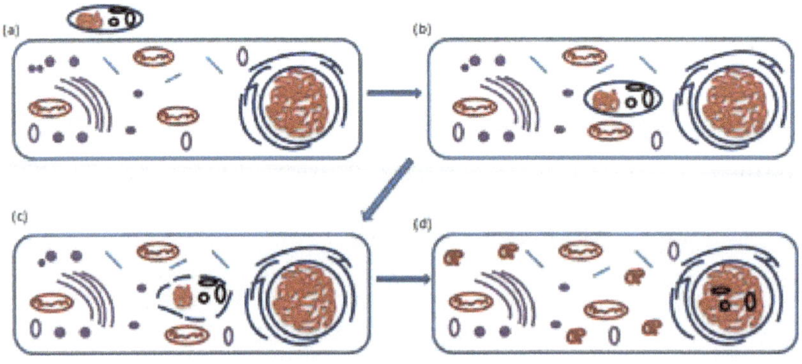

The process of bactofection (a) the transformed bacterial strain with plasmid containing transgene is transferred to target cell (b) genetically engineered bacteria penetrates into the cell (c) In the cytoplasm, the vector undergoes lysis and get destructed releasing plasmids (d) The released plasmids enter into the nucleus where the transgene is expressed by eukaryotic transcription and translation machinery.

Uses

- Bactofection can be used for DNA vaccination against various microbial agents such as viruses, fungi, protozoans and other bacteria.
- It can be used in the treatment of several tumours like melanoma, lung carcinoma and colon carcinoma in mice.

Advantages

- Simple, selective and efficient transfection.
- Low synthesis cost and can be administered easily.

Disadvantages

- Unwanted side effects associated with host-bacteria interaction. This can be reduced by using genetically modified bacteria which contain suicide genes that ease the bacterial destruction and thus reduces the risk of clinical infections.

Transduction

This method involves the introduction of genes into host cell's genome using viruses as carriers. The viruses are used in gene transfer due to following features-

- Efficiency of viruses to deliver their nucleic acid into cells.
- High level of replication and gene expression.

The foreign gene is packaged into the virus particles to enter the host cell. The entry of virus particle containing the candidate gene sequences into the cell and then to the nuclear genome is a receptor- mediated process. The vector genome undergoes complex processes ending up with ds-DNA depending on the vector that can persist as an episome or integrate into the host genome followed by the expression of the candidate gene.

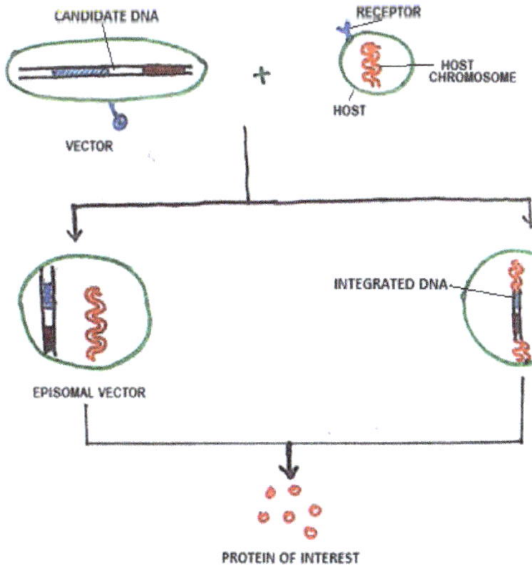

Transduction of a host cell.

Viral Vectors as Therapeutic Agents

Viruses have paved a way into clinical field in order to treat cancer, inherited and infectious diseases. They can be used as vectors to deliver a therapeutic gene into the infected cells. They can be genetically engineered to carry therapeutic gene without having the ability to replicate or cause disease.

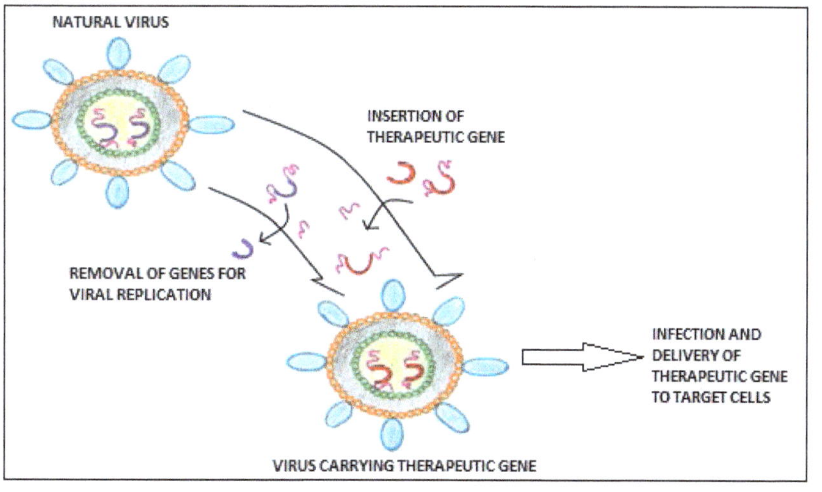

Virus as therapeutic agent.

Strategy for engineering a virus into a vector: The strategy for engineering a virus into a vector requires the following-

Helper Virus

It contains all the viral genes essential for replication but lack the sequence coding for packaging domain (ψ) making it less probable to be packaged into a virion. It can be delivered as helper virus or can stably integrate into the host chromosomal DNA of packaging cell. Some vectors also possess the helper DNA lacking additional transfer functions to increase safety.

Vector DNA

It contains non-coding *cis* -acting viral elements, therapeutic gene sequences (up to 28–32 kb) and the normal packaging recognition signal allowing the selective packaging and release from cells. Some vectors comprise relatively inactivated viral genes as a wide type infection due to lack of other viral genes. The viral proteins essential for replication of the vector DNA are produced which then synthesize multiple copies of the vector genome (DNA or RNA, depending upon the type of vector). These structural proteins recognize the vector (psi plus) but not the helper (psi negative) nucleic acid resulting in the packaging of the vector genome into viral particles.

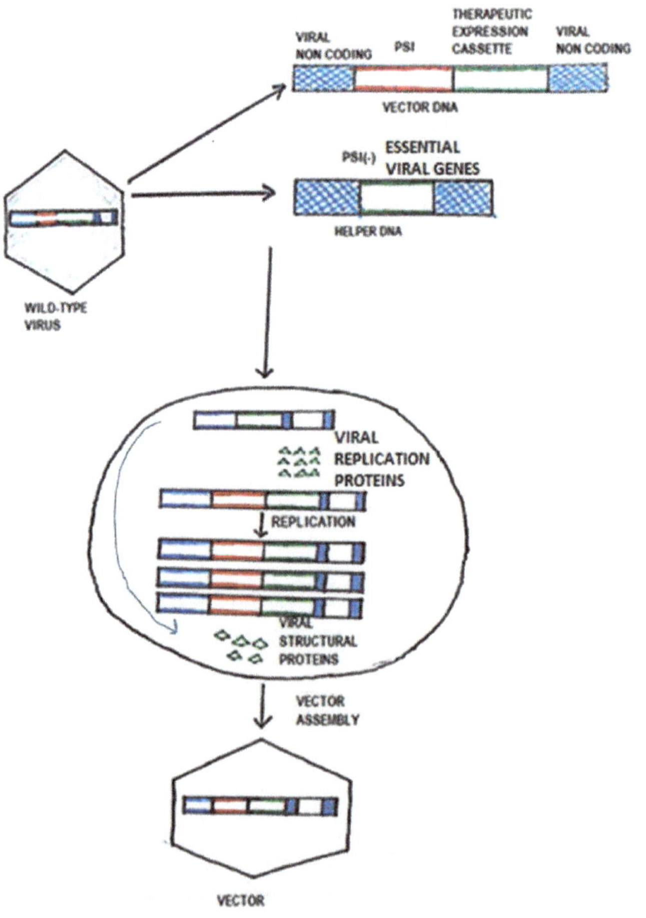

(a): Strategy for engineering a virus into a vector.

Transgene may be incorporated into viral vectors either by addition to the whole genome or by replacing one or more viral genes which can be generally achieved by ligation or homologous recombination.

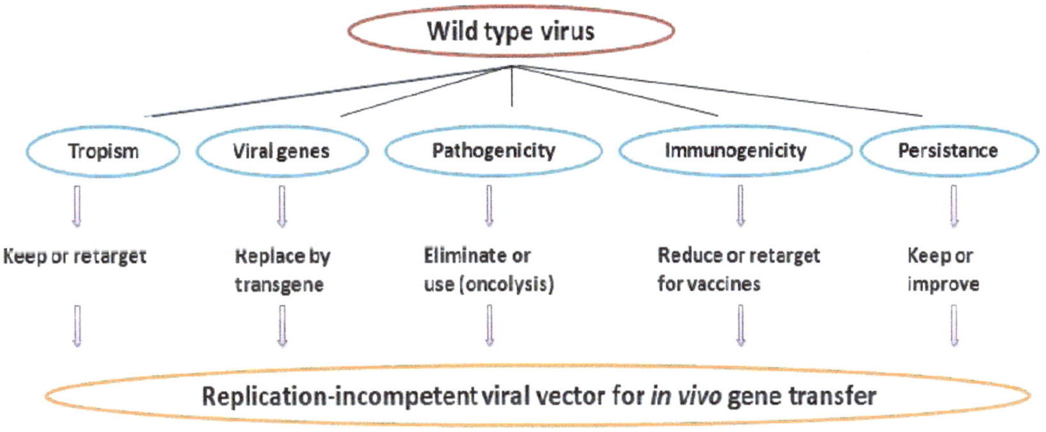

(b): Modifications required for the generation of replication-defective viral vectors from wild type virus for *in vivo* gene transfer.

- If the transgene is added to the genome or replaces one or more non-essential genes for the infection cycle in the expression host, the vector is described as replication-competent or helper-independent, as it can propagate independently *e.g.* helper independent adenoviral vectors.
- However, if the transgene replaces an essential viral gene, this renders the vector replication-defective or helper-dependent, so that missing functions must be supplied *in trans*. This can be accomplished by co-introducing a helper virus or transfecting the cells with a helper plasmid, each of which carry the missing genes *e.g.* helper dependent retroviruses.
- An alternative to the co-introduction of helpers is to use a complementary cell line, which is transformed with the appropriate genes called as 'packaging lines'.
- The vectors from which all viral coding sequences have been deleted and depend on a helper virus which can provide viral gene products *in trans* for packaging and vector DNA replication are known as *`gutless vectors'*.

Advantages

- High capacity for foreign DNA
- The vector has no intrinsic cytotoxic effects.

Viral Vectors

Various kinds of viruses can be used as viral vectors, but five classes of viral vector are used in human gene therapy-

1. Adenovirus
2. Adeno- associated virus (AAV)
3. Herpes virus

4. Retrovirus
5. Lentivirus

Vector	Host cells	Entry pathway	Vector genome forms	Transgene expression	Uses
Retrovirus	Actively dividing cells	Receptor-binding, membrane fusion	Integrated	Long term (years)	SCID, Hyperlipedemia, solid tumors
Lentivirus	Dividing and non-dividing cells	Receptor-binding, membrane fusion	Episomal	Stable	Hematopoetic cells, muscles, neuron, hepatocytes
Adeno virus	Most cells	CAR (Coxsackie and Adenovirus Receptor)-mediated endocytosis endosomal escape	Episomal	Transient (short term for weeks)	CNS, hepatocytes, pancreas
Adeno- associated virus	Most cells	Receptor-mediated endocytosis endosomal escape	Episomal (90%) Integrated (10%)	Medium to long term (year)	lung , muscle, heart, CNS
Herpes virus	Most cells	Endocytotic or membrane fusion	Episomal	Transient	Suitable particularly for nervous system

Table: Viral vectors and their properties.

Adenoviruses

- Adenoviruses are medium-sized (90-100 nm), non-enveloped, icosahedral viruses containing linear, double-stranded DNA of approximately 36 kb.
- 57 immunologically distinct types (7 subgenera) of adenoviruses cause human infections.
- They are unusually stable to physical or chemical agents and adverse pH conditions for long-term survival outside the body.
- There are six early-transcription units, most of which are essential for viral replication, and a major late transcript that encodes capsid.
- They result in transient expression in dividingcells as they do not integrate efficiently into the genome, but prolonged expression can be achieved in post-mitotic cells, like neurons.
- Adenoviruses are mostly attractive as gene therapy vectors, because the virions are taken up efficiently by cells *in vivo*. Adenovirus-derived vaccines have been used in humans with no reported side-effects.

The adenovirus infection cycle comprises two phases-early and late phase,separated by viral DNA replication. The first or "early" phase involves the entry of the virus into the host cell and virus genome to the nucleus. The late genes are transcribed from the major late promoter. The "late" phase involves the formation of gene products related to production and assembly of capsid proteins.

Wild Type Adenovirus Genome

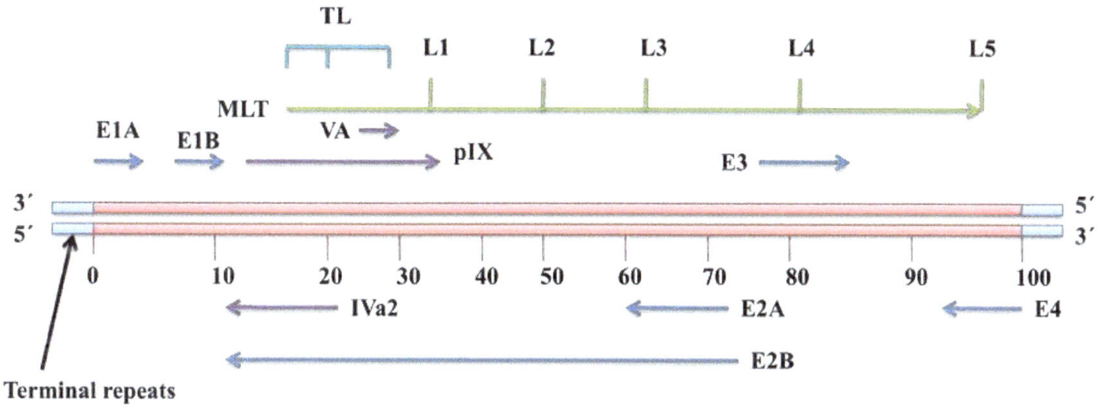

(a): A wild type adenovirus genome. (E1A, E1B, E2A, E2B, E3, E4- early genes; L1 to 5- late genes; MLT- major late transcript; TL-tripartite leader; other genes are represented by pIX, IVa2, VA)

Adenoviral genes	Function
Early genes: E1A, E1B, E2, E3, E4	Transcription, replication, host immune suppression, inhibition of host cell apoptosis.
Delayed early genes: pIX, IVa2	Packaging
Major late gene (L)	Assembly

Table: Different types of adenoviral genes and their function.

Construction of Adenoviral Vectors

First generation adenoviral vectors were *replication deficient* , lacking the essential E1A and E1B genes and often the non-essential gene E3 and were called 'E1 replacement vectors'. They had a maximum capacity of about 7 kb and were propagated in the cell lines transfected with DNA containing E1 genes e.g. human embryonic kidney line 293 (HEK 293).

Drawback

These vectors caused cytotoxic effect due to low-level expression of the viral gene products, and chances of recombination between the vector and the integrated portion of the genome, resulting in the recovery of replication-competent viruses.

Higher-capacity vectors lacking the E2 or E4 regions in addition to E1 and E3 provide a maximum cloning capacity of about 10 kb but still allow low level of transgene expression. These must be propagated on complementary cell lines providing multiple functions. The use of E1/E4 deletions is a sound strategy as the E4 gene is responsible for many of the immunological effects of the virus.

To overcome the above limitations, an alternative strategy employs insertion of 'stuffer DNA' into the nonessential E3 gene as part of the vector backbone so to maintain optimum vector size. Helper dependent adenoviral vectors (HDAd) are favoured for *in vivo* gene transfer due to deletion of all viral coding sequences.

Advantages of HDAd

- Large cloning capacity (up to 37 kb)
- High transduction efficiency
- Long term transgene expression
- Lack of immune response and cytotoxicity.

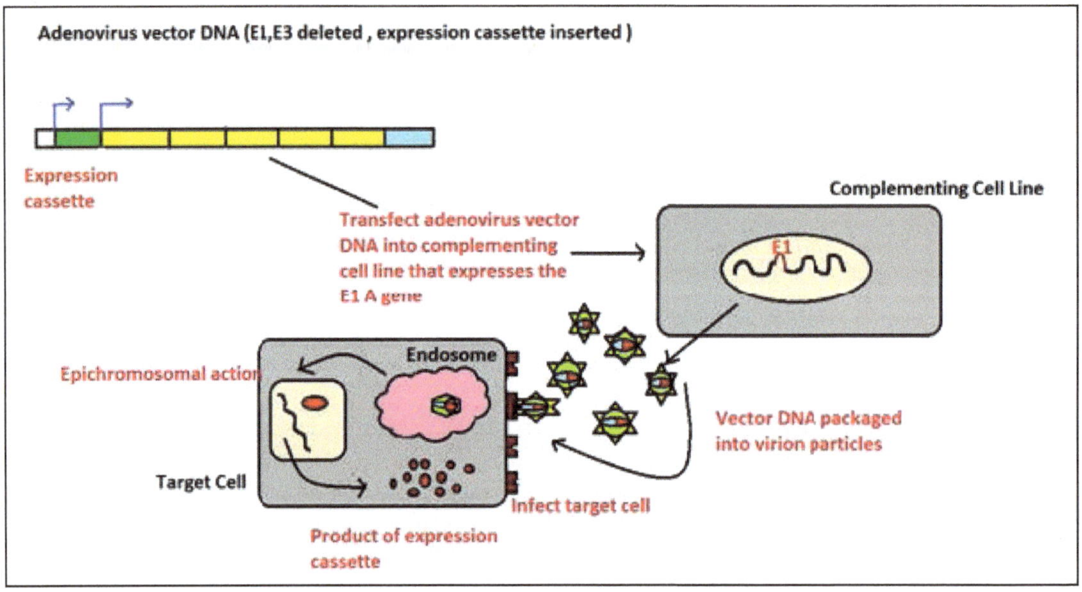

(b):. Adenovirus vectors in gene transfer.

Role in Gene Therapy

- Mainly used for cancer treatment. Gendicine, a recombinant adenovirus, is the first gene therapy product to be licensed for cancer treatment.

Advantages of Adenoviral Vectors

- High transduction efficiency
- Insert size up to 8 kilobases
- Generation of high virus titres
- High level of expression in a wide variety of cell types
- No mutagenic effects due to lack of random integration into the host genome.

Disadvantages of Adenoviral Vectors

- Transient expression due to lack of integration into the host.
- Pathogenic to humans.

Adeno-associated Virus

- It was first discovered as a contaminant in an adenoviral isolate in 1965.
- It is a small, non-enveloped virus packaging a linear single stranded DNA belonging to Parvovirus family.
- It is naturally replication defective thus requiring a helper virus (usually adenovirus or herpes virus) for productive infection.
- In human cells, the provirus integrates predominantly into a 4-kb region (AAVS1) on chromosome 19. Subsequent infection by adenovirus or herpes virus can 'rescue' the provirus and induce lytic infection.

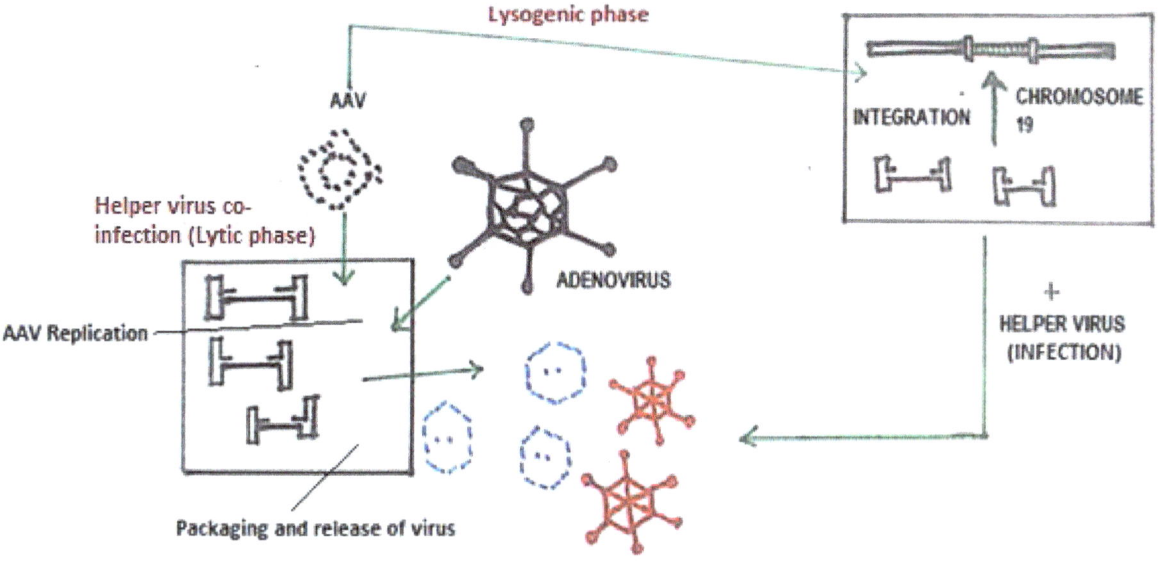

(a): AAV life cycle.

AAV Life Cycle

- AAV life cycle comprises two phases-lytic and lysogeny.
- In the presence of helper virus, AAV undergoes lytic phase comprising genome replication, expression of viral genes and production of virions.
- In the absence of helper virus, it undergoes lysogenic phase and integrates into the host cell's genome as a latent provirus. This latent genome undergoes replication by subsequent infection with helper virus.
- Both the stages of life cycle of AAV are controlled by complex interactions between the AAV genome andhelper virus, AAV and host proteins.

Adeno-associated Viral Genome

The AAV genome is small (about 5 kb) and comprises a central region containing rep (replicase) and cap(capsid) genes flanked by 145 base inverted terminal repeats (ITRs). The rep gene is involved in viral replication and integration whereas cap gene encodes viral capsid proteins. ITRs are required for replication, transcription, proviral integration and rescue.

In earlier AAV vectors, foreign DNA replaced the cap region and was expressed under the control of an endogenous AAV promoter. The transgene expression was inefficient using heterologous promoters due to inhibition of their activity by Rep protein.

Rep interference with endogenous promoters resultedin cytotoxic effects of the virus. To overcome the above limitations, such vectors in which both genes were deleted and the transgene was expressed from either an endogenous or a heterologous promoter, were developed.

In vitro manipulation of AAV is facilitated by cloning the inverted terminal repeats in a plasmid vector and inserting the transgene between them. Transfection of this construct into cells along with a helper plasmid produced recombinant viral particles.

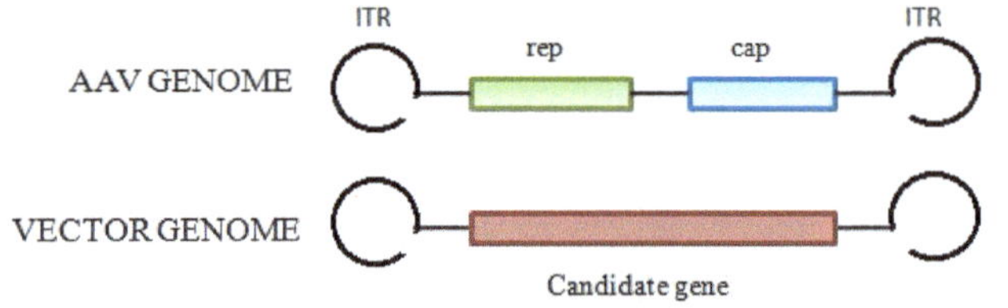

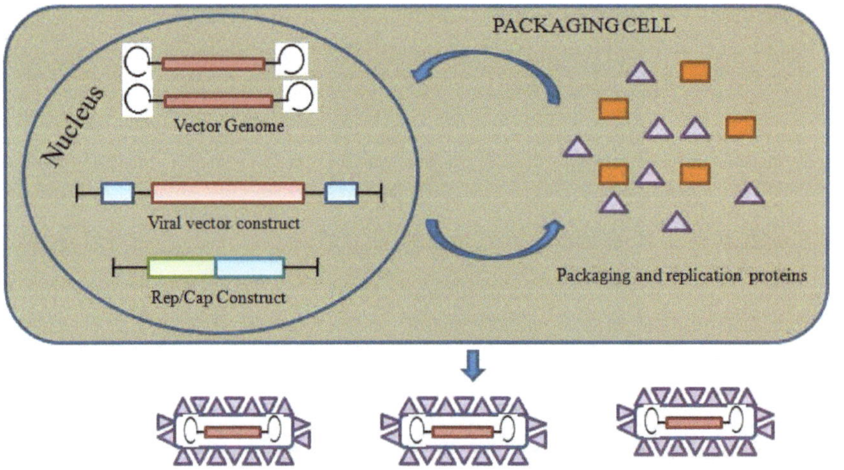

(b): AAV Genome, Vector genome and Packaging coil.

Recombinant AAV (rAAV) is used as an expression cassette containing a reporter or candidate gene of interest. The foreign gene replaces all of the viral genes present in a wild type virus. Only the inverted terminal repeats are left to function as the essential replication/packaging signal.

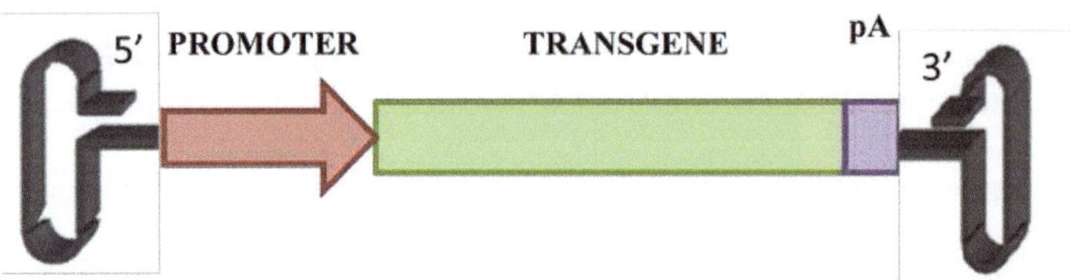

(c): Organization of a typical recombinant AAV (rAAV) genome. pA represents Poly A tail.

Advantages

- Stable and have a wide host range
- Lack of initiating an immune response
- The dependence of AAV on a heterologous helper virus provides higher control over vector replication, making AAV vectors safer for use in gene therapy
- Potential of targeted/site-specific integration
- Non-pathogenic

Disadvantages

- AAV uses concatemeric replication intermediates
- They must be closely screened as they are often contaminated with adenovirus or Herpes Virus.
- Insert size is limited (4Kb)
- Difficult generation of high virus titres

Herpes Virus Vectors

- The herpes viruses are linear ds-DNA viruses of approximately 150 kb size *e.g.* EBV (Epstein–Barr virus) and the HSVs (Hepatitis B virus, e.g. HSV-I, varicella zoster).
- Most HSVs are transmitted without symptoms (varicella zoster virus is exceptional) and cause prolonged infections.
- With the help of two viral glycoproteins, gB and gD, the virus binds to cells through an interaction with heparan sulphate moieties on the cell surface.
- Unlike EBV as a replicon vector (contains both *cis* and *trans* acting genetic elements required for replication), HSV-I have been developed as a transduction vector for purpose of gene transferand can efficiently transduce a wide range of cell types.
- HSV virus is remarkably neurotropic and thus HSV vectors are particularly suitable for gene therapy in the nervous system. HSV can also be transmitted across neuronal synapses during lytic infections which can be used to trace axon pathways.
- Generation of recombinants in transfected cells takes place by homologous recombination. These viral vectors may be replication competent or helper dependent.
- The plasmid based amplicon vectors carrying only the *cis* -acting elements required for replication and packaging can be constructed. These vectors require packaging systems to provide the missing functions in *trans*.

Role in Gene Therapy

Most promised use of HSV vectors involves gene transfer to neural cells where it can cause a latent infection (e.g. spinal cord, brain, and peripheral nervous system).

Advantages

- Infects a wide range of cell types
- Insert size up to 50 kb due to large viral genome size
- Natural tropism to neuronal cells
- Stable viral particles allow generation of high virus titres (10^{12}pfu/ml)

Disadvantages

- No viral integration into host genome and transient transgene expression
- High level of pre-existing immunity
- Cytotoxicity effects
- Risk of recombination with latently HSV-infected cells

Retroviral Vectors

Retroviruses are RNA viruses that replicate via a ds-DNA intermediate. The infection cycle begins with the interaction between viral envelope and the host cell's plasma membrane, delivering the particle into the cell. The capsid contains two copies of the RNA genome, as well as reverse transcriptase/integrase. After infection, the RNA genome is reverse transcribed to produce a cDNA copy, a DNA intermediate, which integrates into the genome randomly.

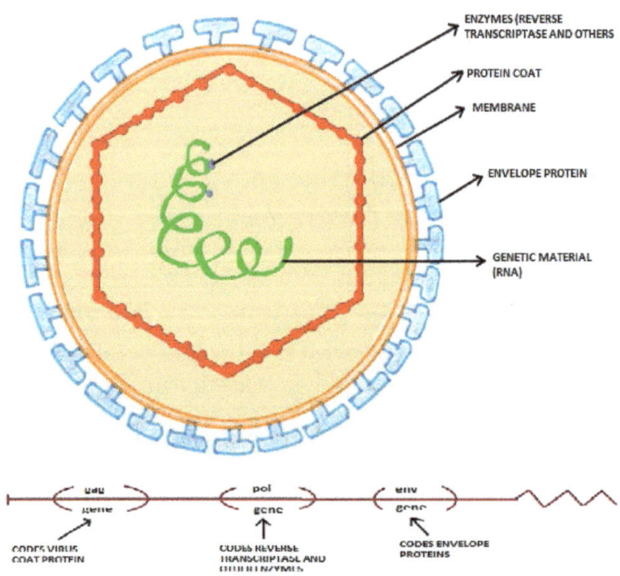

(a): Structure of a Retrovirus vector. RNA showed in the figure is single stranded.

Life Cycle of Retroviruses

A retrovirus, on binding to a cell surface receptor, enters the cell where it reverse transcribes the RNA into double-stranded DNA. Viral DNA gets integrated into the cell chromosome to form a provirus. Cellular machinery transcribes, processes the RNA and undergoes translation into viral proteins. The viral RNA and proteins are then assembled to form new viruses which are released from the cell by budding.

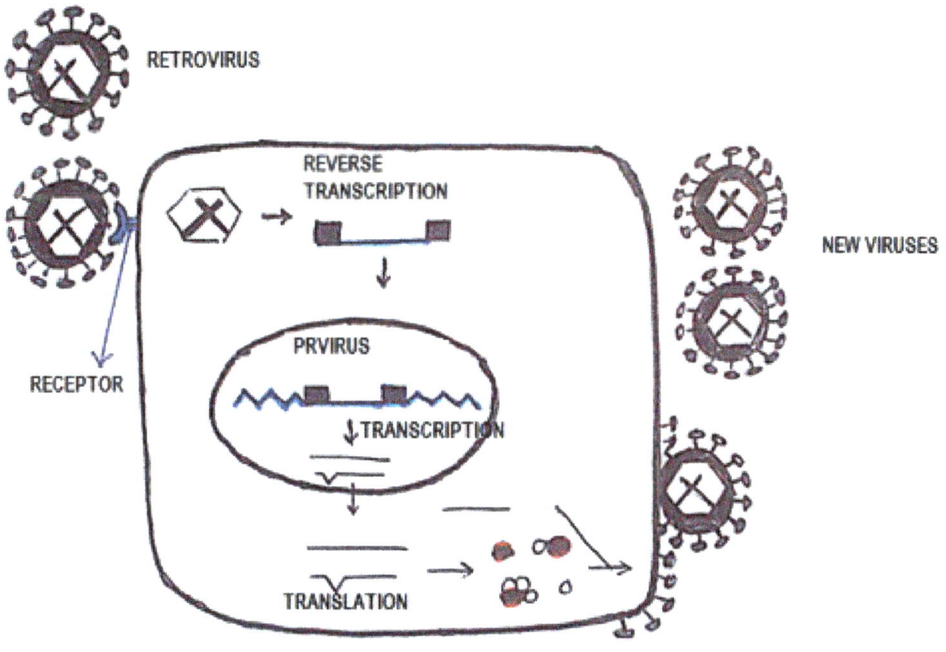

(b): Replication cycle of retroviruses.

Retroviral Genome

The integrated provirus comprises three genes (*gag, pol* and *env*). The *gag* gene encodes a viral structural protein, *pol* encodes the reverse transcriptase and integrase and *env* gene encodes viral envelope proteins. Retrovirus can be classified as oncoviruses, lentiviruses, and spuma-viruses. Oncoviruses are simple whereas lentiviruses and spuma-viruses are complex retroviruses.

Viral genomic RNA is synthesized by transcription from a single promoter located in the left LTR and ends at a poly-A site in the right LTR. Thus, the full-length genomic RNA is shorter than the integrated DNA copy and lacks the duplicated LTR structure. The genomic RNA is capped and polyadenylated, allowing the *gag* gene to be translated. The *pol* gene is also translated by read through, producing a Gag–Pol fusion protein, which is further processed into several distinct polypeptides. Some of the full-length RNA also undergoes splicing, eliminating the *gag* and *pol* genes and allowing the downstream *env* gene to be translated. Two copies of the full-length RNA genome are incorporated into each capsid requiring a specific *cis* -acting packaging site termed ψ. The reverse transcriptase/ integrase are also packaged.

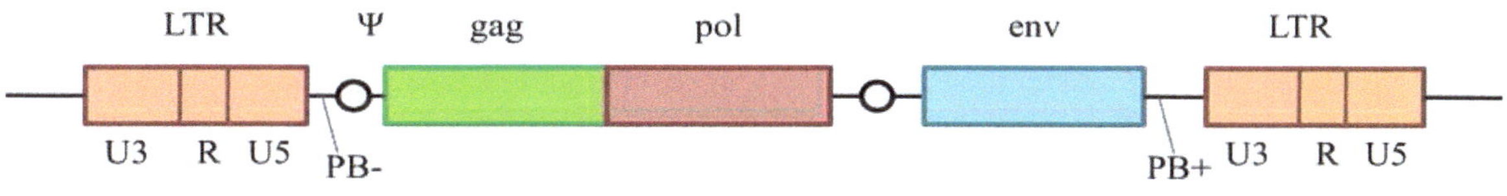

(c): An oncoretrovirus genome comprising long terminal repeats (LTRs) enclosing the three open reading frames *gag, pol* and *env*. PB represents primer binding sites in the viral replication cycle, ψ is the packaging signal and small circles represent splice sites.

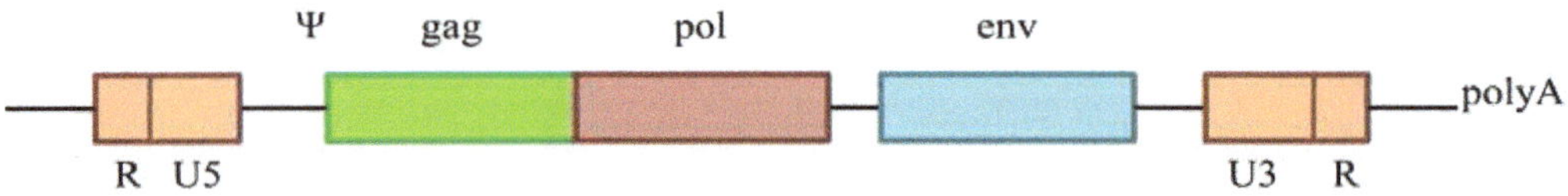

(d): Structure of a packaged RNA genome having a poly (A) tail but lacking the LTRs.

Construction of a Retroviral Vector and Propagation in Helper Cell

The retroviral construct involved in gene delivery comprises two constructs-

- A vector consisting of all *cis* -acting elements required for gene expression and replication
- A helper cell expressing all the viral proteins (*gag, pol, env*) lacking in vector and support the replication of vector. Helper cell lacks RNA containing packaging signal which is required for formation and release of infectious particles but not for non-infectious viral particles.

When the vector DNA is introduced into a helper cell, helper cell produces the viral proteins which help in the assembly of viral particles containing RNA transcribed from the viral vector. These viral particles on infecting the target cell, reverse transcribe the vector RNA into ds-DNA which gets integrated into the host genome forming a provirus which encodes the gene of interest. Target cells do not express viral proteins and cannot generate infectious viral particles containing the vector RNA and thus cannot infect other target cells.

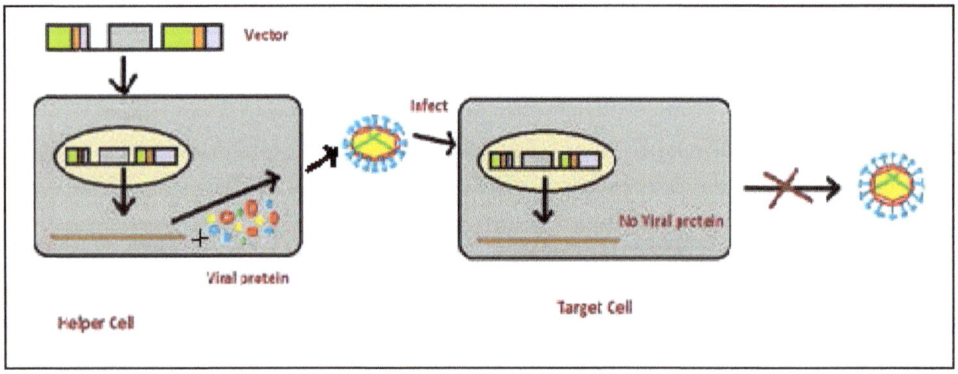

(e): Propagation of retroviral vectors in helper cells.

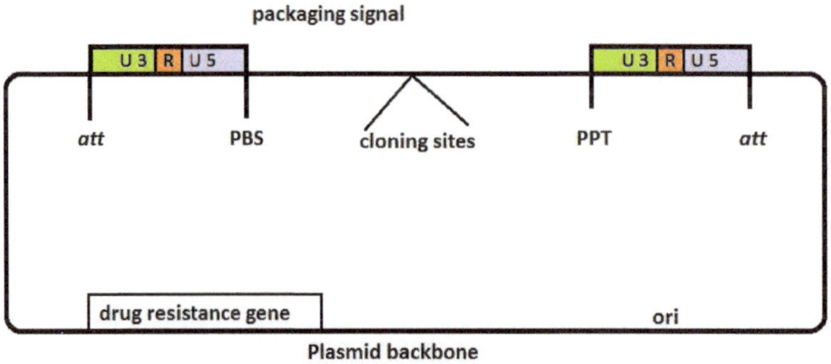

(f): *cis* -acting elements required by a prototypical retroviral vector. The plasmid backbone contains a bacterial origin of replication (ori) and a drug resistance gene.

Advantages

- Well studied system having high transduction efficiency
- Insert size up to 8 kb
- Integration into host genome resulting in sustained expression of the vector
- Vector proteins are not expressed in host

Disadvantages

- Infection by retrovirus requires dividing cells
- Low titres (10^6 -10^7)
- Random integration
- Poor *in vivo* delivery

Lentivirus

- They are subclass of retroviruses.
- They are more efficient and advantageous for gene transfer than other vectors due to following reasons-
    - ◇ Unlike retroviruses which can infect only dividing cells, lentiviruses can be used as vectors due to their ability to infectboth dividing and non-dividing cells.
    - ◇ Low immunogenicity
    - ◇ Stable, long term transgene expression

Properties of Viralvectors Usedin Gene Transfer

The properties of different viral vectors are listed in Table.

Viral vectors	Genome	Insert capacity (kb)	Specific integration	Long-term maintenance	RNA intermediate
Retroviruses	ssRNA with DNA intermediate	1-7	Y	Y	RNA with DNA intermediate
Adenovirus	dsDNA	2-38	N	N	N
Adeno-associated virus	ssDNA	4.5	Y	Y	N
Herpes simplex virus	dsDNA	50	N	Y	N
Lentivirus	RNA with DNA intermediate	7-18	Y	Y	RNA with DNA intermediate

Y-Yes; N-No.

Table: Various viral vectors used in gene transfer

Chemical Methods

Cell membrane is a sheet like assembly of amphipathic molecules that separate cells from their environment. These physical structures allow only the controlled exchange of materials among the different parts of a cell and with its immediate surroundings. DNA is an anionic polymer, larger molecular weight, hydrophilic and sensitive to nuclease degradation in biological matrices. They cannot easily cross the physical barrier of membrane and enter the cells unless assisted.

Various charged chemical compounds can be used to facilitate DNA transfer directly to the cell. These synthetic compoundsare introduced near the vicinity of recipient cells thereby disturbing the cell membranes, widening the pore size and allowing the passage of the DNA into the cell. An ideal chemical used for DNA transfer should have the ability to-

- Protect DNA against nuclease degradation.
- Transport DNA to the target cells.
- Facilitate transport of DNA across the plasma membrane.
- Promote the import of DNA into the nucleus.

The commonly used methods of chemical transfection use the following,

- Calcium phosphate
- DEAE dextran
- Cationic Lipid
- Other polymers - poly-L-lysine (PLL), polyphosphoester, chitosan, dendrimers

Calcium Phosphate Mediated DNA Transfer

Historical Perspective

The ability of mammalian cells to take up exogenously supplied DNA from their culture medium was first reported by Szybalska and Szybalski (1962). They used total uncloned genomic DNA to transfect human cells deficient for the enzyme hypoxanthine guanine phosphoribosyl transferase (HPRT). Rare HPRT-positive cells with fragments of DNA containing the functional gene were identified by selection on HAT medium.Till then, the actual mechanism of DNA uptake was not understood. It was later found that successful DNA transfer takes place by the formation of a fine DNA/calcium phosphate co-precipitate, which first settles onto the cells and is then internalized. This technique was first applied by Graham and Van Der Eb in 1973 for the analysis of the infectivity of adenoviral DNA.

Calcium Phosphate Transfection

This method is based on the precipitation of plasmid DNA and calcium ions by their interaction. In this method,the precipitates of calcium phosphate and DNAbeing small and insolublecan be easily adsorbed on the surface of cell. This precipitate is engulfed by cells through endocytosis and the DNA gets integrated into the cell genome resulting in stable or permanent transfection.

Uses

- This method is mainly used in the production of recombinant viral vectors.
- It remains a choice for plasmid DNA transfer in many cell cultures and packaging cell lines. As the precipitate so formed must coat the cells, this method is suitable only for cells growing in monolayer and not for suspension cultures.

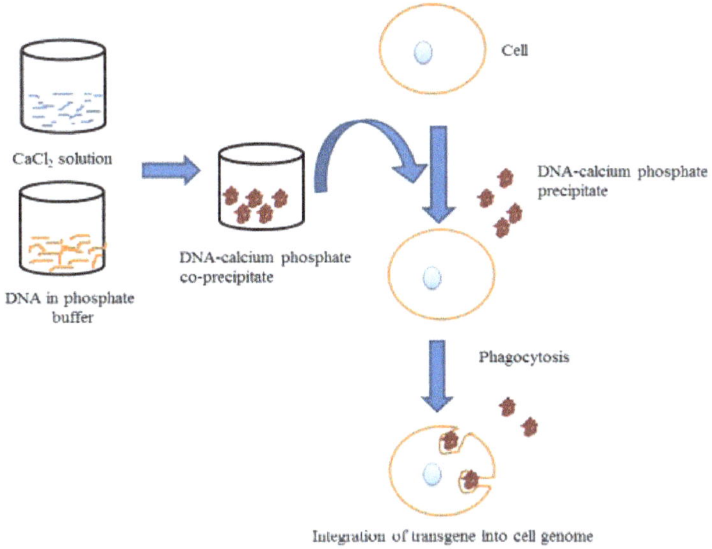

A schematic representation of transfection by Calcium Phosphate Precipitation.

Advantages

- Simple and inexpensive
- Applicability to generate stably transfected cell lines
- Highly efficient (cell type dependent) and can be applied to a wide range of cell types.
- Can be used for stable ortransient transfection

Disadvantages

- Toxic especially to primary cells
- Slight change in pH, buffer salt concentration and temperature can compromise the efficacy
- Relatively poor transfection efficiency compared to other chemical transfection methods like lipofection.
- Limited by the composition and size of the precipitate.
- Random integration into host cell.

Optimal factors (amount of DNA in the precipitate, the length of time for precipitation reaction and exposure of cells to the precipitate) need to be determined for efficient transfection of the cells.

This technique is simple, expensive and has minimal cytotoxic effect but the low level of transgene expression provoked development of several other methods of transfection.

DEAE-Dextran (Diethylaminoethyl Dextran) Mediated DNA Transfer

- This method was initially reported by Vaheri and Pagano in 1965 for enhancing the viral infectivity of cell but later adapted as a method for plasmid DNA transfer.
- Diethylaminoethyl dextran (DEAE-dextran) is a soluble polycationic carbohydrate that promotes interactions between DNA and endocytotic machinery of the cell.
- In this method, the negatively charged DNA and positively charged DEAE – dextran form aggregates through electrostatic interaction and form apolyplex. A slight excess of DEAE – dextran in mixture results in net positive charge in the DEAE – dextran/ DNA complex formed. These complexes, when added to the cells, bind to the negatively charged plasma membrane and get internalized through endocytosis. Complexed DNA delivery with DEAE-dextran can be improved by osmotic shock using DMSO or glycerol.
- Several parameters such as number of cells, polymer concentration, transfected DNA concentration and duration of transfection should be optimized for a given cell line.

Advantages

- Simple and inexpensive
- More sensitive
- Can be applied to a wide range of cell types
- Can be used for transient transfection.

Disadvantages

- Toxic to cells at high concentrations
- Transfection efficiency varies with cell type
- Can only be used for transient transfectionbut not forstable transfection
- Typically produces less than 10% delivery in primary cells.

Another polycationic chemical, the detergent Polybrene, has been used for the transfection of Chinese hamster ovary (CHO) cells, which are not amenable to calcium phosphate transfection.

Lipofection

- Lipofection is a method of transformation first described in 1965 as a model of cellular membranes using liposomes.
- Liposomes areartificial phospholipid vesicles used for the deliveryof a variety of molecules

into the cells. They may be multi-lamellar or unilamellar vesicles with a size range ofo.1 to 10 micrometer or 20-25 nanometers respectively.

- They can be preloaded with DNA by two common methods- membrane-membrane fusion and endocytosisthus forming DNA- liposome complex. This complexfuses with the protoplasts to release the contents into the cell. Animal cells, plant cells, bacteria, yeast protoplasts are susceptible to lipofection method.
- Liposomes can be classified as either cationic liposome or pH-sensitive.

Cationic Liposomes

- Cationic liposomes are positively charged liposomes which associate with the negatively charged DNA molecules by electrostatic interactions forming a stable complex.

Neutral liposomes are generally used as DNA carriers and helpers of cationic liposomes due to their non-toxic nature and high stability in serum.A positively charged lipid is often mixed with a neutral co-lipid, also called helper lipid to enhance the efficiency of gene transfer by stabilizing the liposome complex (lipoplex). Dioleoylphosphatidyl ethanolamine (DOPE) or dioleoylphosphatidyl choline (DOPC) are some commonly used neutral co-lipids.

- The negatively charged DNA molecule interacts with the positively charged groups of the DOPE or DOPC. DOPE is more efficient and useful than DOPC due to the ability of its inverted hexagonal phase to disrupt the membrane integrity.
- The overall net positive charge allows the close association of the lipoplex with the negatively charged cell membrane followed by uptake into the cell and then into nucleus.
- The lipid: DNA ratio and overall lipid concentration used in the formation of these complexes is particularly required for efficient gene transfer which varies with application.

Negatively Charged Liposomes

- Generally pH-sensitive or negatively-charged liposomes are not efficient for gene transfer. They do not form a complex with it due to repulsive electrostatic interactions between the phosphate backbone of DNA and negatively charged groups of the lipids. Some of the DNA molecules get entrapped within the aqueous interior of these liposomes.
- However, formation of lipoplex, a complex between DNA and anionic lipidscan occur by using divalent cations (e.g. Ca^{2+}, Mg^{2+}, Mn^{2+}, and Ba^{2+}) which canneutralize the mutual electrostatic repulsion. These anionic lipoplexes comprise anionic lipids, divalent cations, and plasmid DNA which are physiologically safe components.
- They are termed as pH sensitive due to destabilization at low pH.

The efficiency of both *in vivo* and *in vitro*gene delivery using cationic liposomes is higher thanthat of pH sensitive liposomes. But the cationic liposomes get inactivated and unstable in the presence of serum and exhibit cytotoxicity. Due to reduced toxicity and interference from serum proteins, pH-sensitive liposomes are considered as potential gene delivery vehicles than the cationic liposomes.

Liposome Action

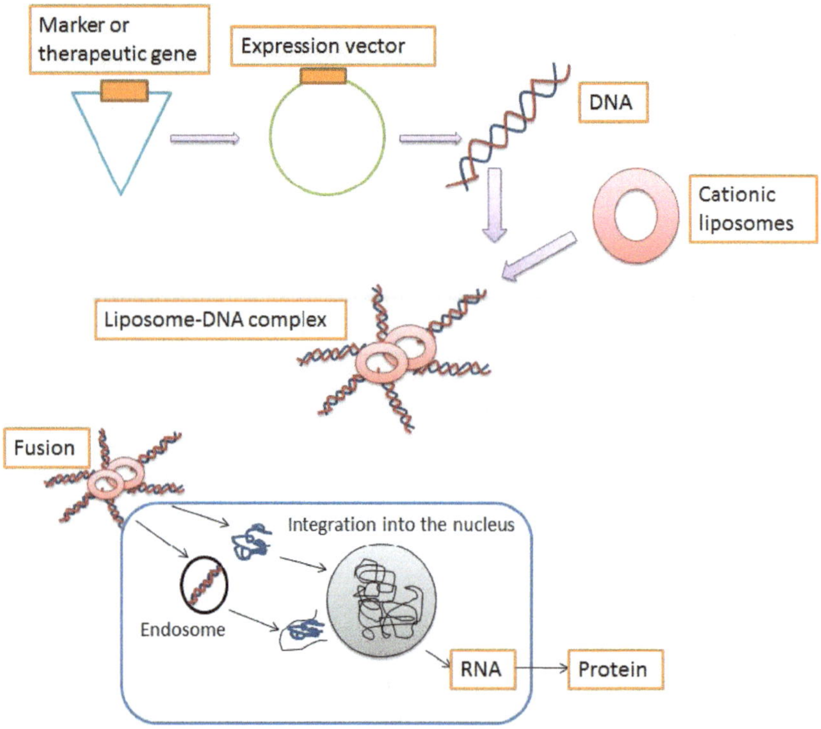

Schematic representation of liposome action in gene transfer.

In addition, liposomes can be directed to cells using monoclonal antibodies which recognize and bind to thespecific surface antigens of cells along with the liposomes. Liposomes can be prevented from destruction by the cell's lysosomes by pre- treating the cells with chemicals such as chloroquine, cytochalasin B, colchicine etc. Liposome mediated transfer into the nucleusis still not completely understood.

Advantages

- Economic
- Efficient delivery of nucleic acids to cells in a culture dish.
- Delivery of the nucleic acids with minimal toxicity.
- Protection of nucleic acids from degradation.
- Measurable changes due to transfected nucleic acids in sequential processes.
- Easy to use, requirement of minimal steps and adaptable to high-throughput systems.

Disadvantages

- It is not applicable to all cell types.
- It fails for the transfection of some cell lines with lipids.

Other Methods

Other methods of chemical transfection involve the use of chemicals such as polyethylenimine, chitosan, polyphosphoester, dendrimers.

Polyethylenimine

- Polyethylenimine (PEI) is a non-degradable, high molecular weight polymer which may accumulate in the body.
- PEI, due to its polycationic nature, condenses with the DNA molecule resulting in the formation of PEI-DNA complex which enters the cell by endocytosis, thus mediating gene transfer.
- PEI exhibit cytotoxicity due to its ability to permeabilize and disrupt cell membranes leading to necrotic cell death.
- The cytotoxicity may be reduced using various methods e.g. PEGylation and conjugation of low molecular weight polyethylenimine with cleavable cross-links such as disulfide bondsin the reducing environment of the cytoplasm.

Chitosan

- Chitosan, a biodegradable polysaccharide is composed of D-glucosamine repeating units and can be used as a non-viral gene carrier.
- It can efficiently bind and protect DNA from nuclease degradation.
- The biocompatibility and low toxicity profile makes it a safe biomedical material for clinical applications.
- Chitosan DNA nanoparticles can transfect several different cell types with relatively low transfection efficiency.
- Modified chitosans such as trimethylated chitosan and chitosan conjugated with deoxycholic acid have been developed to increase the solubility of chitosan at neutral pH which can efficiently transfect COS-1 cells.
- Chitosans with different molecular weights exhibit different DNA binding affinities. The efficiency of transfection is determined by the particle stability which is one of the rate-limiting steps in the overall transfection process.

Polyphosphoester

- Polyphosphoesters (PPE) are biocompatible and biodegradable, particularly those having a backbone analogous to nucleic acids and teichoic acids and used in several biomedical applications.They may result in extracellular persistent release of the DNA molecules thus enhancing the expression of transgene in the muscle as compared to naked DNA intake.
- Several polyphosphoesters with positive charges both in the backbone and in the side chain can be used as non-viral gene carriers.

- They can efficiently bind and protect DNA from nuclease degradation.
- They exhibit a significantly lower cytotoxicity than Poly-L-Lysine or polyethylenimine both *in vitro* and *in vivo*.
- It is a cell type dependent transfection method the efficiency of which can be enhanced using chloroquine.
- The transfection using polyphosphoestersis found to be effective in many cell lines, with some of them comparable to Liposome-mediated transfection.

Dendrimers

- Dendrimers are a new class of polymeric materials that are highly branched and monodisperse macromolecules. Due to their unique behaviour, they are suitable for a wide range of biomedical applications.
- They have positively charged amino groups (termini) on their surface which interact with the negatively charged phosphate groups of the DNA molecule to form a DNA-dendrimer complex.
- This DNA-dendrimer complex has an overall net positive charge and interacts with negatively charged surface molecules of the cell membrane thus allowing the entry of complex into the cell through non-specific endocytosis.
- Once inside the cell, these complexes are then transported to the endosomes where these are protected from nuclease degradation by being highly condensed within the DNA-dendrimer complex.
- The unprotonated amino groups on the dendrimers at neutral pH can become protonated in the acidic environment of the endosome leading to buffering of the endosome and thus inhibiting pH-dependent endosomal nucleases.

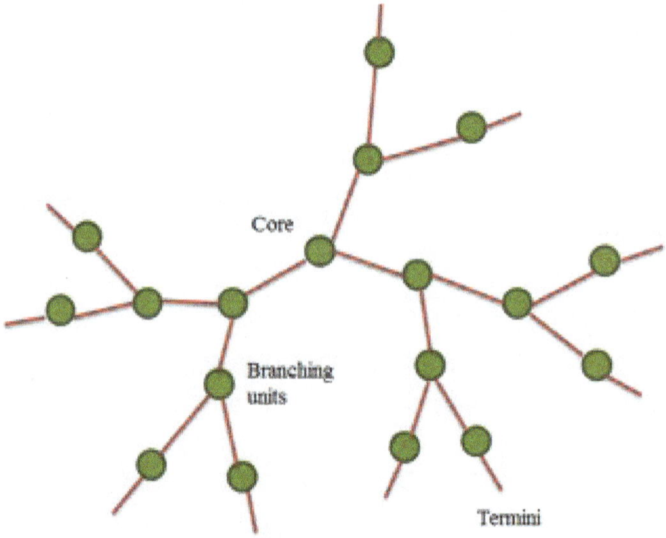

Structure of a dendrimer.

Physical Method

It has been discussed earlier that due to amphipathic nature of the phospholipid bilayer of the plasma membrane, polar molecules such as DNA and protein are unable to freely pass through the membrane. Various physical or mechanical methods areemployed to overcome this and aid ingene transfer as listed below-

1. Electroporation
2. Microinjection
3. Particle Bombardment
4. Sonoporation
5. Laser induced
6. Bead transfection

Electroporation

- Electroporation is a mechanical method used for the introduction of polar molecules into a host cell through the cell membrane.
- This method was first demonstrated by Wong and Neumann in 1982 to study gene transfer in mouse cells.
- It is now a widely used method for the introduction of transgene either stably or transiently into bacterial, fungal, plant and animal cells.
- It involves use of a large electric pulse that temporarily disturbs the phospholipid bilayer, allowing the passage of molecules such as DNA.

The basis of electroporation is the relatively weak hydrophobic/hydrophilic interaction of the phospholipids bilayer and ability to spontaneously reassemble after disturbance. A quick voltage shock may cause the temporary disruption of areas of the membrane and allow the passage of polar molecules. The membrane reseals leaving the cell intact soon afterwards.

(a). Procedure

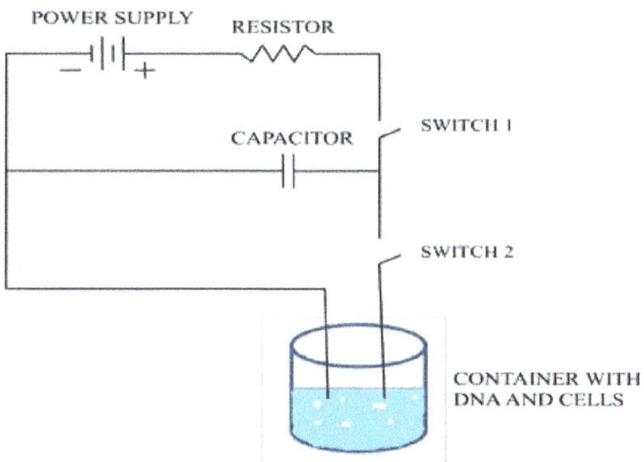

(a): The basic circuit setup of the electroporation apparatus.

The host cells and the DNA molecules to be transported into the cells are suspended in a solution. The basic process inside an electroporation apparatus is represented in a schematic diagram.

- When the first switch is closed, the capacitor charges up and stores a high voltage which gets discharged on closing the second switch.
- Typically, 10,000-100,000 V/cm in a pulse lasting a few microseconds to a millisecond is essential for electroporation which varies with the cell size.
- This electric pulse disrupts the phospholipid bilayer of the membrane causing the formation of temporary aqueous pores.
- When the electric potential across the cell membrane is increased by about 0.5-1.0 V, the charged molecules e.g. DNA migrate across the membrane through the pores in a similar manner to electrophoresis.
- The initiation of electroporation generally occurs when the transmembrane voltage reaches at 0.5-1.5 V. The cell membrane discharges with the subsequent flow of the charged ions and molecules and the pores of the membrane quickly close reassembling the phospholipid bilayer.

(b). Applications

Electroporation is widely used in many areas of molecular biology and in medical field. Some applications of electroporation include:

DNA Transfection or Transformation

Electroporation is mainly used in DNA transfection/transformation which involves introduction of foreign DNA into the host cell (animal, bacterial or plant cell).

Direct Transfer of Plasmids Between Cells

It involves the incubation of bacterial cells containing a plasmid with another strain lacking plasmids but containing some other desirable features. The voltage of electroporation creates pores, allowing the transfer of plasmids from one cell to another. This type of transfer may also be performed between species. As a result, a large number of plasmids may be grown in rapidly dividing bacterial colonies and transferred to yeast cells by electroporation.

Gene Transfer to a Wide Range of Tissues

Electroporation can be performed in vivo for more efficient gene transfer in a wide range of tissues like skin, muscle, lung, kidney, liver, artery, brain, cornea etc. It avoids the vector-specific immune-responses that are achieved with recombinant viral vectors and thus are promising in clinical applications.

(c). Advantages

- It is highly versatile and effective for nearly all cell types and species.

- It is highly efficient method as majority of cells take in the target DNA molecule.
- It can be performed at a small scale and onlya small amount of DNA is required as compared to other methods.

(d). Disadvantages

- Cell damage is one of the limitations of this method caused by irregular intensity pulses resulting in too large pores which fail to close after membrane discharge.
- Another limitation is the non-specific transport which may result in an ion imbalance causing improper cell function and cell death.

Microinjection

- DNA microinjection was first proposed by Dr. Marshall A. Barber in the early of nineteenth century.
- This method is widely used for gene transfection in mammals.
- It involves delivery of foreign DNA into a living cell (e.g. a cell, egg, oocyte, embryos of animals) through a fine glass micropipette. The introduced DNA may lead to the over or under expression of certain genes.
- It is used to identify the characteristic function of dominant genes.

(a). Procedure

- The delivery of foreign DNA is done under a powerful microscope using a glass micropipette tip of 0.5 mm diameter.
- Cells to be microinjected are placed in a container. A holding pipette is placed in the field of view of the microscope thatsucks and holds a target cell at the tip. The tip of micropipette is injected through the membrane of the cell to deliver the contents of the needle into the cytoplasm and then the empty needle is taken out.

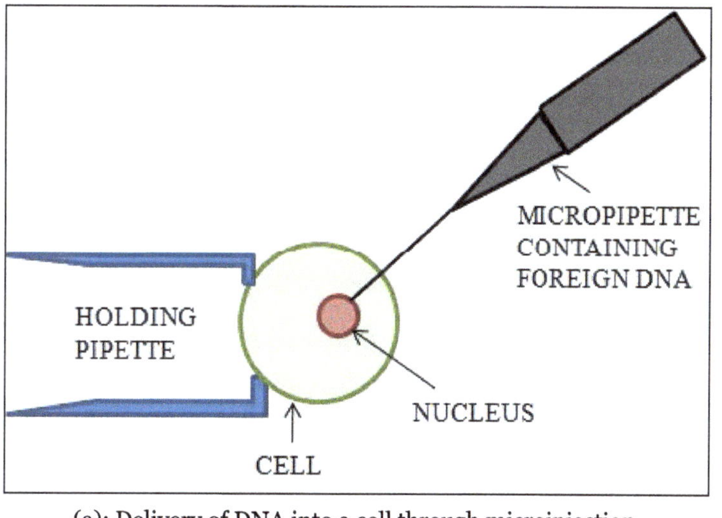

(a): Delivery of DNA into a cell through microinjection.

(b). Advantages

- No requirement of a marker gene.
- Introduction ofthe target gene directly into a single cell.
- Easy identification of transformed cells upon injection of dye along with the DNA.
- No requirement of selection of the transformed cells using antibiotic resistance or herbicide resistance markers.
- It can be used for creating transgenic organisms, particularly mammals.

Particle Bombardment

- Prof Sanford and colleagues at Cornell University (USA) developed the original bombardment concept in 1987 and coined the term "biolistics" (short for "biological ballistics") for both the process and the device.
- Also termed as particle bombardment, particle gun, micro projectile bombardment and particle acceleration.
- It employs high-velocity micro projectiles to deliver substances into cells and tissues.

(a). Uses

- This method is commonly employed for genetic transformation of plants and many organisms.
- This method is applicable for the plants having less regeneration capacity and those which fail to show sufficient response to *Agrobacterium*- mediated gene transfer in rice, corn,wheat, chickpea,sorghum and pigeon-pea.

(b). Apparatus

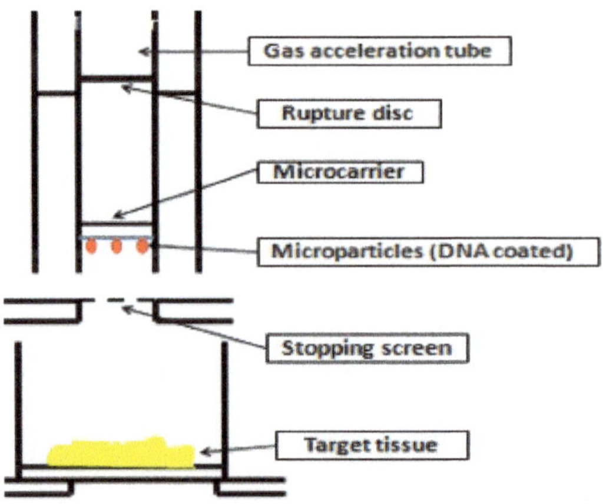

(b): Working system of particle bombardment gun.

The biolistic gun employs the principle of conservation of momentumand uses the passage of helium gas through the cylinder with arrange of velocitiesrequired for optimal transformation of various cell types. It consists of a bombardment chamber which is connected to an outlet for vacuum creation. The bombardment chamber consists of a plastic rupture disk below which macro carrier is loaded with micro carriers. These micro carriers consist of gold or tungsten micro pellets coated with DNA for transformation.

The apparatus is placed in laminar flow while working to maintain sterile conditions. The target cells/tissue is placed in the apparatus and a stopping screen is placed between the target cells and micro carrier assembly. The passage of high pressure helium ruptures the plastic rupture disk propelling the macro carrier and micro carriers. The stopping screen prevents the passage of macro projectiles but allows the DNA coated micro pellets to pass through it thereby, delivering DNA into the target cells.

(c). Advantages

- Simple and convenient method involving coating DNA or RNA on to gold microcarrier, loading sample cartridges, pointing the nozzle and firing the device.
- No need to obtain protoplast as the intact cell wall can be penetrated.
- Manipulation of genome of sub-cellular organelles can be done.
- Eliminates the use of potentially harmful viruses or toxic chemical treatment as gene delivery vehicle.
- This device offers to place DNA or RNA exactly where it is needed into any organism

(d). Disadvantages

- The transformation efficiency may be lower than *Agrobacterium- mediated transformation.*
- Specialized equipment is needed. Moreover the device and consumables are costly.
- Associated cell damage can occur.
- The target tissue should have regeneration capacity.
- Random integration is also a concern.
- Chances of multiple copy insertions could cause gene silencing.

Sonoporation

- Sonoporation involves the use of ultrasound for temporary permeabilization of the cell membrane allowing the uptake of DNA, drugs or other therapeutic compounds from the extracellular environment.
- This method leaves the compound trapped inside the cell after ultrasound exposure.
- It employs the acoustic cavitation of micro bubbles for enhancing the delivery of large molecules like DNA.The micro bubbles form complex with DNA followed by injection and ultrasound treatment to deliver DNA into the target cells.

- Unlike other methods of transfection, sonoporation combines the capability to enhance gene and drug transfer.

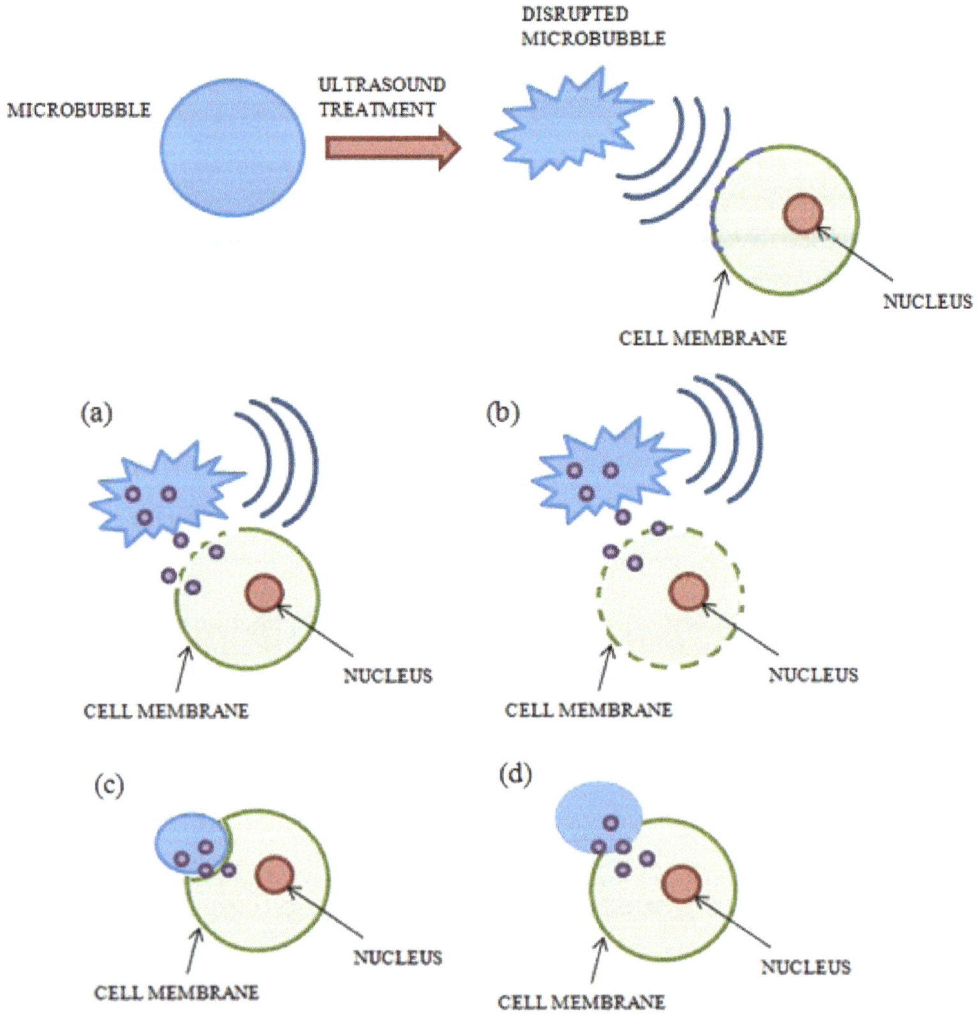

Rupture of microbubbles by ultrasound resulting in enhanced membrane permeability caused by shear stress, increased temperature and activation of reactive oxygen species. Drug delivery by microbubbles by (a) transient holes induced by shear stress for drug delivery (b) increase in membrane fluidity (c) endocytosis of microbubbles (d) microbubble- cell membrane fusion.

(a). Advantages

- Simpleand highly efficient gene transfer method.
- No significant damage is cause to the target tissue.

(b). Disadvantages

- Not suitable for tissues with open or cavitated structures.
- High exposure to low-frequency (<MHz) ultrasounds result in complete cellular death (rupture of the cell). Thus cellular viability must be taken into consideration while employing this technique.

Laser Induced Transfection

- It involves the use of a brief pulse of focused laser beam.
- In this method, DNA is mixed with the cells present in the culture and then a fine focus of laser beam is passed on the cell surface that forms a small pore sufficient for DNA uptake into the cells. The pore thus formed is transitory and repairs soon.

Bead Transfection

- Bead transfection combines the principle of physically producing breaks in the cellular membrane using beads.
- In this method, the adherent cells are incubated for a brief period with glass beads in a solution containing the DNA.
- The efficiency of this rapid technique depends on:
    - o Concentration of DNA in a solution.
    - o Timing of the addition of DNA.
    - o Size and condition of the beads and the buffers utilized.

Immunoporation is a recently developed transfection process involving the use of new type of beads, Immunofect™ beads, which can be targeted to make holes in a specific type of cells.

Agrobacterium

Agrobacterium is a genus of Gram-negative bacteria established by H. J. Conn that uses horizontal gene transfer to cause tumors in plants. *Agrobacterium tumefaciens* is the most commonly studied species in this genus. *Agrobacterium* is well known for its ability to transfer DNA between itself and plants, and for this reason it has become an important tool for genetic engineering.

The *Agrobacterium* genus is quite heterogeneous. Recent taxonomic studies have reclassified all of the *Agrobacterium* species into new genera, such as *Ahrensia*, *Pseudorhodobacter*, *Ruegeria*, and *Stappia*, but most species have been controversially reclassified as *Rhizobium* species.

Plant Pathogen

A. tumefaciens causes crown-gall disease in plants. The disease is characterised by a tumour-like growth or gall on the infected plant, often at the junction between the root and the shoot. Tumors are incited by the conjugative transfer of a DNA segment (T-DNA) from the bacterial tumour-inducing (Ti) plasmid. The closely related species, *A. rhizogenes*, induces root tumors, and carries the distinct Ri (root-inducing) plasmid. Although the taxonomy of *Agrobacterium* is currently under revision it can be generalised that 3 biovars exist within the genus, *A. tumefaciens*, *A. rhizogenes*, and *A. vitis*. Strains within *A. tumefaciens* and *A. rhizogenes* are known to be able to harbour either a Ti or Ri-plasmid, whilst strains of *A. vitis*, generally restricted to grapevines, can harbour a Ti-plasmid. Non-*Agrobacterium* strains have been isolated from environmental samples which harbour a Ri-plasmid whilst laboratory studies have shown that non-*Agrobacterium* strains can also harbour a Ti-plasmid. Some environmental strains of *Agrobacterium* possess neither a Ti nor Ri-plasmid. These strains are avirulent.

The large growths on these roots are galls induced by *Agrobacterium* sp.

The plasmid T-DNA is integrated semi-randomly into the genome of the host cell, and the tumor morphology genes on the T-DNA are expressed, causing the formation of a gall. The T-DNA carries genes for the biosynthetic enzymes for the production of unusual amino acids, typically octopine or nopaline. It also carries genes for the biosynthesis of the plant hormones, auxin and cytokinins, and for the biosynthesis of opines, providing a carbon and nitrogen source for the bacteria that most other micro-organisms can't use, giving *Agrobacterium* a selective advantage. By altering the hormone balance in the plant cell, the division of those cells cannot be controlled by the plant, and tumors form. The ratio of auxin to cytokinin produced by the tumor genes determines the morphology of the tumor (root-like, disorganized or shoot-like).

In Humans

Although generally seen as an infection in plants, *Agrobacterium* can be responsible for opportunistic infections in humans with weakened immune systems, but has not been shown to be a primary pathogen in otherwise healthy individuals. One of the earliest associations of human disease caused by *Agrobacterium radiobacter* was reported by Dr. J. R. Cain in Scotland (1988). A later study suggested that *Agrobacterium* attaches to and genetically transforms several types of human cells by integrating its T-DNA into the human cell genome. The study was conducted using cultured human tissue and did not draw any conclusions regarding related biological activity in nature.

Uses in Biotechnology

The ability of *Agrobacterium* to transfer genes to plants and fungi is used in biotechnology, in particular, genetic engineering for plant improvement. A modified Ti or Ri plasmid can be used. The plasmid is 'disarmed' by deletion of the tumor inducing genes; the only essential parts of the T-DNA are its two small (25 base pair) border repeats, at least one of which is needed for plant transformation. The genes to be introduced into the plant are cloned into a plant transformation vector that contains the T-DNA region of the disarmed plasmid, together with a selectable marker (such as antibiotic resistance) to enable selection for plants that have been successfully transformed. Plants are grown on media containing antibiotic following transformation, and those that do not have the T-DNA integrated into their genome will die. An alternative method is agroinfiltration.

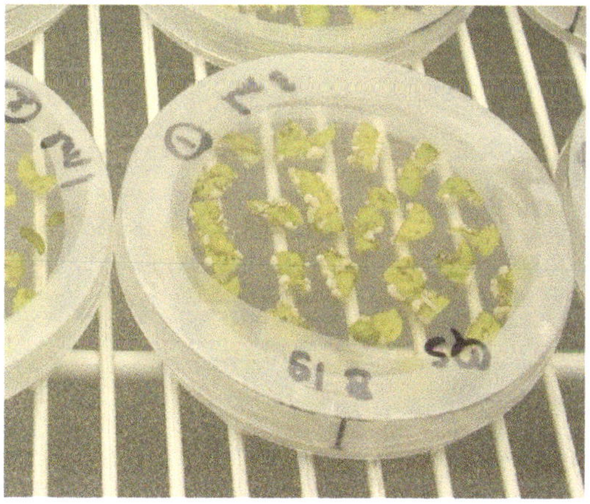

Plant (*S. chacoense*) transformed using *Agrobacterium*. Transformed cells start forming calluses on the side of the leaf pieces

Transformation with *Agrobacterium* can be achieved in two ways. Protoplasts or alternatively leaf-discs can be incubated with the *Agrobacterium* and whole plants regenerated using plant tissue culture. A common transformation protocol for *Arabidopsis* is the floral-dip method: inflorescence are dipped in a suspension of *Agrobacterium*, and the bacterium transforms the germline cells that make the female gametes. The seeds can then be screened for antibiotic resistance (or another marker of interest), and plants that have not integrated the plasmid DNA will die when exposed to the correct condition of antibiotic.

Agrobacterium does not infect all plant species, but there are several other effective techniques for plant transformation including the gene gun.

Agrobacterium is listed as being the vector of genetic material that was transferred to these USA GMOs:

- Soybean
- Cotton
- Corn
- Sugar Beet
- Alfalfa
- Wheat
- Rapeseed Oil (Canola)
- Creeping bentgrass (for animal feed)
- Rice (Golden Rice)

Genomics

The sequencing of the genomes of several species of *Agrobacterium* has permitted the study of the evolutionary history of these organisms and has provided information on the genes and systems involved in pathogenesis, biological control and symbiosis. One important finding is the possibility that chromosomes are evolving from plasmids in many of these bacteria. Another discovery is that the diverse chromosomal structures in this group appear to be capable of supporting both symbiotic and pathogenic lifestyles. The availability of the genome sequences of *Agrobacterium* species will continue to increase, resulting in substantial insights into the function and evolutionary history of this group of plant-associated microbes.

History

Marc Van Montagu and Jozef Schell at the University of Ghent (Belgium) discovered the gene transfer mechanism between *Agrobacterium* and plants, which resulted in the development of methods to alter *Agrobacterium* into an efficient delivery system for gene engineering in plants. A team of researchers led by Dr Mary-Dell Chilton were the first to demonstrate that the virulence genes could be removed without adversely affecting the ability of *Agrobacterium* to insert its own DNA into the plant genome (1983).

Chloroplast Transformation

Chloroplast transformation is an important tool for biotechnological applications and has gained much interest in the regulation of gene expression and plant physiology.

Chloroplast

Chloroplasts are sub cellular organelles (plastids) of plant cells generally considered to be derived from the symbiotic cyanobacteria. They are present in shoots and leaves of green plants and contain pigment called chlorophyll. They are also present in several forms as colorless plastids (amyloplasts) in roots and as colored plastids (chromoplasts) in fruits. A cell comprises variable number of plastids, each plastid containing many copies of genome (50 to 100).

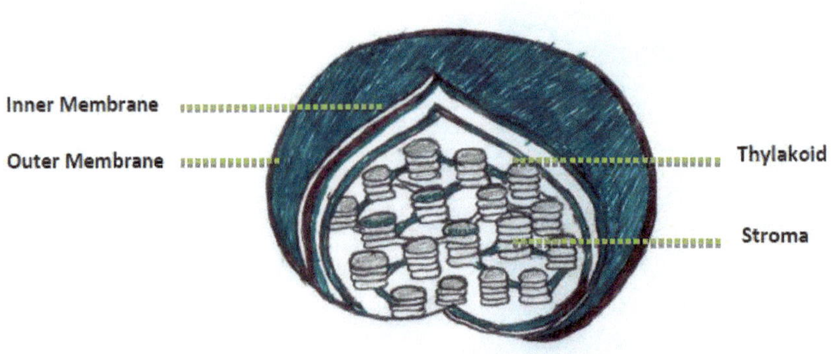

Schematic diagram of Chloroplasts

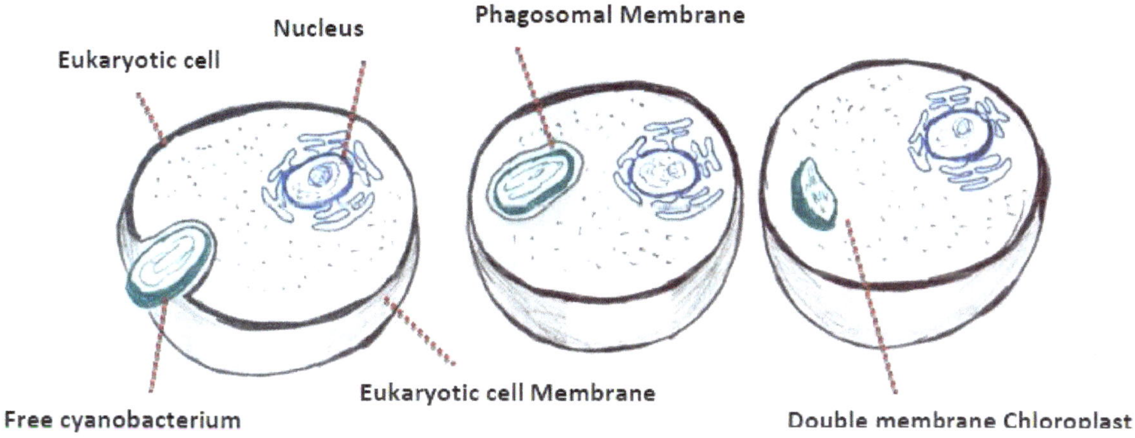

Chloroplasts derived from endosymbiotic cyanobacteria.

Plastid genomes resemble bacterial genomes in many aspects and also contain some features of multicellular organisms, such as RNA editing and split genes. Most of the proteins are encoded in the plant nucleus, synthesized and then imported into the chloroplast. However, chloroplast genome encodes for transcription and translation machinery and numerous proteins.

Chloroplast Transformation: Historical Perspective

First stable chloroplast transformation was achieved in the alga *Chlamydomonas reinhardtii*. In addition, the *aadA* marker and methods for removal of marker were first demonstrated. In higher plants, Tobacco due to its ease of culture and regeneration, gained significant attention for chloroplast transformation.

Tobacco protoplasts were co-cultivated with *Agrobacterium* but the resulted transgenic lines showed the unstable integration of foreign DNA into the chloroplast genome.

The candidate genes were introduced in isolated intact chloroplasts and then into protoplasts resulting in transgenic plants. Gene gun, a transformation device, was developed by John Sanford to enable the transformation of plant chloroplasts without using isolated plastids.

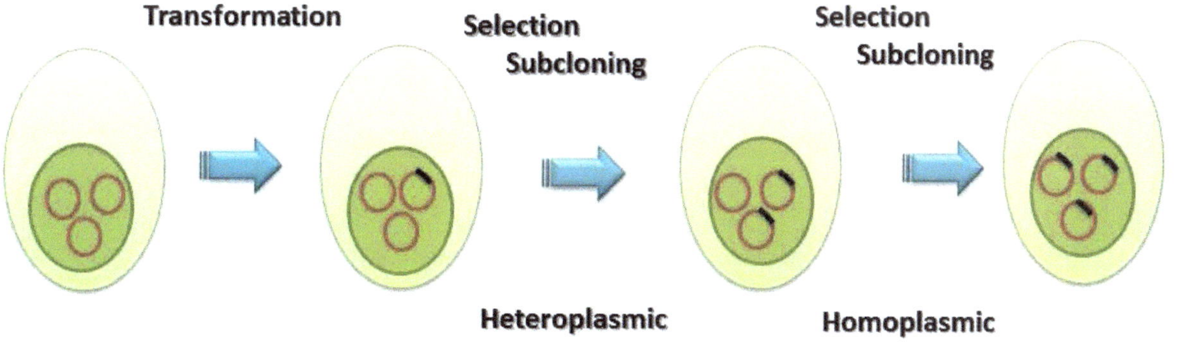

Chloroplast transformation in *Chlamydomonas*: *Chlamydomonas* comprises a single large chloroplast with about hundreds of copies of its genome. Initial integration occurs in only one copy of the polyploid plastome resulting in heteroplasmic. Repeated sub-cloning and selection result in recovery of homoplasmic clones.

Advantages of Chloroplast Transformation

Chloroplast transformation offer several advantages compared with nuclear transformation which are as follows.

Risk of Transgene Escape

Chloroplast genome is maternally inherited and there is rare occurrence of pollen transmission. It provides a strong level of biological containment and thus reduces the escape of transgene from one cell to other.

Expression Level

- It exhibits higher level of transgene expression and thus higher level of protein production due to the presence of multiple copies of chloroplast transgenes per cell and
- Remains unaffected by phenomenon such as pre or post-transcriptional silencing.

Homologous Recombination

- Chloroplast transformation involves homologous recombination and is therefore precise and predictable.
- This minimizes the insertion of unnecessary DNA that accompanies in nuclear genome transformation.
- This also avoids the deletions and rearrangements of transgene DNA, and host genome DNA at the site of insertion.

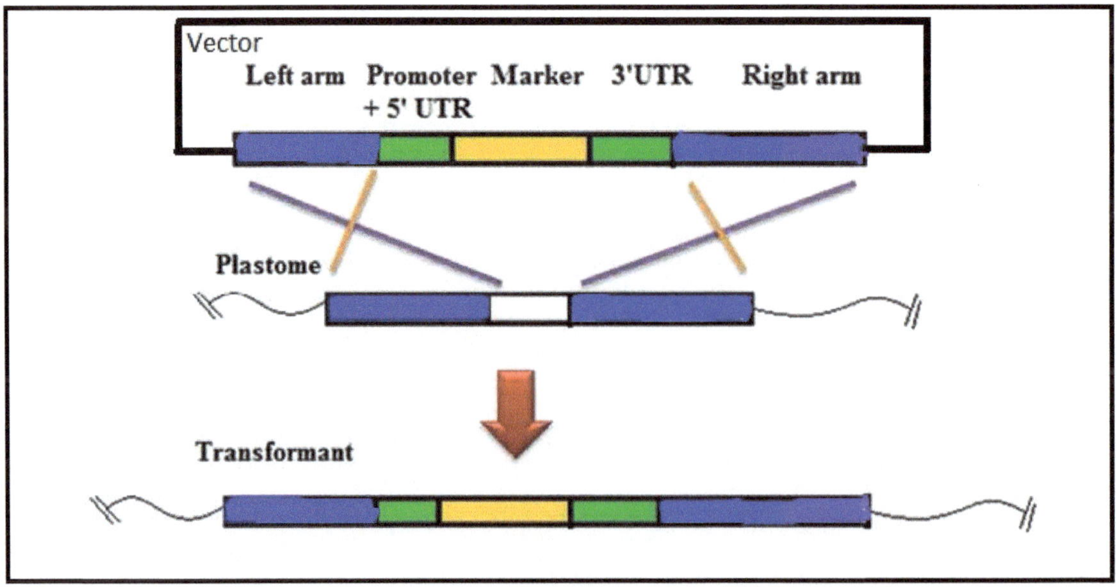

A vector containing a selectable marker (yellow) under the control of plastid expression signals (promoter, 5′ UTR and 3′ UTR , shown in green) flanked by chloroplast sequences (shown in purple). Homologous recombination takes place between the flanking targeting arms and recipient plastid genome (plastome).

Gene Silencing/ RNA Interference

Gene silencing or RNA interference does not occur in genetically engineered chloroplasts.

Position Effect

- Absence of position effect due to lack of a compact chromatin structure and efficient transgene integration by homologous recombination.
- Avoids inadvertent inactivation of host gene by transgene integration.

Disulphide Bond Formation

Ability to form disulfide bonds and folding human proteins results in high-level production of biopharmaceuticals in plants.

Multiple Gene Expression

Multiple transgene expression is possible due to polycistronic mRNA transcription.

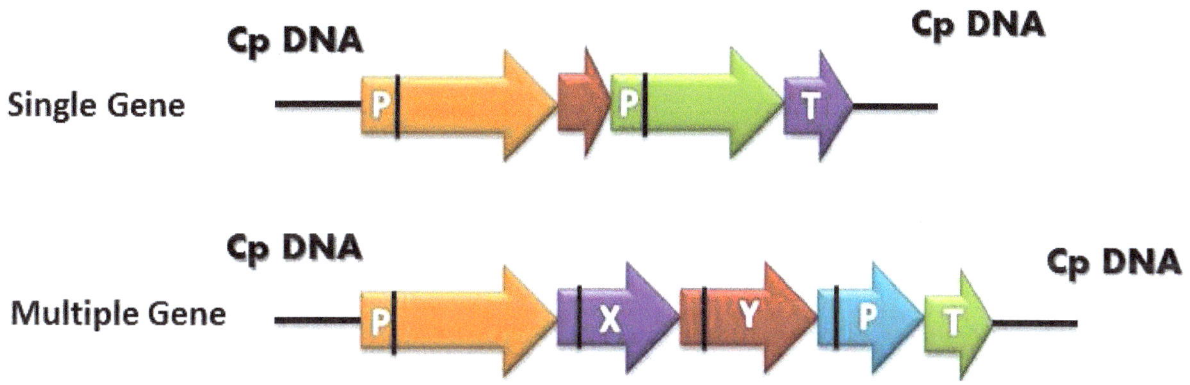

Multigene expression

Expression of Edible Vaccine

High level of expression and engineering foreign genes without the use of antibiotic resistant genes makes this compartment ideal for the development of edible vaccines.

Codon Usage

Chloroplast is originated from cyanobacteria through endosymbiosis. It shows significant similarities with the bacterial genome. Thus, any bacterial genome can be inserted in chloroplast genome.

Expression of Toxic Proteins

Foreign proteins observed to be toxic in the cytosol are non-toxic when accumulated within transgenic chloroplasts as they are compartmentalized inside chloroplast.

Comparison of Chloroplast and Nuclear Genetic Engineering

Transgenic	Chloroplast Genome	Nuclear Genome
Transgene copy number	10-100 plastid genome	Species - Specific
Level of gene expression	High	Limited
Gene arrangement	Occurs aften	Transgene is independently inserted
Position effect	No	Yes
Gene silencing	No	Yes
Gene Containment	Yes	No
Disulphide bond formation	Yes	No
Toxicity of foreign proteins	Low	High
Transgenic lines	Uniform gene expression	Variable gene expression
Homogenecity	Homoplasmic	Heterozygous or homozygous

Table: Differences between chloroplast and nuclear genetic engineering

Transformation Methods for Chloroplast Transformation

- Biolistic/Particle bombardment method involves the introduction of *Escherichia coli* plasmids containing a gene of interest and marker gene into chloroplasts or plastids. The insertion of foreign genes into plasmid DNA occurs by homologous recombination via the sequences flanking at the insertion site. First successful chloroplast transformation was performed in *Chlamydomonas reinhardtii* by particle bombardment method. Simple operation and high transformation efficiency makes it a favorable way for plastid or chloroplast transformation.
- PEG-mediated and *Agrobacterium- mediated* transformation method was also employed in the early days.

After the first chloroplast transformation in *Chlamydomonas reinhardttii,* the stable plastid transformation has also been established in higher plants, *Nicotiana tobacum, Arabidopsis,* rape, *Lesquerella,* rice, potato, lettuce, soybean, cotton, carrot and tomato.

However, plastid transformation is routinely performed only in tobacco because of higher efficiency of transformation in tobacco than in other plants.

Vector Design for Chloroplast Transformation

Selectable Marker Genes

Due to the presence of multi-copy plastid DNA, selectable marker genes are very important to achieve uniform transformation of all genome copies. The non-transformed plastids are gradually left out during an enrichment process using a selective medium. Initially plastid *16S rRNA* (*rrn16*) gene was used as a selection marker in chloroplast transformation. The transgenic lines were selected by spectinomycin resistance but with low efficiency.

Various selectable markers for chloroplast transformation have been developed based on various features like dominance, cell-autonomy or portability.

- Dominant markers confer high transformation efficiency due to the expression at early stages although present in a minority of the plastomes. e. g. *aadA* (aminoglycoside 3′adenylyltransferase) gene confers resistance to streptomycin and spectinomycin by inactivation of antibiotics.
- Recessive markers confer lower transformation efficiency. They confer resistance only when enough transformed plastome copies are produced by random segregation resulting in a selectable phenotype. e. g. point mutation in the ribosomal RNA, *rrnS* and *rrnL* genes, confer antibiotic resistance by relieving the sensitivity of individual ribosome.
- Plastid- or cell-autonomous markers confer their phenotype only to the organelle or the cell in which they reside. Some markers may integrate at a specific locus of the plastid genomee.g. *rrnS* or *rrnL* genes.
- Autonomous and portable markers can be inserted virtually in any locus of the plastome e.g. *aadA* gene.
- Some markers have a property to confer a phenotype strong enough for direct selection of transformants.

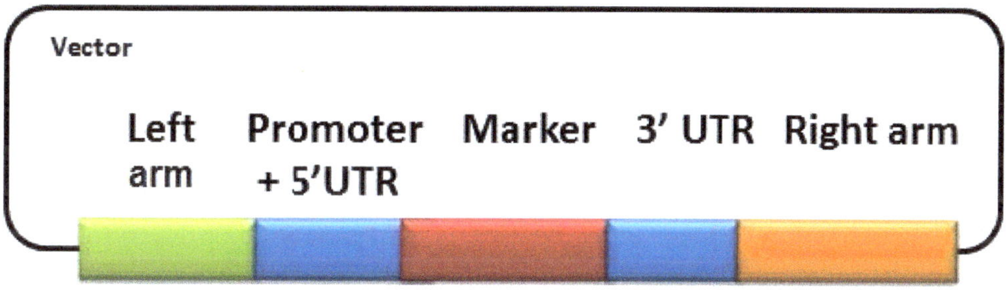

Selectable marker (in case when it is a foreign gene to be inserted into plastid genome shown in red) is under the control of expression signals such as promoter, a 3′ UTR and a 5′ UTR (shown in blue).

Marker	Selection	Organism
Photosynthesis		
atpB	Photo autotrophy	*Chlamydomonas*
tscA	Photo autotrophy	*Chlamydomonas*
psaA/B	Photo autotrophy	*Chlamydomonas*
petB	Photo autotrophy	*Chlamydomonas*
rbcL	Photo autotrophy	Tobacco
Drug resistance		
rrnS	Spectinomycin, Streptomycin	*Chlamydomonas*, Tobacco, Tomato
rrnL	Erythromycin	*Chlamydomonas*
nptII	Kanamycin	Tobacco (low transformation efficiency), Cotton
aphA-6	Kanamycin, Amikacin	*Chlamydomonas*, Tobacco, Cotton
aad (preferred for chloroplast transformation)	Trimethoprim, Spectinomycin, Streptomycin, Sulphonamides	*Solanum tuberosum* (Potato)

ble	Bleomycin	Tobacco
dhfr	Methotraxate	Rice
hpt	Hygromycin	Rice
gat	Gentamycin	Tobacco
Herbicide resistance		
psbA	DCMU, metribuzin	*Chlamydomonas*
bar (not good enough for transformation) and *pat*	Phosphinothricin	Tobacco
AHAS	Sulfometuron methyl	*Porphyridium* sp.
EPSP	Glyphosate	Tobacco
HPPD	Diketonitrile	Tobacco
Csr 1-1	Chlorsulphuron	Rice
dhps	Sulphonamides	Tobacco
Metabolism		
BADH	Betaine aldehyde	Tobacco
coda	5-fluorocytosine	Tobacco
ARG9	Arg autotrophy	*Chlamydomonas*
ASA2	Trp analogues	Tobacco

Table. Various selectable markers used for chloroplast transformation

Insertion Sites

- Plastid expression vectors possess left and right flanking sequences each with 1–2 kb in size from the host plastid genome, which facilitates foreign gene insertion into plastid DNA via homologous recombination.
- The insertion site in the plastid genome is determined by the choice of plastid DNA segment flanking the marker gene and the gene of interest. The foreign DNA is inserted in intergenic regions of the plastid genome.
- Most commonly used insertion sites are *trnV-3'rps12'trnI-trnA and trnfM-trnG.*

The *trnV-3'rps12* and *trnI-trnA* sites are located in the *25 kb* inverted repeat *(IR)* region of plastid DNA and insertion of gene into these sites results in the rapid copying into two copies in the IR region.

The *trnfM-trnG* site is located in the large single copy region of the plastid DNA. The insertion of gene between *trnfM* and *trnG* should contain only one copy per plastid DNA. *pSBL- CTV2* was the first vector developed in the Daniell laboratory for expression of several proteins. This vector inserts the foreign gene in *trnI-trnA* intergenic region.

Regulatory Sequences

- The level of gene expression in plastids is predominately determined by regulatory sequences such as promoter as well as 5′UTR elements.
- Strong promoter is required to ensure high mRNA level for high-level of protein accumulation e.g. rRNA operon *(rrn)* promoter *(Prrn)*. Most commonly used promoter is *CaMV* 35S

promoter of cauliflower mosaic virus which drives high level of transgene expression in dicots.

- In plastid expression vectors, a suitable 5′ untranslated region (5′-UTRs) containing a ribosomal binding site (RBS) is an important element.
- Stability of the transgenic mRNA is ensured by the 5′ UTR and 3′ UTR sequences flanking the transgene. Protein accumulation from the transgene depends on the 5′-UTR inserted upstream of the open reading frame encoding the genes of interest.

Applications of Chloroplast Engineering

Chloroplast transformation can be used in the production of transgenic plants with herbicide resistance, insect resistance, viral resistance, fungal resistance, abiotic and biotic stress tolerance, production of biopharmaceuticals etc.

References

- Uchino, Yoshihito; Yokota, Akira; Sugiyama, Junta (1997). "Phylogenetic position of the marine subdivision of Agrobacterium species based on 16S rRNA sequence analysis". The Journal of General and Applied Microbiology. 43 (4): 243–247. PMID 12501326. doi:10.2323/jgam.43.243
- Kunik, T.; Tzfira, T; Kapulnik, Y; Gafni, Y; Dingwall, C; Citovsky, V (2001). "Genetic transformation of HeLa cells by Agrobacterium". Proceedings of the National Academy of Sciences. 98 (4): 1871–6. Bibcode:2001PNAS...98.1871K. JSTOR 3054968. PMC 29349. PMID 11172043. doi:10.1073/pnas.041327598
- Farrand, S. K.; Van Berkum, P. B.; Oger, P (2003). "Agrobacterium is a definable genus of the family Rhizobiaceae". International Journal of Systematic and Evolutionary Microbiology. 53 (5): 1681–7. PMID 13130068. doi:10.1099/ijs.0.02445-0
- Schell, J.; Van Montagu, M. (1977). "The Ti-Plasmid of Agrobacterium Tumefaciens, A Natural Vector for the Introduction of NIF Genes in Plants?". In Hollaender, Alexander; Burris, R. H.; Day, P. R.; Hardy, R. W. F.; Helinski, D. R.; Lamborg, M. R.; Owens, L.; Valentine, R. C. Genetic Engineering for Nitrogen Fixation. Basic Life Sciences. 9. pp. 159–79. ISBN 978-1-4684-0882-9. PMID 336023. doi:10.1007/978-1-4684-0880-5_12
- Joos, H; Timmerman, B; Montagu, M. V.; Schell, J (1983). "Genetic analysis of transfer and stabilization of Agrobacterium DNA in plant cells". The EMBO Journal. 2 (12): 2151–60. PMC 555427. PMID 16453483
- Thomson JA. "Genetic Engineering of Plants" (PDF). Biotechnology. Encyclopedia of Life Support Systems. 3. Retrieved 17 July 2016
- Hulse, M.; Johnson, S.; Ferrieri, P. (1993). "Agrobacterium Infections in Humans: Experience at One Hospital and Review". Clinical Infectious Diseases. 16 (1): 112–7. PMID 8448285. doi:10.1093/clinids/16.1.112
- Clough, Steven J.; Bent, Andrew F. (1998-12-01). "Floral dip: a simplified method forAgrobacterium-mediated transformation ofArabidopsis thaliana". The Plant Journal. 16 (6): 735–743. ISSN 1365-313X. PMID 10069079. doi:10.1046/j.1365-313x.1998.00343.x
- Cain, John Raymond (1988). "A case of septicaemia caused by Agrobacterium radiobacter". Journal of Infection. 16 (2): 205–6. PMID 3351321. doi:10.1016/s0163-4453(88)94272-7
- Setubal, Joao C.; Wood, Derek; Burr, Thomas; Farrand, Stephen K.; Goldman, Barry S.; Goodner, Brad; Otten, Leon; Slater, Steven (2009). In Jackson, Robert W. Plant Pathogenic Bacteria: Genomics and Molecular Biology. Caister Academic Press. pp. 91–112. ISBN 978-1-904455-37-0

6

Current Developments in Genetic Engineering

Transgenic science is used for both animals and plants. In plants, genes are introduced normally to strengthen them against diseases, pests and stress. Transgenic animals, on the other hand, are used for clinical research. Knockout and knockin animals are two types of transgenic animals. The aspects elucidated in this chapter are of vital importance, and provide a better understanding of the advancements made in genetic engineering.

Transgenic Science

Transgenic plants are developed by transferring or modifying genes from another organism by a diverse technique like physical, chemical and biological methods. Transgenic plants are obtained by introducing a gene into its genome with the help of vectors in order to develop a plant with new characteristics. This process of recombinant DNA technology is used for developing genetically modified plants to embolden the variety against pests, diseases and stress or to improve the quality of the product. These can also be achieved by conventional breeding if the genes for the desired characters are present in close or wild relatives else transgenesis is the only viable alternative. Apart from this, genetic engineering can overcome the other limitations of conventional breeding such as large space requirement, time consuming, and uncertainty of the result. It is not mandatory to opt for transgenics if the problem can be addressed by conventional breeding. Rational decision making can be done by following figure.

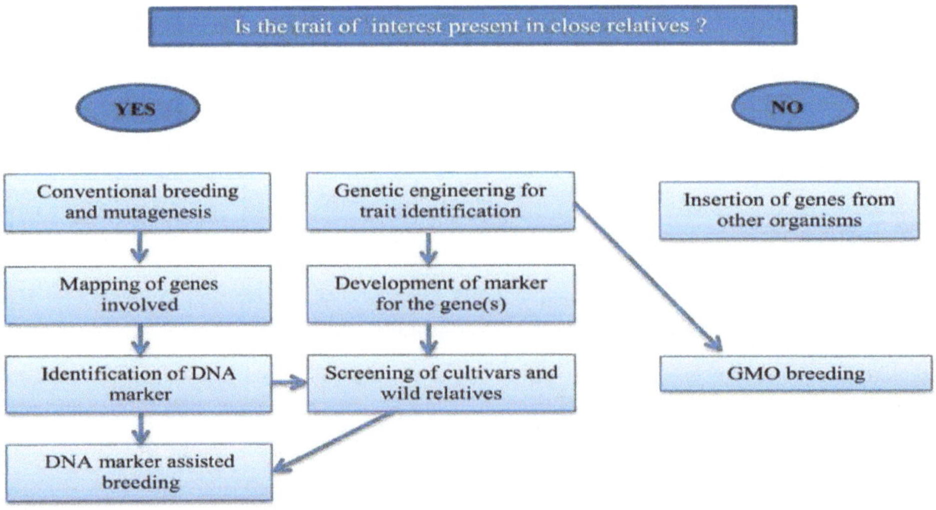

Decision making box by using both conventional and modern biotechnology approaches for crop breeding

Modern Plant Breeding

Modern plant breeding uses techniques of molecular biology to select the variations. This approach which combines biotechnology and molecular biology is known as molecular breeding.

Steps Involved in Plant Breeding

Following steps are involved in both traditional as well as modern plant breeding-

1. Generation (or identification) of variant may be done simply by collection of wild type varieties or traditional varieties from farmers or by hybridization (crossing 2 or more plants species) and by induced mutation and induced polyploidy

2. Selection for desired characteristics like grains for large seed size, seed dormancy, and non-shattering seed heads. It is possible now to select variants based on molecular markers.

These two steps are still followed for plant breeding but biotechnology has changed the breeding techniques. In this chapter we are going to focus more on the new technology of introducing characters to the plant called Transgenic Technology.

Transgenic Technology results into genetic variation across kingdoms, rather than within a species or genus. Gene transfer is more precise than previous methods.

Steps Involved in Production of Transgenic Plants

There are several steps involved in production of transgenic plants which is given in figure.

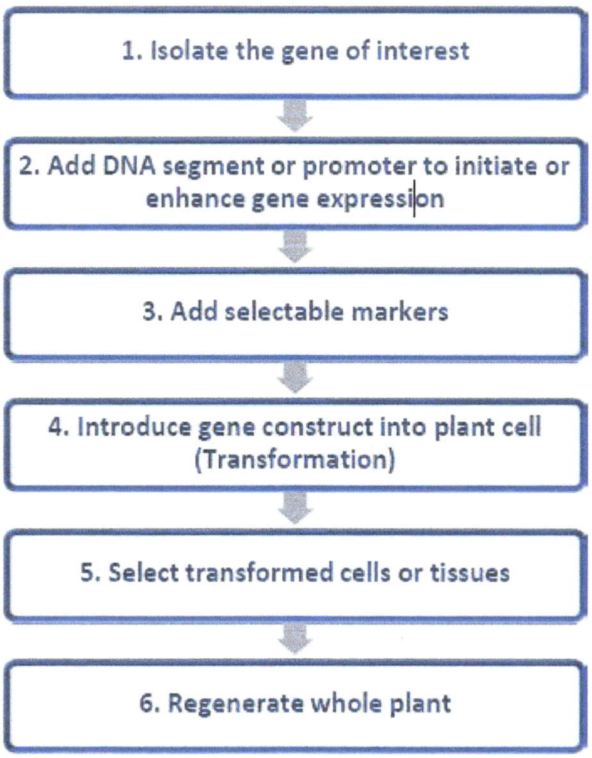

Steps Involved in production of transgenic plants

Isolation and Cloning the Gene of Interest

Identification and cloning of the gene of interest is a first limiting step in the transgenic development process. Locating , identifying, characterizing, and cloning genes of agricultural importance requires, a huge effort both in terms of human and financial capital. One of the earliest development is the introduction of insect resistance by transgenic technology. The discovery of Bt Genes has revolutionized plant transgenics.

Spores of the soil bacterium *Bacillus thuringiensis* (Bt) contain a crystal (cry) protein (δ-endotoxin). Inside insect gut, the crystals break apart and release a toxin that binds to and creates pores in the intestinal lining. Instead of whole gene truncated cry gene is used in Bt crops. Figure shows the truncated cry gene structure.

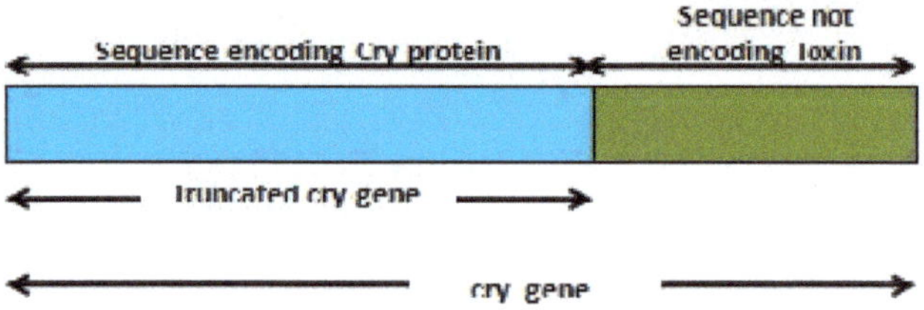

Gene sequence showing the truncated cry gene

Mechanism of Toxicity

- *Bt* gene (also known as *cry* gene) was found in a gram positive bacteria *Bacillus thuringiensis* . The structure of Bt gene is shown in Figure.
- This *Bt* gene is used in the production of insect resistant crops (genetically modified) and biological insecticides as well.
- *Bacillus thuringiensis*, during sporulation produces a toxic protein which possesses insecticidal activity against Lepidoptera, Coleoptera, Hymenoptera, Diptera and Nematode.
- This crystal protein is called as Cry protein, encoded by cry gene present on the plasmid (non-chromosomal gene).
- As soon as the Cry protein crystals reach the digestive tract of the insect, the prevailing alkaline condition there causes denaturation of the insoluble crystals.

This denaturation makes the crystal soluble and prone to proteases activity in the gut of insect.

- Proteolysis of Cry crystal leads to release of cry toxin, which forms pore in the cell membrane of the gut by inserting themselves into it.

The pore causes cell lysis and ultimately death of insects.

Control of Gene Expression

- The level of gene expression is determined by regulatory sequences such as promoters as well as 5' UTR elements.

- Transgene Promoters: Most commonly used is the CaMV 35S promoter of cauliflower mosaic virus. It is a constitutive promoter (turned on all the time in all tissues), that gives high levels of expression in plants.

Most commonly used terminator sequence is the nopaline synthase (*nos*) gene from *Agrobacterium tumefaciens*.

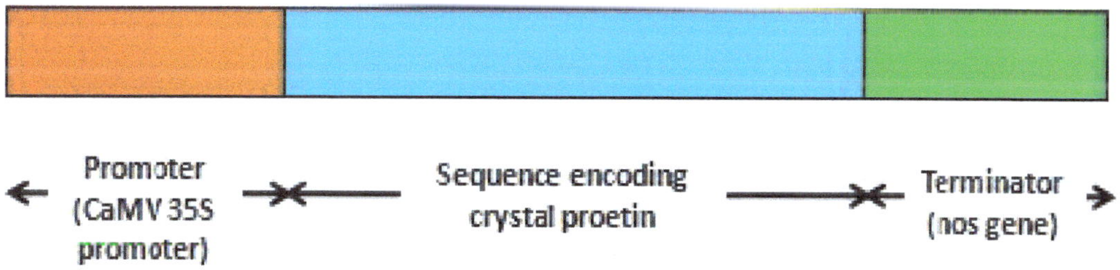

Shows expression cassette of Bt gene along with promoter and terminator.

Selectable Markers

Various selectable markers are used for the selection of transgenic plants.

Plant Tissues used for Transformation

The tissue must be capable of generating callus (undifferentiated tissue), from which the complete plant can be produced. The choice of tissue depends on the species. Some commonly used tissues are immature embryos, leaf disks, and apical meristems.

Introduction of Gene Construct into Plant Cells (Transformation)

The basic methods and techniques used for plant cell transformation are as follows:

i. *Agrobacterium* mediated transformation
ii. Agro-infection
iii. Chloroplast transformation
iv. Indirect gene transfer
v. Electroporation
vi. Biolistic (Gene gun) Method
vii. Chemical method of gene transfer
viii. Microinjection
ix. Pollen Transformation
x. Direct DNA uptake by mature zygotic embryos

Selection of Transformants

- Cells/tissues in which new genes are incorporated into plant's DNA are identified and then grown in media containing antibiotics or herbicides.
- So during transformation, selectable marker gene (antibiotic resistant etc.) is incorporated such that, cells which are devoid of this plasmids get killed or their growth is arrested. Kanamycin, an antibiotic capable of killing plant cells.
- Since transformed plants contain kanamycin resistance gene, they only can survive.
- In the *Agrobacterium* infiltration method, the seed of the infiltrated plants are plated on agar containing kanamycin – the plant seed containing the introduced DNA will germinate (less in number) and will grow on agar plates containing Kanamycin.

Regeneration of Transformant

- During regeneration, whole plants with inserted genes are developed through tissue culture.
- Cultured transformed plant cells are regenerated under appropriate conditions.
- Auxin/cytokinin ratio is important for plant regeneration.
- In the culture media high concentration of auxin (plant growth regulator), causes rooting of the cultured transformed cells. Further, shooting is initiated by increasing the concentration of cytokinin (plant growth regulator-phytohormones) and new plantlet develops from the culture.
- After that plantlets are transferred to soil for proper anatomical and physiological developments, and for better acclimatization to environment.

Confirmation of Transformed Plants

The presence and activity of introduced gene is confirmed by observation of phenotype and by advanced methods as listed below.

1. Southern blot: In Southern blotting separated DNA fragments obtained after electrophoresis, are transferred to a filter membrane and subsequent fragment detection is accompanied by probe hybridization.
2. Northern blot: It is used to study gene expression by detection of mRNA (or isolated mRNA) in a plant sample.
3. Western blot: It is used to study the gene expression by detection of the protein produced by the transformed regenerated plants.

Evaluation of Transformed Plants

- Plant is evaluated for the presence and activity of introduced gene.
- The effect of various environmental factors on the transgenic plant is also evaluated.

- Evaluation for food or feed safety.
- Evaluation of basic containment level is also very important.

Chloroplast Transformation

Here we are going to see some of the application of chloroplast engineering.

Improvement of Plant Traits

Plastid genome can be used to engineer many agronomic traits such as herbicide resistance, insect resistance, and antibiotic resistance.

Biotic Stress

Expression of insect resistance genes such as *cry* genes can be well performed in the plastid genome without modifications of codon usage or other sequence manipulations.

Abiotic Stress

Chloroplast engineering can be executed for the successful development of plants suffering from abiotic stresses like salinity, high or low temperature, drought etc. Some examples are listed below-

- The over-expression of enzymes required for Glycine betaine (GlyBet) biosynthesis in transgenic plants improve tolerance to various abiotic stresses.
- Chloroplast transformation has been used to transfer choline monooxygenase (BvCMO), an enzyme that catalyzes the conversion of choline into betaine aldehyde from beet (*Beta vulgaris)* into the plastid genome of tobacco.
- Transplastomic carrot plants expressing betaine aldehyde dehydrogenase (BADH) gene have also been developed using chloroplast engineering showing highest level of salt tolerance.

Production of Biopharmaceuticals

Chloroplast engineering can also be employed for the production of biopharmaceutical proteins, antigens listed in Tables.

Table: Depicts biopharmaceutical proteins expressed via chloroplast transformation

Biopharmaceutical Proteins	Gene	Site of Integration	Promoter	5'/3' Regulatory Elements
Elastin Derived Polymer	*EG 121*	*tm|/trnA*	*Prrn*	*ggagg/TpsbA*
Antimicrobial Peptide	*MSI-99*	*tm|/trnA*	*Prrn*	*ggagg/TpsbA*
Insulin-like growth factor	*IGF-1*	*tm|/trnA*	*Prrn*	*PpsbA/TpsbA*
Interferon alpha 5	*INFa5*	*tm|/trnA*	*Prrn*	*PpsbA/TpsbA*
Interferon alpha 2b	*INFa2b*	*tm|/trnA*	*Prrn*	*PpsbA/TpsbA*
Human serum Albumin	*hsa*	*tm|/trnA*	*Prrn, PpabA*	*ggagga,TpsbA /TpsbA*
Monoclonal Antibodies	*Guy's 13*	*tm|/trnA*	*Prrn*	*ggagg/TpsbA*
Interferon gamma	*IFN-g*	*rbcL/accD*	*PpsbA*	*PpsbA/TpsbA*
Human somatotropin	*hST*	*trnV/rps 12/7*	*Prrn, PpabA*	*T7gene 10psbA/ trps16*

Table: Depicts vaccine antigens expressed via chloroplast transformation

Vaccine Antigens	Gene	Site of Integration	Promoter	5'/3' Regulatory Elements
Cholera toxin	*ctxB*	*trnI/trnA*	*Prrn*	*ggagg/ TpsbA*
Canine Parvovirus (CPV)	*ctxB-2L21 gfp-2L21*	*trnI/trnA*	*Prrn*	*PpsbA/ TpsbA*
Anthrax Protective antigen	*pag*	*trnI/trnA*	*Prrn*	*PpsbA/ TpsbA*
Plague Vaccine	*caF1~LcrV*	*trnI/trnA*	*Prrn*	*PpsbA/ TpsbA*
Tetanus toxin	*tetC (bacterial and synthetic)*	*trnV/rps 12/7*	*Prrn*	*T7 gene 10, atpB/Trbcl*

Metabolic Pathway Engineering

- Chloroplast engineering represents an attractive alternative to conventional nuclear transgene expression for metabolic engineering. The strong transgene containment and possibility to stack multiple transgenes by linking them in operon makes this process widely exploited in chloroplast genome.

  Synthesis of polyhydroxybutyrate (PHB), a bioplastic has been achieved by introduction of the most complex metabolic pathway into the chloroplast genome so far. Three enzymes of PHB biosynthesis are co-transcribed into the tobacco plastid genome. Significant accumulation of PHB in chloroplasts causes male sterility and severe growth retardation. The application of plastid transformation to metabolic pathway engineering is still restricted to few model species like agrobacterium.

- Metabolic engineering has emerged as a promising technology in food crops. Recently, plastid transformation has been performed in tomato to alter carotenoid biosynthesis for the production of fruits with elevated contents of provitamin A (β- carotene), an important antioxidant and essential vitamin for human nutrition.

Examples of GM (Genetically Modified) Crops or Transgenic Plants

Table: Crops from which transgenes might introgress into wild relatives

Crop	Main species	Progenitor species	Genomic structure, chromosome number (2n)	Transgenic varieties	Main wild relatives (2n) (distribution)	References*
Sorghum	*Sorghum bicolor*	*S.bicolor*	Two subgenomes of five chromosomes (2n=10)	None	*S. bicolor* (2n=20) (Old and New World) *Sorghum halepense* (2n=40) (Old and New World) *Sorghum propinquum* *Sorghum almum*	106,107 65,66 66 66
Sunflower	*Helianthus annuus*	*H. annuus*	(2n=34)	Herbicide and insect tolerance in development	*H. annuus* (2n=34) (New World) *Helianthus petiolaris* (2n=34) (New World)	62,63 16,64
Canola	*Brassica napus*	*Brassica rapa* and *Brassica oleracea*	AACC (2n=38)	Herbicide tolerance commercially released, insect and disease stress tolerance	*B. rapa* (AA, 2n=20) (widespread in most canola growing areas)	57‡
	B. rapa	*B. rapa*	AA (2n=20)	Commercial release withdrawn	*B. rapa* (AA, 2n=20) (widespread in most canola growing areas)	–
Wheat	*Triticum aestivum*	Wild *Triticum* and *Aegilops* species	AABBDD (2n=42)	Herbicide tolerance developed, not yet commercially released	*Aegilops cylindrica* (DD, 2n=28) (North America)	54, 55
Sugar beet	*B.vulgaris* ssp. *vulgaris*	*B. vulgaris ssp. maritima*	(2n=18/27)	Herbicide and disease (virus) tolerance commercially released	*B. vulgaris* ssp. *maritima* (2n=18) (Old World)	52,53,108,109
Alfalfa	*Medicago sativa* ssp. *sativa*	*M.sativa*	(2n=32)	Herbicide and stress tolerance developed but not commercially released	*M. sativa* ssp. *sativa* (spp. *caerulea*) (2n=16) (Old and New World) *M. sativa* ssp. *falcata* (2n=16/32) (Old and New World)	50,51

Pharming (Genetics)

Pharming, a portmanteau of "farming" and "pharmaceutical", refers to the use of genetic engineering to insert genes that code for useful pharmaceuticals into host animals or plants that would otherwise not express those genes, thus creating a genetically modified organism (GMO). Pharming is also known as molecular farming, molecular pharming or biopharming.

The products of pharming are recombinant proteins or their metabolic products. Recombinant proteins are most commonly produced using bacteria or yeast in a bioreactor, but pharming offers the advantage to the producer that it does not require expensive infrastructure, and production capacity can be quickly scaled to meet demand, at greatly reduced cost.

History

The first recombinant plant-derived protein (PDP) was human serum albumin, initially produced in 1990 in transgenic tobacco and potato plants. Open field growing trials of these crops began in the United States in 1992 and have taken place every year since. While the United States Department of Agriculture has approved planting of pharma crops in every state, most testing has taken place in Hawaii, Nebraska, Iowa, and Wisconsin.

In the early 2000s, the pharming industry was robust. Proof of concept has been established for the production of many therapeutic proteins, including antibodies, blood products, cytokines, growth factors, hormones, recombinant enzymes and human and veterinary vaccines. By 2003 several PDP products for the treatment of human diseases were under development by nearly 200 biotech companies, including recombinant gastric lipase for the treatment of cystic fibrosis, and antibodies for the prevention of dental caries and the treatment of non-Hodgkin's lymphoma.

Several proteins were brought to market as research and bioproduction reagents, mostly by Sigma-Aldrich. ProdiGene struck agreements with Sigma to distribute ProdiGene's corn-produced aprotinin, trypsin, beta-glucuronidase (GUS), and avidin. Large Scale Biology and SIgma agreed that Sigma would distribute LSBC's tobacco-produced aprotinin. Sigma also agreed to distribute Ventria's rise-produced Lactoferrin and Lysozyme.

However, in late 2002, just as ProdiGene was ramping up production of trypsin for commercial launch it was discovered that volunteer plants (leftover from the prior harvest) of one of their GM corn products were harvested with the conventional soybean crop later planted in that field. ProdiGene was fined $250,000 and ordered by the USDA to pay over $3 million in cleanup costs. This raised a furor and set the pharming field back, dramatically. Many companies went bankrupt as companies faced difficulties getting permits for field trials and investors fled. In reaction, APHIS introduced more strict regulations for pharming field trials in the US in 2003. In 2005, Anheuser-Busch threatened to boycott rice grown in Missouri because of plans by Ventria Bioscience to grow pharm rice in the state. A compromise was reached, but Ventria withdrew its permit to plant in Missouri due to unrelated circumstances.

The industry has slowly recovered, by focusing on pharming in simple plants grown in bioreactors and on growing GM crops in greenhouses. Some companies and academic groups have continued with open-field trials of GM crops that produce drugs. In 2006 Dow AgroSciences received USDA approval to market a vaccine for poultry against Newcastle disease, produced in plant cell culture – the first plant-produced vaccine approved in the U.S.

Pharming in Mammals

Historical Development

Milk is presently the most mature system to produce recombinant proteins from transgenic organisms. Blood, egg white, seminal plasma, and urine are other theoretically possible systems, but all have drawbacks. Blood, for instance, as of 2012 cannot store high levels of stable recombinant proteins, and biologically active proteins in blood may alter the health of the animals. Expression in the milk of a mammal, such as a cow, sheep, or goat, is a common application, as milk production is plentiful and purification from milk is relatively easy. Hamsters and rabbits have also been used in preliminary studies because of their faster breeding.

One approach to this technology is the creation of a transgenic mammal that can produce the biopharmaceutical in its milk (or blood or urine). Once an animal is produced, typically using the pronuclear microinjection method, it becomes efficacious to use cloning technology to create additional offspring that carry the favorable modified genome. In February 2009 the US FDA granted marketing approval for the first drug to be produced in genetically modified livestock. The drug is called ATryn, which is antithrombin protein purified from the milk of genetically modified goats. Marketing permission was granted by the European Medicines Agency in August 2006.

Patentability Issues Regarding Pharming

As indicated above, some mammals typically used for food production (such as goats, sheep, pigs, and cows) have been modified to produce non-food products, a practice sometimes called pharming. Use of genetically modified goats has been approved by the FDA and EMA to produce ATryn, i.e. recombinant antithrombin, an anticoagulant protein drug. These products "produced by turning animals into drug-manufacturing 'machines' by genetically modifying them" are sometimes termed biopharmaceuticals.

The patentability of such biopharmaceuticals and their process of manufacture is uncertain. Probably, the biopharmaceuticals themselves so made are unpatentable, assuming that they are chemically identical to the preexisting drugs that they imitate. Several 19th century United States Supreme Court decisions hold that a previously known natural product manufactured by artificial means cannot be patented. An argument can be made for the patentability of the process for manufacturing a biopharmaceutical, however, because genetically modifying animals so that they will produce the drug is dissimilar to previous methods of manufacture; moreover, one Supreme Court decision seems to hold open that possibility.

On the other hand, it has been suggested that the recent Supreme Court decision in *Mayo v. Prometheus* may create a problem in that, in accordance with the ruling in that case, "it may be said that such and such genes manufacture this protein in the same way they always did in a mammal, they produce the same product, and the genetic modification technology used is conventional, so that the steps of the process 'add nothing to the laws of nature that is not already present. If the argument prevailed in court, the process would also be ineligible for patent protection. This issue has not yet been decided in the courts.

Pharming in Plants

Plant-made pharmaceuticals (PMPs), also referred to as pharming, is a sub-sector of the biotechnology industry that involves the process of genetically engineering plants so that they can produce certain types of therapeutically important proteins and associated molecules such as peptides and secondary metabolites. The proteins and molecules can then be harvested and used to produce pharmaceuticals.

Recently, several non-crop plants such as the duckweed *Lemna minor* or the moss *Physcomitrella patens* have shown to be useful for the production of biopharmaceuticals. These frugal organisms can be cultivated in bioreactors (as opposed to being grown in fields), secrete the transformed proteins into the growth medium and, thus, substantially reduce the burden of protein purification in preparing recombinant proteins for medical use. In addition, both species can be engineered to cause secretion of proteins with human patterns of glycosylation, an improvement over conventional plant gene-expression systems. Biolex Therapeutics developed a duckweed-based expression platform; it sold that business to Synthon and declared bankruptcy in 2012.

Additionally, an Israeli company, Protalix, has developed a method to produce therapeutics in cultured transgenic carrot or tobacco cells. Protalix and its partner, Pfizer, received FDA approval to market its drug, taliglucerase alfa (Elelyso), treatment for Gaucher's disease, in 2012.

Arabidopsis is often used as a model organism to study gene expression in plants, while actual production may be carried out in maize, rice, potatoes, tobacco, flax or safflower. The advantage of rice and flax is that they are self-pollinating, and thus gene flow issues are avoided. However, human error could still result in pharm crops entering the food supply. Using a minor crop such as safflower or tobacco, avoids the greater political pressures and risk to the food supply involved with using staple crops such as beans or rice.

Regulation

The regulation of genetic engineering concerns the approaches taken by governments to assess and manage the risks associated with the development and release of genetically modified crops. There are differences in the regulation of GM crops - including those used for pharming - between countries, with some of the most marked differences occurring between the USA and Europe. Regulation varies in a given country depending on the intended use of the products of the genetic engineering. For example, a crop not intended for food use is generally not reviewed by authorities responsible for food safety.

Controversy Over Pharming

There are controversies around GMOs generally on several levels, including whether making them is ethical, issues concerning intellectual property and market dynamics; environmental effects of GM crops; and GM crops' role in industrial agricultural more generally. There are also specific controversies around pharming.

Advantages

Plants do not carry pathogens that might be dangerous to human health. Additionally, on the level of pharmacologically active proteins, there are no proteins in plants that are similar to human proteins. On the other hand, plants are still sufficiently closely related to animals and humans that they are able to correctly process and configure both animal and human proteins. Their seeds and fruits also provide sterile packaging containers for the valuable therapeutics and guarantee a certain storage life.

Global demand for pharmaceuticals is at unprecedented levels. Expanding the existing microbial systems, although feasible for some therapeutic products, is not a satisfactory option on several grounds. Many proteins of interest are too complex to be made by microbial systems or by protein synthesis. These proteins are currently being produced in animal cell cultures, but the resulting product is often prohibitively expensive for many patients. For these reasons, science has been exploring other options for producing proteins of therapeutic value.

These pharmaceutical crops could become extremely beneficial in developing countries. The World Health Organization estimates that nearly 3 million people die each year from vaccine preventable disease, mostly in Africa. Diseases such as measles and hepatitis lead to deaths in countries where the people cannot afford the high costs of vaccines, but pharm crops could help solve this problem.

Disadvantages

While molecular farming is one application of genetic engineering, there are concerns that are unique to it. In the case of genetically modified (GM) foods, concerns focus on the safety of the food for human consumption. In response, it has been argued that the genes that enhance a crop in some way, such as drought resistance or pesticide resistance, are not believed to affect the food itself. Other GM foods in development, such as fruits designed to ripen faster or grow larger, are believed not to affect humans any differently from non-GM varieties.

In contrast, molecular farming is not intended for crops destined for the food chain. It produces plants that contain physiologically active compounds that accumulate in the plant's tissues. Considerable attention is focused, therefore, on the restraint and caution necessary to protect both consumer health and environmental biodiversity.

The fact that the plants are used to produce drugs alarms activists. They worry that once production begins, the altered plants might find their way into the food supply or cross-pollinate with conventional, non-GM crops. These concerns have historical validation from the ProdiGene incident, and from the StarLink incident, in which GMO corn accidentally ended up in commercial food products. Activists also are concerned about the power of business. According to the Canadian Food Inspection Agency, in a recent report, says that U.S. demand alone for biotech pharmaceuticals is expanding at 13 percent annually and to reach a market value of $28.6 billion in 2004. Pharming is expected to be worth $100 billion globally by 2020.

List of Originators (Companies and Universities), Research Projects and Products

- Dow AgroSciences - poultry vaccine against Newcastle disease virus (first PMP to be approved for marketing by the USDA Center for Veterinary Biologics Dow never intended to market the vaccine. "'Dow Agrosciences used the animal vaccine as an example to completely run through the process. A new platform needs to be approved, which can be difficult when authorities get in contact with it for the first time', explains the plant physiologist Stefan Schillberg, head of the Molecular Biology Division at the Fraunhofer Institute for Molecular Biology and Applied Ecology Aachen."

- Fraunhofer Institute for Molecular Biology and Applied Ecology, with sites in Germany, the US, and Chile is the lead institute of the Pharma Planta consortium of 33 partner organizations from 12 European countries and South Africa, funded by the European Commission. Pharma Planta is developing systems for plant production of proteins in greenhouses in the European regulatory framework. It is collaborating on biosimilars with Plantform and Pharma Praxis.
- Genzyme - antithrombin III in goat milk
- GTC Biotherapeutics - ATryn (recombinant human antithrombin) in goat milk
- Icon Genetics produces therapeutics in transiently infected *Nicotiana benthamiana* (relative of tobacco) plants in greenhouses in Halle, Germany or in fields. First product is a vaccine for a cancer, non-Hodgkin's lymphoma.
- Iowa State University - immunogenic protein from *E. coli* bacteria in pollen-free corn as a potential vaccine against *E. coli* for animals and humans
- Kentucky Bioprocessing took over Large Scale Biology's facilities in Owensboro, Kentucky, and offers contract biomanufacturing services in tobacco plants, grown in greenhouses or in open fields.
- Medicago Inc. - Pre-clinical trials of Influenza vaccine made in transiently infected *Nicotiana benthamiana* (relative of tobacco) plants in greenhouses Medicago has a system for pharming in alfalfa that their website says is "not suited for the production of vaccines"
- PharmaPraxis - Developing biosimilars in collaboration with PlantForm and Fraunhofer.
- Pharming - C1 inhibitor, human collagen 1, fibrinogen (with American Red Cross), and lactoferrin in cow milk The intellectual property behind the fibrinogen project was acquired from PPL Therapeutics when PPL went bankrupt in 2004.
- Phyton Biotech uses plant cell culture systems to manufacture active pharmaceutical ingredients based on taxanes, including paclitaxel and docetaxel
- Planet Biotechnology - antibodies against Streptococcus mutans, antibodies against doxorubicin, and ICAM 1 receptor in tobacco
- Plant Form Corporation - biosimilar trastuzumab in tobacco - It is developing biosimilars in collaboration with PharmaPraxis and Fraunhofer.
- Prodi Gene - was developing several proteins, including aprotinin, trypsin and a veterinary TGE vaccine in corn. Was in process of launching trypsin product in 2002 when later that year its field test crops contaminated conventional crops. Unable to pay the

$3 million cost of the cleanup, it was purchased by International Oilseed Distributors in 2003 International Oilseed Distributors is controlled by Harry H. Stine, who owns one of the biggest soybeans genetics companies in the US. ProdiGene's maize-produced trypsin, with the trademark TrypZean is currently sold by Sigma-Aldritch as a research reagent.

- Syngenta - Beta carotene in rice (this is "Golden rice 2"), which Syngenta has donated to the Golden Rice Project
- University of Arizona - Hepatitis C vaccine in potatoes
- Ventria Bioscience - lactoferrin and lysozyme in rice
- Washington State University - lactoferrin and lysozyme in barley
- European COST Action on Molecular Farming - COST Action FA0804 on Molecular Farming provides a pan-European coordination centre, connecting academic and government institutions and companies from 23 countries. The aim of the Action is to advance the field by encouraging scientific interactions, providing expert opinion and encouraging commercial development of new products. The COST Action also provides grants allowing young scientists to visit participating laboratories across Europe for scientific training.
- Mapp Biopharmaceutical in San Diego, California was reported in August 2014 to be developing ZMapp, an experimental cure for the deadly Ebola virus disease. Two Americans who had been infected in Liberia were reported to be improving with the drug. ZMapp was made using antibodies produced by GM tobacco plants.

Projects Known to be Abandoned

- Agragen, in collaboration with University of Alberta - docosahexaenoic acid and human serum albumin in flax
- Chlorogen, Inc. - cholera, anthrax, and plague vaccines, albumin, interferon for liver diseases including hepatitis C, elastin, 4HB, and insulin-like growth factor in tobacco chloroplasts. Went out of business in 2007.
- Dow Chemical Company made a deal with Sunol Molecular in 2003 to develop antibodies against tissue factor in plants and in mammalian cell culture and to compare them. In 2005 Sunol sold all its tissue factor antagonists to Tanox, which in turn was bought by Genentech in 2003. Genentech licensed the tissue factor program to Altor in 2008 Altor is itself a spinout from Sunol. The product under development, ALT-836, formerly known as TNX-832 and Sunol-cH36, is the not the plant-produced antibody, but rather is mammalian antibody, more specifically, a chimeric antibody produced in a hybridoma.
- Epicyte - spermicidal antibodies in corn Epicyte was purchased by Biolex in 2004 at which time Epicyte's portfolio was described as "focused on the discovery and development of human monoclonal antibody products as treatments for a wide range of infectious and inflammatory diseases."

- Large Scale Biology Corporation (LSBC) (*bankrupt*) - used Tobacco mosaic virus to develop reagents and patient-specific vaccines for Non-Hodgkin's lymphoma, Papillomavirus vaccine, parvovirus vaccine, alpha galactosidase for Fabry disease, lysosomal acid lipase, aprotinin, interferon Alpha 2a and 2b, G-CSF, and Hepatitis B vaccine antigens in tobacco. In 2004, LSBC announced an agreement with Sigma-Aldritch under which LSB would produce recombinant aprotinin in plants of the tobacco family and Sigma-Aldrich would commercially distribute LSBC's recombinant product to its customers in the R&D, cell culture and manufacturing markets. As of October 2012 SIgma still has the protein in stock.
- Meristem Therapeutics - Lipase, lactoferrin, plasma proteins, collagen, antibodies (IgA, IgM), allergens and protease inhibitors in tobacco. Liquidated in 2008.
- Novoplant GmgH - therapeutic proteins in tobacco and feed peas. Conducted field trials in US of feed peas for pigs that produced anti-bacterial antibodies. Former CSO is now with another company; appears that Novoplant is out of business.
- Monsanto Company - abandoned development of pharmaceutical producing corn
- PPL Therapeutics - Alpha 1-antitrypsin for cystic fibrosis and emphysema in sheep milk. This is the company that created Dolly the Sheep, the first cloned animal. Went bankrupt in 2004. Assets were acquired by Pharming and an investment group including University of Pittsburgh Medical Center.
- SemBioSys - insulin in safflower. In May 2012, SemBioSys terminated its operations.

Transgenic science offers the methods for in vitro introduction of DNA fragment(s) into a cell and alters the genome in order to achieve a desired character. Animals whose genomes have been altered by transfer of a gene or introducing an exogenous gene are called transgenic animals. Transgenic animals serve as important model for clinical research as well as bioreactors in pharmaceutical manufacturing sector. The term "transgenic" was first coined by Gordon and Ruddle in 1981 and later adopted to describe broad range of experimental animal models having a specific targeted modification in their genome.

Types of Transgenic Animals

Transgenic animals can be broadly classified into two categories depending on the modification in the genome.

Knockout Animals

Transgenic animals produced by replacing the endogenous functional gene with mutated non-functional gene are called knockout animals. In laboratory, knockout mice are generated for clinical studies.

Knockin Animals

Transgenic animals having exogenous functional gene incorporated to the genome by targeted insertion are termed as knockin animals. Several animal transgenic models have been developed by this approach to use as disease models or in biofarming technology.

Methods for creating Transgenic Animals

With the advances of recombinant DNA technology, different methods have been developed for generating transgenic animals. The examples of commonly used methods applied in transgenic animal production are-

- Gene knockdown using RNA interference (RNAi),
- Retroviral mediated gene transfer
- Embryonic stem cell technology
- DNA microinjection
- Nuclear transfer
- Sperm as vector

Gene Knockdown using RNA Interference (RNAi)

RNA interference (RNAi) induced in mammalian cells by introducing antisense RNA to generate double stranded RNA (dsRNA) structure (or hairpin) which are cleaved to generate small fragments of RNA (~22bp length). Double stranded RNA are indentified and cleaved by Dicer enzyme to generate small interfering RNA (siRNA) or micro RNA (miRNA). miRNA is single stranded RNA molecule which inhibits translation of mRNA whereas siRNA are double stranded RNA which cleave RNA. A riboprotein complex is generated by the small RNA fragments called RISC (RNA induced silencing complex) which binds to homologous mRNA molecule and initiate the cleavage, thus silencing the gene expression.

In mammalian cells, RNAi effect can be introduced by generating small interfering RNA (siRNA) or small hairpin RNA (shRNA). Using this approach, several transgenic mice and rat models have been developed for potential medical and pharmaceutical applications.

RNAi technology has several advantages over other approaches. Once stably established RNAi can be synthesized directly inside the mammalian cells thus avoiding any other cloning steps. RNAi approach can be designed to regulate gene expression either transcription or translation stage.

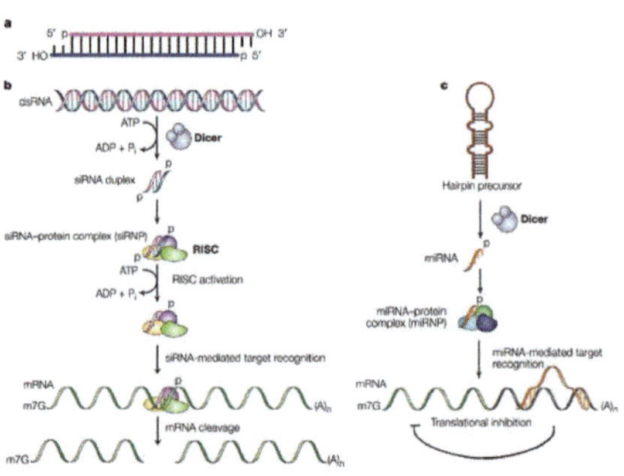

Gene silencing by RNAi: a. Double stranded RNA construct formed by introducing antisense RNA into the system; b. Mechanism of generation of siRNA from dsRNA by Dicer in presence of ATP; c. Generation of miRNA from hairpin precursor RNA (shRNA) by Dicer.

Retroviral Mediated Gene Transfer

Retrovirus is a virus of family "Retroviridae" that has RNA as its genetic material and replicate inside the host cell using reverse transcription machinery (to convert RNA to DNA). Retroviruses are commonly used as vectors to transfer genetic material into the host cell because of its high efficiency to transfer RNA to the host. Recombinant retrovirus RNA genome is then copied by reverse transcriptase (encoded by retrovirus) to yield a DNA copy, which then becomes integrated into the host genome at random sites. The offspring results from such transformation are chimeric, an organism consisting of tissues of diverse genetic constitution. To transmit transgene to next generation, any gene transfer technique including retroviral mediated should target germline cells. Generally, early 4-16 celled embryos are used to get maximum of retrovirus infection resulting in chimeric animal. Although homozygosity is attained just after several generations of selfing, these chimeric animals are inbred for 10-20 generations to obtain homozygous transgenic animals.

Although retroviral vectors have high efficiency of transgene integration into host cell, there are certain limitations of this approach.

Limitations

- Low copy number integration.
- Additional steps needed for construction of recombinant retrovirus.
- Limited size of DNA insert (~<15kb).
- Mosaicism (same individual having two different types of genotypes) of the recovered animals.
- Possible interference of the proviral LTR sequences with the expression of transgene.

Embryonic Stem Cell Technology

Embryonic stem (ES) cells are pluripotent stem cells isolated from inner cell mass (ICM) of blastocyst stage of embryo. Use of ES cell mediated transgenic animal production is quite effective specially to generate targeted gene modification at precise location.

The procedure to generate transgenic animals using modified ES cell technology is described in brief below:

- Isolation of pluripotent ES cells from ICM of blastocyst and culturing of the cells *in vitro*.
- Transgene introduction to ES cells using methods like DNA micro-injection, electroporation, precipitation reaction, transfection etc. DNA transfection to ES cells can be carried out using different vectors like liposome, retroviral vectors etc.
- Selection of transformed ES cells for either knockin or knockout constructs.
- Transformed cells are injected into blastocyst followed by implantation in a surrogate female.

After birth, the offsprings are screened for chimerism. Inbreeding of the genetic chimeras is performed to obtain homozygous transgenic animal carrying both the mutated alleles for that character.

Limitation

- Difficulty in *in vitro* culturing and maintenance of pluripotent ES cells.
- Germ line transfection with low frequency.

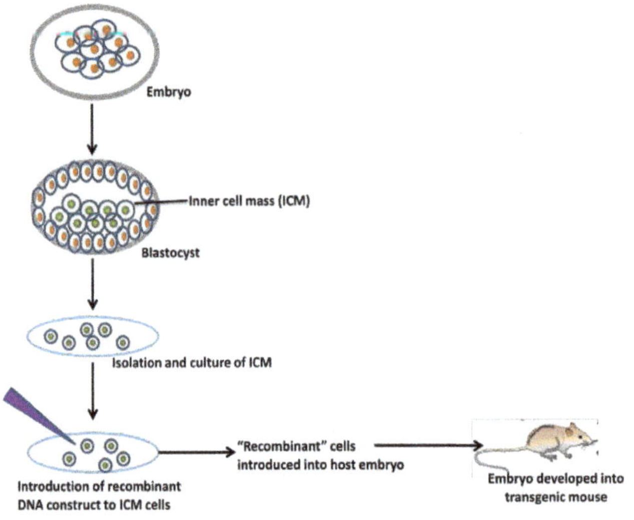

Embryonic stem cell technology

DNA Microinjection

DNA microinjection technique was first described in 1966 by T. P. Lin, which paved the way of generating transgenic animal by direct introduction of foreign DNA into the host cell. Linearized DNA construct is injected into the cell which finally integrates to the host genome at random sites by homologous recombination. DNA microinjection can be targeted to two sites: Pronuclear injection and cytoplasmic injection. Following steps are involved in production of transgenic animal by microinjection:

1. Transgene DNA Construct: cDNA of a gene of interest along with the regulatory elements (promoter, terminator etc) are used to design the DNA construct. Promoter and 3' UTR are necessary for proper gene regulation. Some evidence shows that intron splicing plays role in gene expression in mammalian system. Hence introns are also included in the construct sometimes. Construct should be linearized before injection.

2. Embryo Collection: Donor parental strain for the production of embryo are selected considering factors like response to super ovulation (ability to produce mature ova at large number), frequencies of embryo survival following microinjection, size of pronuclei and incidence of specific pathologies inherent to strain.

Successful super ovulation depends on the strain, age and weight of the animal. Breeding should be monogamous. After post-mating, embryos are collected at single cell stage and proceed for microinjection.

3. DNA Microinjection: The purified double stranded linear DNA construct containing the transgene DNA sequence is introduced to host cell by microinjection. The foreign DNA must be integrated (although random) into the host genome prior to cell division. To facilitate this transgene DNA is introduced into zygote at the earliest possible stage (pronuclear period) immediately after fertilization. Usually male pronucleus is preferred because it is larger and easier to inject. The host chromosome at the site of integration generally undergoes duplication, deletion or re-arrangements due to transgene incorporation. This may lead to insertional mutagenesis and thus producing detectable phenotypic trait.

Note: Critical points for successful DNA microinjection technique-

- Careful collection of relatively large group of accurately single cell embryo.
- Embryo transfer to suitable recipient female (standardized in each case).
- Construction and preparation of transgene DNA fragment to be injected.

Advantage

- No size limitation of insert DNA.
- Applicable to wide range of species.
- Less time consuming and short generation time.

Limitation

- Low success rate
- Random integration of transgene constructs to host chromosome, resulting in undesirable effect or phenotype.
- Potential undesired insertional mutagenesis

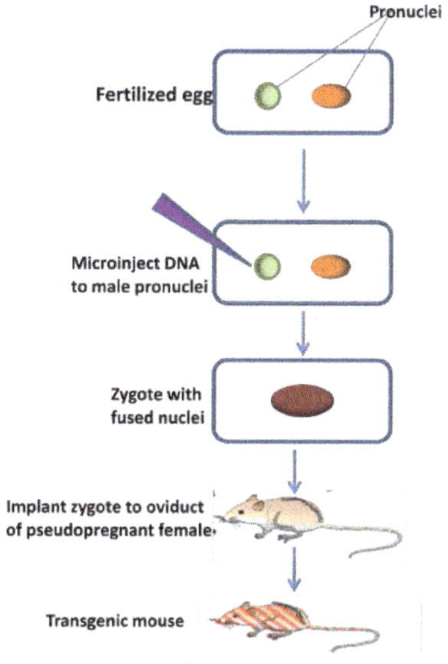

Transgenic mice by DNA microinjection

Nuclear Transfer

Nuclear transfer involves transfer of nuclei from a donor cell (somatic cell or stem cell) to an enucleated oocyte, thereby reprogramming development. This process is also referred as somatic cell nuclear transfer (SCNT). In early 1980, first mammalian nuclear transfer experiments were conducted in mice. In 1997, a team lead by Ian Wilmut created Dolly the sheep became the first vertebrate clone to generate by somatic cell nuclear transfer. The technique however was not full proof as Dolly was reported to have progressive lung disease and premature arthritis and died prematurely at the age of 6 in 2003.

Following steps are associated with generation of transgenic animal by SCNT procedure:

1. Mature oocyte is arrested at metaphase-II phase and enucleated.
2. Somatic donor cells are cultured in vitro and somatic cell derived nucleus is transferred to enucleated oocytes.
3. The resulting nuclear transfered oocytes are activated and embryogenesis is activated.
4. The embryo is generally allowed to develop in vitro till blastocyst stage.
5. The blastcyst embryo is then implanted into the uterus of surrogate female where the further development of the embryo takes place.

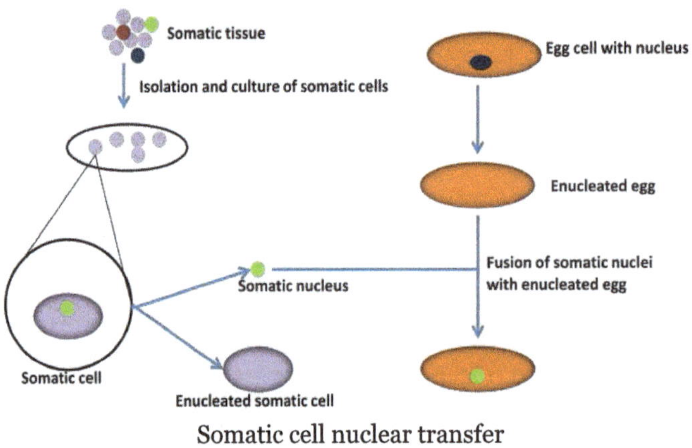

Somatic cell nuclear transfer

Sperm Mediated Gene Transfer (SMGT)

Sperm mediated gene transfer was first reported in 1989 which is based on ability of sperm cells to internalize exogenous DNA and transfer it to the egg cells. This method utilizes a "natural vector" (sperms) to deliver foreign DNA. The exogenous DNA binds to sperm's head at sub-acrosomal region and DNA molecules are taken up by sperm cells. After DNA-sperm interaction and internalization, the exogenous DNA gets integrated to the genome.

Criteria for Successful SMGT Technique

- Quality of semen sample
- DNA uptake rate which depends on viability and motility of sperm cells.

Limitation

- Sperm cells take up any DNA from the environment which reduces the specificity of the vector
- High amount of foreign DNA inhibits sperm movement.
- Large amount of DNA may get fragmented during the activation of sperm nucleus.

Characterization of Transgenic Animals

After successful introduction of transgene into the host animal, the resulting transgenic animal should be kept under observation to study the stability and expression pattern of transgene inside host system. Continued expression of the product depends on two aspects of stability of the transgene:

- The structural stability or integrity of the transgene.
- The stability of transgene expression.

A. Genetic Stability

In presence of multiple integration sites in the host genome, there may be occurrence of rearrangement or deletion of integrating transgene as well as host genome.To overcome this, stability of the transgene should be regularly monitored by tests like Southern blots, sequencing etc.

B. Stability of Expression:

The stability of expression of a transgene product can vary depending on interaction of the genetic background of the host animals or imprinting effects due to maternal or paternal inheritance. Stability of expression of the transgene product should be monitored over the productive life of the transgenic animal. Methods may include Northern blots, RT-PCR, DNase protection assay or other appropriate techniques.

- Detailed characterization of the original gene intended for introduction into the animals should be provided.
- The natural protein and its functions should be described along with a description of its pattern of expression.
- Both the original vector sequence and the transgene construct should be extensively characterized by restriction maps and nucleotide sequences.

C. Characterization of the Transgenic Founder (Fo) Animal:

- The medical history of the animals which donate gametes or embryonic stem cells (donor) and foster or recipient animals should be well known.
- Animals should undergo detailed veterinary evaluations of health, including breed-related and specific tests for species related disease problems.

Importance of Transgenic Animals:

1. Targeted Production of Pharmaceutical Proteins: Transgenic animals are used for the production of valuable human enzymes, hormones, antibodies and growth factors.Transgenic animals (eg. larger mammals) as bioreactor ("Pharmaceutical pharming") is cost effective compared to cell culture methods as animals can supplement their body fluid with fresh nutrients, remove waste products, internally regulate the temperature and pH and resist pathogens on their own. The transgene products can be collected by targeting its expression in the secretory cells of the liver, lactating mammary gland and kidney. e.g. - Sheep secreting human Factor IX in milk used for the treatment of hemophilia B.

2. Models of Human Disease Processes: To study different human disease patterns and their mechanisms, different lower animals having evolutionary resemblance to human are used. Due to stringent regulatory and ethical issues, direct experiment on human is not possible. The phenotypic and regulatory parameters of a particular human disease may be evaluated in an animal model with relatively short generation time. By expressing human proteins in a transgenic animal, the regulation and activity can be studied. Also by creating a condition similar to human diseased condition (eg. diabetic mice), one can study various parameters and their interaction. Developing diverse transgenic phenotypes permits the innovative testing of diagnostics and therapeutic agents with a reduced population of experimental animals. Utilizing novel cell lines from transgenic organs can lead to huge reduction in the number of research animals required for the evaluation. It is possible to create a transgenic genome in which more than one transgene may interact or in which a transgene may interact with endogenous normal or mutated gene. e.g. - transgenic rodent model are developed for numerous human disease like cardiovascular disease, cancer, etc.

3. Analysis of a Developmental Pathway: Many transgenic animals are non-viable, either prenatally or perinatally. This can give insights into genes essential for during development. Correlation of time of embryonic lethality to known animal embryonic gene expression patterns provides knowledge of insight into the pathways and interaction of the mutated gene. Information on gene expression can easily be obtained by using a reporter gene, which combines the regulatory regions of the gene under study with an easily detectable marker protein. For e.g. – Using Cre-lox system both temporal and tissue specific control of expression of knockout gene construct can be studied.

4. Modification of Animal Anatomy and Physiology: Transgenic technology can be used for genetic improvement of an animal. A foreign gene can be introduced by selective deletion of endogenous gene or specific region. A direct approach is to add a gene for a growth factor or hormone to the genome instead of altering the complex multigenic physiology of the mammal. For developing animals with novel disease resistance, higher yield of meat or milk, decreased body fat, etc.

5. Production of Organs for Xenotransplantation: Transplantation of living cells, tissues or organs from one donor organism to another receipient organism is known as xenotransplantation. Such cells, tissues or organs are called xenografts or xenotransplants. The shortage of healthy donor organ can be mitigated by developing animals whose organs would be suitable for xenotransplantation. Transgenic animals have aided the understanding of

molecular basis of graft rejection, which includes foreign antigens, coagulation defects and interspecies ligand-receptor interactions. e.g. - Transgenic pigs have been developed that do not express the surface glycoproteins highly immunogenic to humans.

As discussed earlier, transgenic animals are genetically modified by stable incorporation of foreign DNA into its genome by artificial gene transfer technology in order to introduce, silence or delete specific phenotypic characteristic. Transgenic animals can be used to express recombinant proteins in milk, urine, blood, sperm, or eggs, or even to grow organs for xenotransplantation. "Biopharming" in animals encompass the production of biologically active recombinant proteins using transgenic animals as bioreactor.

Although production of recombinant human therapeutic proteins in microbial bioreactors is extremely cost-efficient, it suffers from many limitations like:

1. Prokaryotic cells lack post translational modification machinery (glycosylation, phosphorylation etc) essential for proper folding and functioning of complex eukaryotic proteins.
2. The lack of post-translational modifications in prokaryotic system can be overcome by replacing them with large-scale animal cell cultures. But the long generation time and the requirement for rich culture media make animal cell bioreactors more expensive.
3. In both the above cases, the upstream production operations are relatively inexpensive but the downstream processing of the final product is sometimes extremely complex and costly.

These drawbacks can be overcome by using transgenic animals for production of human or veterinary pharmaceuticals. The use of targeted gene transfer for the expression of the transgene in the mammary gland of large farm animals like cow, pigs, goats and sheep offers advantages like easy isolation of the product secreted into the milk and low cost-large scale production.

Transgenic have been produced in a variety of animal species e.g., mice, rabbits, swine, sheep, goat, cattle, poultry, fish, amphibians, insects and nematodes to produce therapeutic proteins against osteoporosis, cystic fibrosis, hemophilia, malaria, arthritis, HIV etc. Transgenic animals have also been developed to produce monoclonal antibodies (therapeutic antibodies against disease proteins) which are used in vaccine development. Other than being used as bioreactor, transgenic animals are also used as disease models and for improving livestock.

Advantages of using Transgenic Animal in Biopharming

- Protein function and activity depends on the three dimensional structure achieved by post-translational modification. The post-translational modification machinery is absent in bacteria (like *E. coli*) and plants have significantly different machinery than animals. To produce complex human therapeutic proteins, nothing can be better than animal livestock.
- Animal pharming does not require constant monitoring and sampling.
- Animal pharming is cost effective as it does not require substantial plant machinery which has to be purchased and maintained otherwise.

- Purification of proteins from an animal's secretary or excretory gland products (milk, body fluid, urine etc) is simpler.
- Scale-up of production of a recombinant protein is easier with animal livestock.

Purification and Characterization of Transgenic Product

Product Recovery From the Transgenic Animals

Recovery can be done by milking, exsanguinations (process of blood collection), surgical excision of tissues, or other methods as applicable. The recovery procedure should be designed to maximize the safety, sterility, potency, and purity of the product. The collection facilities should be maintained at the best possible hygiene condition to reduce any contamination.

Quality of a Production Batch

Monitoring the quality of the product and the manufacturing processes require the purity of the product to be determined at multiple points during the production process. The tests found to be most sensitive to changes in product structure and/or potency should be validated and then incorporated as part of the batch release protocol. Conditions and criteria by which a production batch is defined (e.g., animals used, batch size, collection storage times, the time before sterile filtering and acceptance criteria for source material and product pooling) should be provided.

Endogenous and Adventitious Agents Detection

1. Choice of host animal:

Many animal species have been proposed to develop as hosts for production of therapeutic products. The lack of experience with many of these hosts raises potential safety concerns about adventitious agents are considered on a case-by-case basis by the FDA. Adventitious agents are defined as micro-organisms which are unintentionally introduced to the manufacturing process of a biological product. There is a possibility that adventitious agents or chemical contaminants could enter into the host animals and be concentrated in the product by the purification procedure without signaling their presence. Zoonotic diseases (diseases caused by infectious agents that can be transmitted from vertebrate animals to humans and vice-versa) should be taken care of extremely. Therefore, health monitoring of the livestock is necessary, although not sufficient alone to guarantee absence of these contaminants. The rigor of infection control in the animal host and validation of elimination of adventitious agents from the product will depend on several factors like:

- The intended use of the product;
- The tissue from which the product is derived;
- The method by which product is collected;
- The purification process;
- Animal husbandry practices used during production of the founder and production animals.

2. Source tissue:

Currently therapeutic products are produced in transgenic animals by utilizing systems in which the product will be isolated from bodily fluids (milk, blood, or urine) where significant batch-to-batch variation in both microbiological and virological loads may arise in unpurified products. The extent of such variability should be documented for each product. The downstream manufacturing procedures should have full capability for providing a safe and consistent product.

3. Pathogen testing and elimination:

Transgenic animal source materials have the potential to contain a variety of human pathogens (e.g., viruses, bacteria, mycoplasma, and transmissible spongiform encephalopathy agents). Analytical methods and purification processes should be standardized to ensure the safety of the product. After consulting with veterinarians and other experts in the field, a specific list of pathogenic agents and accepted testing methods should be prepared and prioritized with regard to the risk of human infection. In cases where recombinant DNA was introduced by viral vector administration, or similar methods, the possibility of viral load in the product should be monitored.

Analysis of Product Identity and Purity

Before initiating studies in humans, the product from the transgenic source material should be characterized with respect to activity, safety, purity, and potency. Product similarities (e.g., potency, pharmacokinetics) and differences (e.g., glycosylation, antigenicity) with well characterized natural or recombinant molecules should also be performed.

Biochemical Identity, Purity and Potency

Physicochemical, immunological and biological characterizations should be carried out on the transgene product to assess its identity, purity and activity which include chemical structure, amino acid composition, disulfide linkages etc. The anatomical site of production in the transgenic animal may differ from the natural site of production and result in biochemical modifications that affect the activity of biological product.

Product Heterogeneity

Variations in final product activity should be examined with respect to variables associated with maintaining the transgenic herd and generating source material. For example, the identification of any variation in product structure, purity or activity when the source material is obtained: a) from different animals, b) during different seasons, c) from animals facing other sources of stress, d) from source material that has been stored for extended periods of time etc should be ascertained.

Characterization of the transgenic production process should initially focus on:

1. The presence of known and potential human pathogens, (both adventitious and endogenous agents) in the source material.
2. The amount of immunogenic and toxic materials in the final product,
3. The batch-to-batch consistency of the products physicochemical propreties.

Batch Release Testing of the Final Product

The purified product prior to final formulation should be characterized in a manner similar to other recombinant products. Contaminant levels in the final product that are acceptable for product release will be determined by preclinical studies, the dose and route of product administration, frequency and duration of product administration and the proposed patient population for the clinical study.

Preclinical safety Evaluation

If there is similarity between the product and the naturally occurring molecules, it may still require limited toxicology studies. *In vitro* studies in animal models may be used to examine product efficacy and safety.

Application of Transgenic Animal Biopharming

Biomedical Application

To overcome the limitations of conventional recombinant production system of human therapeutic proteins, transgenic animals have been developed as bioreactor.

Therapeutic Protein Production

Several human therapeutic proteins were produced using transgenic livestock as bioreactor. Low cost, higher level of production and easy downstream processing make transgenic animal a better alternative to conventional production system. Table shows examples of some therapeutic proteins produced using transgenic animals.

Table: Human therapeutic proteins in transgenic animals

Animal	Drug/protein	Use
sheep	alpha1 anti trypsin	deficiency leads to emphysema
sheep	CFTR (Cystic fibrosis transmembrane conductance regulator)	treatment of cystic fibrosis
sheep	tissue plasminogen activator	treatment of thrombosis
sheep	factor VIII & IX (blood clotting protein)	treatment of hemophilia
sheep	fibrinogen	treatment of wound healing
pig	tissue plasminogen activator	treatment of thrombosis
pig	factor VIII, IX	treatment of hemophilia
goat	human protein C	treatment of thrombosis
goat	antithrombin 3	treatment of thrombosis
goat	glutamic acid decarboxylase	treatment of type 1 diabetes

Antibody Production using Transgenic Animals

Mammary glands of several transgenic animals are targeted for specific production of monoclonal antibodies specific for various diseases. Transgenic goats and cattle are developed for such application.

Blood Replacement

Functional human blood proteins including haemoglobin were produced using transgenic swine. Further improvement of this strategy is required to achieve desirable level of expression without any health risk to the animal and the recipient (human).

Xenotransplantation of Organs

Organ transplantation has tremendous importance in medical science which is a critical life saving option in several cases. The shortage of human organs for transplantation leads to the development of xenograft techniques. There are certain pre-requisites for successful xenotransplantation:

- Overcome immunological rejection (called as "graft rejection") of donor organs in the recipient body.
- Prevent transmission of pathogens from donor to recipient.

Transgenic animals (pigs) have been developed to achieve reduced immunological rejection and better compatibility.

Agricultural Application

Transgenic animal has been developed to provide various biopharming applications in agriculture. Few examples of the applications are discussed below-

Improvement of Economically Important Traits in Livestock

- Transgenic pig bearing human metallothionein promote/porcine growth hormone gene showed better growth rate, feed conversion, body fat/muscle ratio etc, thus improving the productivity.
- Transgenic pigs having human insulin-like growth factor-I had more lean muscle and about ~30% less fat; thus improving the commercial significance of these transgenics.
- Transgenic poultry animals were developed to increase the healthiness and nutrient quotient of meat and other products (less saturated fat, lactose level etc.).

Lactation

Dairy farming is an attractive field in agriculture which has huge commercial potential. Transgenic mice models have been developed to demonstrate the feasibility of such modifications in larger animals but further research is required to develop the concept. Production of milk having modified properties like lactose free milk, modified lipid content etc would have a niche market. Transgenic goat has been developed to secret spider silk in goat milk.

Wool Production

Transgenic sheep has been developed to increase wool production (~6.2% increase) by introducing "keratin-IGF-I" construct. Such modification showed no adverse effect on the growth and development of the transgenic animals.

Eco-friendly Farm Animals

The high phosphorous content in manure of farm animals causes environmental pollution. Transgenic pigs have been developed bearing a bacterial "*phytase*" gene which allow digestion of plant phytate. Transgenic pigs showed ~75% lesser fecal phosphate output compared to the wildtype pigs.

ATryn® -First Recombinant Human Antithrombin

The first product derived from a transgenic animal is a recombinant antithrombin called ATryn® was developed by GTC Biotherapeutics and approved by US Food and Drug Administration (FDA) in 2009. It is used for the prophylactic treatment of deep vein thrombosis in patients with hereditary antithrombin deficiencies. The promoter region from goat beta-casein gene was linked to human antithrombin (*hAT*) cDNA and the recombinant construct containing the transgene was introduced into the goat embryos, which were then implanted into surrogate mothers. The resulting offsprings carrying this transgene produced the recombinant human antithrombin (rhAT) in their milk. The rhAT protein is isolated from the milk by conventional purification method using tangential flow filtration, anion exchange chromatography, heparin affinity chromatography, nanofiltration and hydrophobic interaction chromatography with a yield of greater than 50%.

Important Safety Information:

- ATryn® is not advisable to use in patients with known hypersensitivity to goat milk proteins. Allergic-type hypersensitivity reactions, including anaphylaxis may occur in some patients. Use of drug must be discontinued at once if these reactions are reported.
- The alteration of anticoagulant effect of drugs that use antithrombin as anticoagulating agent may be observed when ATryn® is added or withdrawn. To avoid insufficient or excessive anticoagulation, suitable coagulation tests for the anticoagulant should be performed.

Limitation of Transgenic Animals in Pharming

- Both the time and cost to develop a transgenic model with high production level of therapeutic protein is high.
- Highly skilled persons are required for handling transgenic animals.
- Environmental and social impact of using transgenic animals as biopharming tools is complex and uncertain.
- Regulatory processes associated with the commercialization of a given product require description of detailed procedures and for the enforcement of intellectual property rights.
- Other concerns surrounding these technologies include the environmental and ethical aspects of transgenesis. Environmental issues associated with genetically modified organisms that have caused the most public outcry, including food and ecosystem contamination or threats to biodiversity, are generally more problematic in transgenic plants than animals.

Safety Aspects of Biopharming

FDA has developed several regulations and guidelines for safe production and commercialization of recombinant products using transgenic animals. A crucial criterion in animal derived products is prevention of transmission of pathogenic factors from animals to human. All transgenic application of livestock requires compliance to the standards of genetic security and reliability of method applied in genetic modification.

Gene Mapping

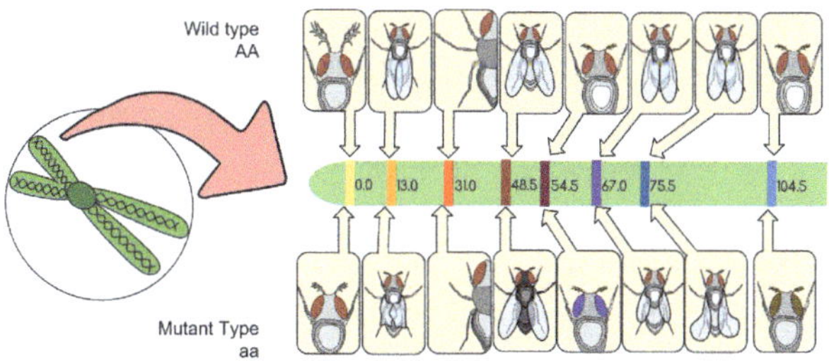

Thomas Hunt Morgan's *Drosophila melanogaster* genetic linkage map. This was the first successful gene mapping work and provides important evidence for the Boveri–Sutton chromosome theory of inheritance. The map shows the relative positions of allelic characteristics on the second Drosophila chromosome. The distance between the genes (map units) are equal to the percentage of crossing-over events that occurs between different alleles.

Gene mapping describes the methods used to identify the locus of a gene and the distances between genes.

The essence of all genome mapping is to place a collection of molecular markers onto their respective positions on the genome. Molecular markers come in all forms. Genes can be viewed as one special type of genetic markers in the construction of genome maps, and mapped the same way as any other markers.

Gene Mapping vs. Physical Mapping

There are two distinctive types of "Maps" used in the field of genome mapping: genetic maps and physical maps. While both maps are a collection of genetic markers and gene loci, genetic maps' distances are based on the genetic linkage information, while physical maps use actual physical distances usually measured in number of base pairs. While the physical map could be a more "accurate" representation of the genome, genetic maps often offer insights into the nature of different regions of the chromosome, e.g. the genetic distance to physical distance ratio varies greatly at different genomic regions which reflects different recombination rates, and such rate is often indicative of euchromatic (usually gene-rich) vs heterochromatic (usually gene poor) regions of the genome.

Gene Mapping

Researchers begin a genetic map by collecting samples of blood or tissue from family members that carry a prominent disease or trait and family members that don't. Scientists then isolate DNA from the samples and closely examine it, looking for unique patterns in the DNA of the family members who do carry the disease that the DNA of those who don't carry the disease don't have. These unique molecular patterns in the DNA are referred to as polymorphisms, or markers.

The first steps of building a genetic map are the development of genetic markers and a mapping population. The closer two markers are on the chromosome, the more likely they are to be passed on to the next generation together. Therefore, the "co-segregation" patterns of all markers can be used to reconstruct their order. With this in mind, the genotypes of each genetic marker are recorded for both parents and each individual in the following generations. The quality of the genetic maps is largely dependent upon these factors: the number of genetic markers on the map and the size of the mapping population. The two factors are interlinked, as a larger mapping population could increase the "resolution" of the map and prevent the map being "saturated".

In gene mapping, any sequence feature that can be faithfully distinguished from the two parents can be used as a genetic marker. Genes, in this regard, are represented by "traits" that can be faithfully distinguished between two parents. Their linkage with other genetic markers are calculated same way as if they are common markers and the actual gene loci are then bracketed in a region between the two nearest neighbouring markers. The entire process is then repeated by looking at more markers which target that region to map the gene neighbourhood to a higher resolution until a specific causative locus can be identified. This process is often referred to as "positional cloning", and it is used extensively in the study of plant species.

Physical Mapping

Since actual base-pair distances are generally hard or impossible to directly measure, physical maps are actually constructed by first shattering the genome into hierarchically smaller pieces. By characterizing each single piece and assembling back together, the overlapping path or "tiling path" of these small fragments would allow researchers to infer physical distances between genomic features. The fragmentation of the genome can be achieved by restriction enzyme cutting or by physically shattering the genome by processes like sonication. Once cut, the DNA fragments are separated by electrophoresis. The resulting pattern of DNA migration (i.e. its genetic fingerprint) is used to identify what stretch of DNA is in the clone. By analyzing the fingerprints, contigs are assembled by automated (FPC) or manual means (pathfinders) into overlapping DNA stretches. Now a good choice of clones can be made to efficiently sequence the clones to determine the DNA sequence of the organism under study.

In physical mapping, there are no direct ways of marking up a specific gene since the mapping does not include any information that concerns traits and functions. Genetic markers can be linked to a physical map by processes like in situ hybridization. By this approach, physical map contigs can be "anchored" onto a genetic map. The clones used in the physical map contigs can then be sequenced on a local scale to help new genetic marker design and identification of the causative loci.

Macrorestriction is a type of physical mapping wherein the high molecular weight DNA is digested with a restriction enzyme having a low number of restriction sites.

There are alternative ways to determine how DNA in a group of clones overlaps without completely sequencing the clones. Once the map is determined, the clones can be used as a resource to efficiently contain large stretches of the genome. This type of mapping is more accurate than genetic maps.

Genome Sequencing

Genome sequencing is sometimes mistakenly referred to as "genome mapping" by non-biologists. The process of "shotgun sequencing" resembles the process of physical mapping: it shatters the genome into small fragments, characterizes each fragment, then puts them back together (more recent sequencing technologies are drastically different). While the scope, purpose and process are totally different, a genome assembly can be viewed as the "ultimate" form of physical map, in that it provides in a much better way all the information that a traditional physical map can offer.

Use of Gene Mapping

Identification of genes is usually the first step in understanding a genome of a species; mapping of the gene is usually the first step of identification of the gene. Gene mapping is usually the starting point of many important downstream studies.

Disease Association

The process to identify a genetic element that is responsible for a disease is also referred to as "mapping". If the locus in which the search is performed is already considerably constrained, the search is called the *fine mapping* of a gene. This information is derived from the investigation of disease manifestations in large families (genetic linkage) or from populations-based genetic association studies.

Marker-assisted Selection

Marker assisted selection or marker aided selection (MAS) is an indirect selection process where a trait of interest is selected based on a marker (morphological, biochemical or DNA/RNA variation) linked to a trait of interest (e.g. productivity, disease resistance, abiotic stress tolerance, and quality), rather than on the trait itself. This process is used in plant and animal breeding.

For example, using MAS to select individuals with disease resistance involves identifying a marker allele that is linked with disease resistance rather than the level of disease resistance. The assumption is that the marker associates at high frequency with the gene or quantitative trait locus (QTL) of interest, due to genetic linkage (close proximity, on the chromosome, of the marker locus and the disease resistance-determining locus). MAS can be useful to select for traits that are difficult or expensive to measure, exhibit low heritability and/or are expressed late in development. At certain points in the breeding process the specimens are examined to ensure that they express the desired trait.

Marker Types

The majority of MAS work in the present era uses DNA-based markers. However, the first markers that allowed indirect selection of a trait of interest were morphological markers. In 1923, Sax first reported association of a simply inherited genetic marker with a quantitative trait in plants when he observed segregation of seed size associated with segregation for a seed coat color marker in beans (*Phaseolus vulgaris* L.). In 1935, Rasmusson demonstrated linkage of flowering time (a quantitative trait) in peas with a simply inherited gene for flower color.

Markers may be:

- Morphological - These markers are often detectable by eye, by simple visual inspection. Examples of this type of marker include the presence or absence of an awn, leaf sheath coloration, height, grain color, aroma of rice etc. In well-characterized crops like maize, tomato, pea, barley or wheat, tens or hundreds of genes that determine morphological traits have been mapped to specific chromosome locations.
- Biochemical- A protein that can be extracted and observed; for example, isozymes and storage proteins.
- Cytological - The chromosomal banding produced by different stains; for example, G banding.

Positive and Negative Selectable Markers

The following terms are generally less relevant to discussions of MAS in plant and animal breeding, but are highly relevant in molecular biology research:

- Positive selectable markers are selectable markers that confer selective advantage to the host organism. An example would be antibiotic resistance, which allows the host organism to survive antibiotic selection.
- Negative selectable markers are selectable markers that eliminate or inhibit growth of the host organism upon selection. An example would be thymidine kinase, which makes the host sensitive to ganciclovir selection.

A distinction can be made between selectable markers (which eliminate certain genotypes from the population) and screenable markers (which cause certain genotypes to be readily identifiable, at which point the experimenter must "score" or evaluate the population and act to retain the preferred genotypes). Most MAS uses screenable markers rather than selectable markers.

Gene vs Marker

The gene of interest directly causes production of protein(s) or RNA that produce a desired trait or phenotype, whereas markers (a DNA sequence or the morphological or biochemical markers produced due to that DNA) are genetically linked to the gene of interest. The gene of interest and the marker tend to move together during segregation of gametes due to their proximity on the same chromosome and concomitant reduction in recombination (chromosome crossover events) between the marker and gene of interest. For some traits, the gene of interest has been discovered and the presence of desirable alleles can be directly assayed with a high level of confidence. However, if the gene of interest is not known, markers linked to the gene of interest can still be used to select for individuals with desirable alleles of the gene of interest. When markers are used there may be some inaccurate results due to inaccurate tests for the marker. There also can be false positive results when markers are used, due to recombination between the marker of interest and gene (or QTL). A perfect marker would elicit no false positive results. The term 'perfect marker' is sometimes used when tests are performed to detect a SNP or other DNA polymorphism in the gene of interest, if that SNP or other polymorphism is the direct cause of the trait of interest. The term 'marker' is still appropriate to use when directly assaying the gene of interest, because the test of genotype is an indirect test of the trait or phenotype of interest.

Important Properties of Ideal Markers for MAS

An ideal marker:

- Easy recognition of all possible phenotypes (homo- and heterozygotes) from all different alleles
- Demonstrates measurable differences in expression between trait types or gene of interest alleles, early in the development of the organism
- Testing for the marker does not have variable success depending on the allele at the marker locus or the allele at the target locus (the gene of interest that determines the trait of interest).
- Low or null interaction among the markers allowing the use of many at the same time in a segregating population
- Abundant in number
- Polymorphic

Demerits of Morphological Markers

Morphological markers are associated with several general deficits that reduce their usefulness including:

- the delay of marker expression until late into the development of the organism
- dominance
- deleterious effects
- pleiotropy
- confounding effects of genes unrelated to the gene or trait of interest but which also affect the morphological marker (epistasis)
- rare polymorphism
- frequent confounding effects of environmental factors which affect the morphological characteristics of the organism

To avoid problems specific to morphological markers, the DNA-based markers have been developed. They are highly polymorphic, exhibit simple inheritance (often codomimant), are abundant throughout the genome, are easy and fast to detect, exhibit minimum pleiotropic effects, and detection is not dependent on the developmental stage of the organism. Numerous markers have been mapped to different chromosomes in several crops including rice, wheat, maize, soybean and several others, and in livestock such as cattle, pigs and chickens. Those markers have been used in diversity analysis, parentage detection, DNA fingerprinting, and prediction of hybrid performance. Molecular markers are useful in indirect selection processes, enabling manual selection of individuals for further propagation.

Selection for Major Genes Linked to Markers

'Major genes' that are responsible for economically important characteristics are frequent in the plant kingdom. Such characteristics include disease resistance, male sterility, self-incompatibility, and others related to shape, color, and architecture of whole plants and are often of mono- or oligogenic in nature. The marker loci that are tightly linked to major genes can be used for selection and are sometimes more efficient than direct selection for the target gene. Such advantages in efficiency may be due for example, to higher expression of the marker mRNA in such cases that the marker is actually a gene. Alternatively, in such cases that the target gene of interest differs between two alleles by a difficult-to-detect single nucleotide polymorphism, an external marker (be it another gene or a polymorphism that is easier to detect, such as a short tandem repeat) may present as the most realistic option.

Situations that are Favorable for Molecular Marker Selection

There are several indications for the use of molecular markers in the selection of a genetic trait.

In such situations that:

- the selected character is expressed late in plant development, like fruit and flower features or adult characters with a juvenile period (so that it is not necessary to wait for the organism to become fully developed before arrangements can be made for propagation)
- the expression of the target gene is recessive (so that individuals which are heterozygous positive for the recessive allele can be crossed to produce some homozygous offspring with the desired trait)
- there is requirement for the presence of special conditions in order to invoke expression of the target gene(s), as in the case of breeding for disease and pest resistance (where inoculation with the disease or subjection to pests would otherwise be required). This advantage derives from the errors due to unreliable inoculation methods and the fact that field inoculation with the pathogen is not allowed in many areas for safety reasons. Moreover, problems in the recognition of the environmentally unstable genes can be eluded.
- the phenotype is affected by two or more unlinked genes (epistatis). For example, selection for multiple genes which provide resistance against diseases or insect pests for gene pyramiding.

The cost of genotyping (an example of a molecular marker assay) is reducing while the cost of phenotyping is increasing particularly in developed countries thus increasing the attractiveness of MAS as the development of the technology continues.

Steps for MAS

Generally the first step is to map the gene or quantitative trait locus (QTL) of interest first by using different techniques and then using this information for marker assisted selection. Generally, the markers to be used should be close to gene of interest (<5 recombination unit or cM) in order to ensure that only minor fraction of the selected individuals will be recombinants. Generally, not only a single marker but rather two markers are used in order to reduce the chances of an error due to homologous recombination. For example, if two flanking markers are used at same time with an interval between them of approximately 20cM, there is higher probability (99%) for recovery of the target gene.

QTL Mapping Techniques

In plants QTL mapping is generally achieved using bi-parental cross populations; a cross between two parents which have a contrasting phenotype for the trait of interest are developed. Commonly used populations are recombinant inbred lines (RILs), doubled haploids (DH), back cross and F_2. Linkage between the phenotype and markers which have already been mapped is tested in these populations in order to determine the position of the QTL. Such techniques are based on linkage and are therefore referred to as "linkage mapping".

Single Step MAS and QTL Mapping

In contrast to two-step QTL mapping and MAS, a single-step method for breeding typical plant populations has been developed.

In such an approach, in the first few breeding cycles, markers linked to the trait of interest are identified by QTL mapping and later the same information is used in the same population. In this approach, pedigree structure is created from families that are created by crossing number of parents (in three-way or four way crosses). Both phenotyping and genotyping is done using molecular markers mapped the possible location of QTL of interest. This will identify markers and their favorable alleles. Once these favorable marker alleles are identified, the frequency of such alleles will be increased and response to marker assisted selection is estimated. Marker allele(s) with desirable effect will be further used in next selection cycle or other experiments.

High-throughput Genotyping Techniques

Recently high-throughput genotyping techniques are developed which allows marker aided screening of many genotypes. This will help breeders in shifting traditional breeding to marker aided selection. One example of such automation is using DNA isolation robots, capillary electrophoresis and pipetting robots.

One recent example of capllilary system is Applied Biosystems 3130 Genetic Analyzer. This is the latest generation of 4-capillary electrophoresis instruments for the low to medium throughput laboratories.

Use of MAS for Backcross Breeding

A minimum of five or six-backcross generations are required to transfer a gene of interest from a donor (may not be adapted) to a recipient (recurrent – adapted cultivar). The recovery of the recurrent genotype can be accelerated with the use of molecular markers. If the F1 is heterozygous for the marker locus, individuals with the recurrent parent allele(s) at the marker locus in first or subsequent backcross generations will also carry a chromosome tagged by the marker.

Marker Assisted Gene Pyramiding

Gene pyramiding has been proposed and applied to enhance resistance to disease and insects by selecting for two or more than two genes at a time. For example, in rice such pyramids have been developed against bacterial blight and blast. The advantage of use of markers in this case allows to select for QTL-allele-linked markers that have same phenotypic effect.

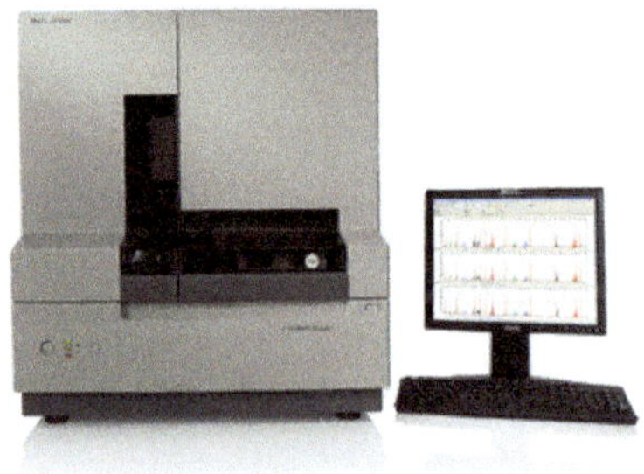

ABI 3130 genetic analyzer.

MAS has also been proved useful for livestock improvement.

A coordinated effort to implement wheat (*Triticum turgidum* and *Triticum aestivum)* marker assisted selection in the U.S. as well as a resource for marker assisted selection exists at the Wheat CAP (Coordinated Agricultural Project).

Marker Assisted Selection Learning Lessons

- Using OptiMAS for Marker-Assisted Recurrent Selection in a Multi-Parental Population
- Video Overview of DNA Extraction and Marker-Assisted Selection
- Marker Assisted Selection
- Background Selection
- Forward Selection
- Marker Assisted Selection in Tomato
- Marker-Assisted Selection for Bacterial Spot Resistance in Tomato Case Study
- Marker-Assisted Selection for PVY Resistance in Potato
- Marker-Assisted Selection for Resistance to Golden Nematode in Potato
- Gene Pyramiding Using Molecular Markers
- Pyramiding Resistance Genes for Bacterial Spot and Bacterial Speck on Tomato Chromosome 5

- The SolCAP Tomato Phenotypic Data: Estimating Heritability and BLUPs for Traits
- rrBLUP Package in R for Genomewide Selection

Traditional breeders have very effectively manipulated the genomes of both crops and livestock species, exploiting the existing natural variations within a species, a breed, and a population. Early breeders lacked knowledge of molecular genetics or selection markers and enhanced the production traits in their herds through selectionof superior individuals as progenitors fornext generations. It required dedication and hard work of combining phenotypic records of individual performance with genealogical information often for many years for achieving enhanced "breeding values".

Increased milk production of 110 kg per animal per year in Holstein dairy cattle or the attainment of 50% decreased feed conversion (the kilograms of feed required to produce a kilogram of pork,) in pig production between the 1960s and 2005 are few examples to illustrate the power of traditional breeding. However, the efficiency of traditional methods decreases in cases where there is difficulty to measure traits, low heritability, or cannot be correctly, quickly and inexpensively measured in a large number of animals (Eggen 2012). These difficult-to-measure traits including fertility, longevity, feed efficiency, and disease resistance are often critically important and the selection of these traits must be achieved through genomic means.

Marker Assisted Selection

The use of DNA markers to target the loci to assist phenotypic screening is known as Marker Assisted Selection (MAS). MAS help to make selection decisions before phenotypes are available. A broader terminology is Marker Assisted Breeding (MAB) which includes apart from MAS, several modern breeding strategies like marker-assisted recurrent selection (MARS), marker-assisted backcrossing (MABC), genome-wide selection (GWS) or genomic selection (GS) etc (Ribaut et al., 2010). MAB is the application of molecular markers along with linkage maps and genomics to alter and improve plant or animal traits based on genotypic assays.

An ideal marker is the one which allows inheritance to be followed across generations. Recent advances in molecular genetics offer methods to dissect the genetic variability of complex traits into quantitative trait loci (QTL). Generally a marker is associated to a QTL. The best scenario is when the marker and QTL are in the same locus. For practical purposes the marker and QTL should be closely located so that recombination between marker and QTL is rare.

Most traits of economic importance are complex quantitative traits controlled by many genes. However, some of these genes might have a larger effect then the others and are called as major genes located at QTL. QTL strictly applies to genes of any effect, however in practice it applies only to major genes, having effectslarge enough for detection and mapping the genome. The pattern of inheritance of such genes may assist in selection.

Thus marker-assisted selection helps in the accurate selection of specific DNA variations that have been associated with a measurable difference or effect on complex traits. It is critical to understand that markers for complex traits like milk yield etc. are associated with only one of the many genes (major gene) that contribute towards that trait. The presence or absence of the numerous other "unmarked" genes both minor and/or other major (if any) and the production environment will ultimately determine whether an animal actually express the trait at the desired intensity. Therefore MAS is considered as an additional tool, and not as a replacement for, traditional selection techniques.

The potential benefits from MAS are maximum for the following traits which:

1. Have low heritability,
2. Are difficult or expensive to measure,
3. Are not measured until the animal had already contributed to next generation,
4. Are currently not selected because they are not routinely measured,
5. Are genetically correlated with a trait that the breeders do not want to increase.

Current MAS schemes consist of the following steps (Meuwissen 2003):

1. Determine the biggest, statistically significant QTL(s) in a genome wide scan for QTL. There are two approaches to locate a QTL (i) Genome scan and (ii) Candidate gene approach.
2. Select for these big QTL next to selecting for polygenes (the remaining often being unidentified smaller genes).

MAS for Genetic Improvement of Crops and Livestock

Markers must be tightly-linked to target loci. Ideally markers must be more than 5 cM (centimorgan) away from a gene or QTL for its linkage confirmation.

Using a pair of flanking markers greatly improve the reliability but increases time and cost.

The markers used for selection of a trait must be polymorphic to distinguish the banding pattern among same or different species.

It is important to locate all markers and protein coding genes in the chromosomes.

For detecting QTL based on genetic markers, development of genetic maps of the species of interest is required.

Development of PCR based techniques and microsatellite markers, increases the efficiency of the mapping procedure in all living organisms.

Prerequisites for a Successful MAS Programme

The crucial requirements for making a MAS programme successful are summarized below (Xu2003).

i. High throughput DNA extraction. Extraction of genomic DNA from plants is a limitation associated with sample processing for MAS. Various methods involving liquid nitrogen, phenol-chloroform, detergents, polyvinylchloride and salts (to remove polysaccharides and polyphenols) are available which are tedious and labor-intensive. A rapid, novel and automated genomic DNA extraction method from leaves of the plant is required that is compatible for PCR (polymerase chain reaction) applications.

ii. Genetic markers. Both DNA and other types of markers (protein, morphological, cytological) can be used in MAS programs. DNA markers are advantageous as they are polymorphic, distributed in large numbers throughout the genome, their presence or absence unaffected by environment and do not directly affect the phenotype.

The commonly used DNA markers like RFLP, SSR, RAPD, AFLP and SNP. Overall for successful MAS programme a marker should have the following attributes,

- Easy to detect
- Have high rate of polymorphism
- Co-dominance
- Occurrence throughout the genome
- Amount of DNA required be small
- Low/null interaction among the markers for use of many at the same time in a segregating population.

iii. Genetic maps. Linkage maps help in detecting marker-trait associations and in choosing markers for a MAS programme. After a marker is found to be associated with a trait in a given population, a dense molecular marker map from a standard reference population will help in the identification of other markers that are closer to or that flank, the target gene.

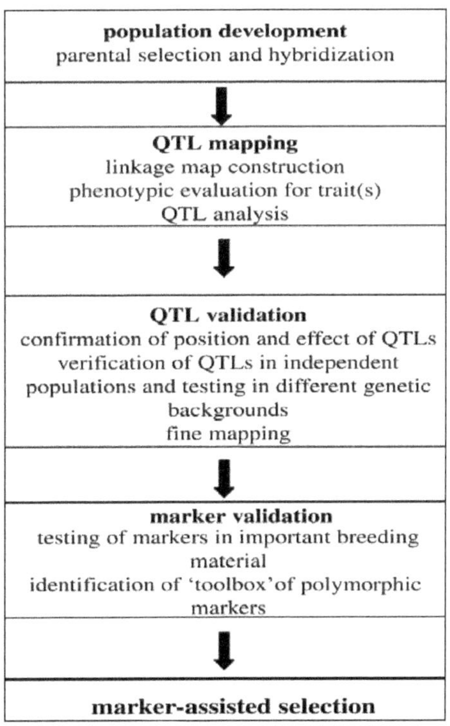

Marker assisted selection

iv. Knowledge of associations between molecular markers and traits of interest. The knowledge of markers associated with traits relevant to a breeding program is the most important ingredient for MAS which is obtained from QTL or gene mapping studies, classical mutant analysis, bulked segregant analysis (Michelmore et al., 1991)etc.

v. Data management system. The number of samples handled in a MAS program is very large, where each sample is evaluated for multiple markers. In such a situation data handling and processing becomes very important. An efficient system for labeling, storing, retrieving, and analyzing large data sets, and producing reports useful to the breeder should be put in place on priority basis.

Steps in QTL Mapping

The fundamental approach for QTL mapping is illustrated in figure by taking into consideration the hair density (i.e. trichomes) that are present on a plant leaf.

Step1. Inbred parental lines having different trichomes density (i.e. Dense vs. Sparse) are crossed. The F1 population thus obtained has intermediate density of trichomes.

Step2. It is then followed by selfing of F1 population to obtain F2 population.

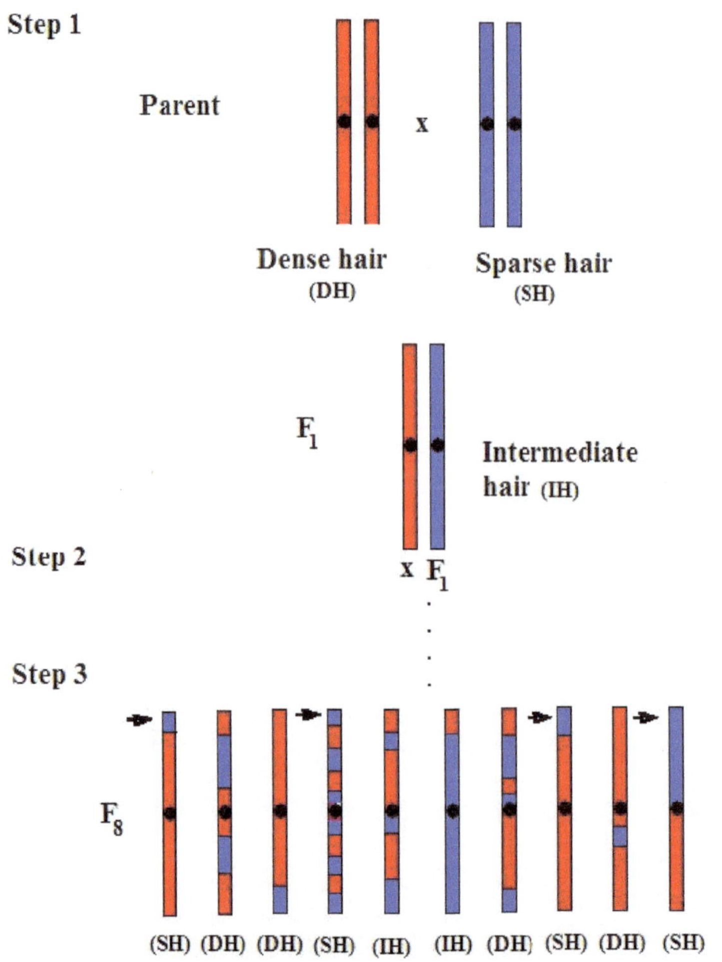

A schematic representation of QTL mapping.

Step3. Each F2 individual is selfed for 6 subsequent generations resulting in the formation of several recombinant inbred lines (RILs). Each RIL thus obtained is homozygous for a section of a parental chromosome. Scoring of these RIL's is done for several genetic markers and the trichome density phenotype.

Step4. The arrow represents a section of chromosome derived from the parent having sparse trichome density. Leaves of all the individuals that have inherited this section also have sparse trichome density, demonstrating that this chromosomal region probably contains a QTL for this trait.

Relevance of QTL Mapping For MAS

QTL mapping is the foundation of the development of markers for MAS. Factors that affect the accuracy of QTL mapping are population size and type, level of replication of phenotypic data, environmental effects and genotyping errors. Following steps are involved in QTL mapping:

i. Marker conversion may be required to make the marker genotyping method technically simpler for the improvement of MAS reliability.

ii. QTL confirmation tests the accuracy of result from the primary QTL mapping studies.

iii. QTL validation talks about the verification of the QTL's effectiveness in different genetic backgrounds.

iv. Marker validation is required to test the level of polymorphism of tightly-linked markers within a very short distance (approx 5-10 cM) covering a target locus and also to test the reliability of markers in phenotype prediction.

MAS Schemes in Plant Breeding

Marker Assisted Backcrossing

Holland (2004) has described three general levels of marker-assisted backcrossing namely fore ground, recombinant and background selection.

In Foreground selection, markers are employed in combination with or for replacing the screening for target gene or QTL (Hospital & Charcosset, 1997). This help in:

a) Screening traits that have laborious or time-consuming phenotypic procedures,

b) To select for reproductive-stage traits in the seedling stage allowing the best plants to be identified for backcrossing,

c) Selection of recessive alleles, which is difficult to achieve by conventional methods.

Recombinant selection involves selection of BC progeny with the target gene and involves the recombination events that occur between the target loci and linked flanking markers (e.g. less than 5 cM on either side). This helps in reducing the size of the donor chromosome fragment comprising the target locus (i.e. size of the introgression) to minimize 'linkage drag' (Hospital, 2005). In conventional breeding, the donor segment can remain very large till many BC generations (e.g. more than 10; Ribaut & Hoisington, 1998; Salina et al., 2003). Due to rare possibility of occurrence of double recombination events on both sides of target loci "recombinant selection" can generally be performed utilizing at least two BC generations (Frisch et al., 1999a).

Background selection involves selection of BC progeny having greatest part of recurrent parent (RP) genome by using background markers. Background markers are those unlinked to the target gene or QTL on all other chromosomes. These are also described as, markers that can be used to select against the donor genome. This accelerates RP recovery as compared to conventional back-crossing which takes a minimum of six BC generations to recover the RP the same can be achieved by BC_4, BC_3 or even BC_2 (Visscher *et al.* 1996; Hospital & Charcosset 1997; Frisch *et al.* 1999*a*,*b*), thus saving two to four BC generations.

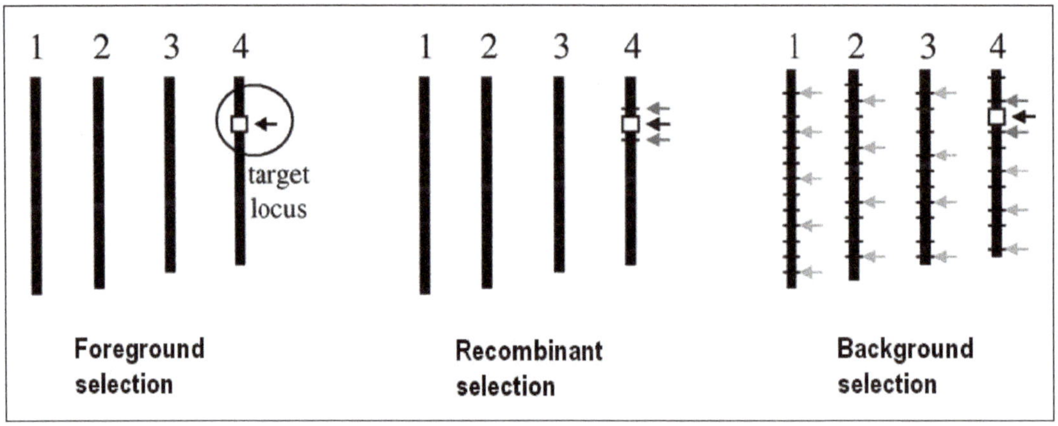

Recombinant and background selection.

Marker Assisted Pyramiding involves combining simultaneously multiple genes/QTLs together into a single genotype. Gene pyramiding is possible through conventional breeding but it is extremely difficult as individual animals must be phenotypically screened for all traits tested and nearly impossible at early generations. DNA markers can facilitate selection as they result in non-destructive assays. Also, with as single DNA sample, markers for multiple specific genes or QTLs can be easily verified without requiring phenotyping. These are mostly used for combining multiple disease resistance genes so as to develop durable disease resistance.

Early generation marker assisted selection: MAS provides a huge advantage that many plants with unwanted gene combinations, especially those lacking essential disease resistance traits and plant height, can be discarded in the early stage, so as to result in more efficient and cheap evaluation of other traits with fewer breeding lines at the later stages. However the major disadvantage of applying MAS at early generations is the high cost of genotyping a larger number of plants.

Combined Approaches: A combination of phenotypic screening and MAS approach can be utilized to maximize genetic gain (when some QTLs have been unidentified from QTL mapping) and to reduce the population sizes for such traits where marker genotyping is easier and cheaper than phenotypic screening.

Advantages of MAS

The efficiency and effectiveness for breeding may be greatly increased by MAS. The main advantages of MAS compared to conventional phenotypic selection are:

- It is simple and save time.
- Selection may be carried out at seedling stage.
- Single plants/animals may be selected with high reliability.
- Useful for traits that cannot be improved easily through phenotypic selection, either due to difficulty in measuring on young animals (i.e. before sexual reproduction), or due to low heritability.

Limitations of MAS

1. It requires prior knowledge of gene alleles or markers associated with the traits of interest as well as the quantitative estimates of these associations in the specific population.
2. It explains only a limited part of the genetic differences between individuals.

Therefore in spite of the initial enthusiasm MAS created the outcome was not that promising. Currently using adequate genetic markers, one can follow the segregation of the entire genome in spite of a set of specific regions of interest, thereby, going from MAS towards genomic selection. Today parental relationships are no longer essential for explaining related performances in animals because, these can now be explained by the fact that animals are sharing identical chromosome regionsusing a dense set of SNPs, (Eggen 2012). The cost-effectiveness of whole-genome SNP panels for major livestock species and genomic selection tools is emerging as a paradigm shift in animal breeding. SNP panels for many species, including several agriculturally relevant species have been developed.

Table: List of whole-genome single nucleotide polymorphism (SNP) chips developed for important crop plants and livestock.

Species	Identification1	Classification	Provider2	Consortium	SNP, no.
Potato	Potato	Public	Illumina	solCAP	8,303
Tomato	Tomato	Public	Illumina	solCAP	7,720
Apple	Apple	Public	Illumina	RosBREED	8,788
Peach	Peach	Public	Illumina	RosBREED	8,144
Cherry	Cherry	Public	Illumina	RosBREED	5,696
Maize	MaizeSNP50	Commercial	Illumina	Commercial	56,110
Rice	Rice44K	Commercial	Affymetrix	Commercial	44,100
Chicken	Chicken	Private: public sale	Illumina	Cobb Vantress-Hendrix-USDA	57,636
Cat	Feline	Private: public sale	Illumina	Morriss Animal Foundation	62,897
Horse	Equine	Private: public sale	Illumina	Neogen (Gene Seek)	65,157
Sheep	Ovine	Private: public sale	Illumina	AgResearch	5,409

Cattle	BovineHD	Commercial	Illumina	Various	777,962
Cattle	BovineSNP	Commercial	Illumina	Various	54,609
Cattle	BOS 1	Commercial	Affymetrix	Various	648,000
Sheep	OvineSNP50	Commercial	Illumina	Various	52,241
Cattle	BovineLD	Commercial	Illumina	Various	6,909
Pig	PorcineSNP60	Commercial	Illumina	Various	62,163
Dog	CanineHD	Commercial	Illumina	Various	173,662

1HD = high density; LD = low density, 2Illumina Inc., San Diego, CA; Affymetrix, Santa Clara, CA.

3. In spite of various advantages the impact of MAS in crop improvement is low due to following reasons:

- Resources (equipment) not available.
- Markers may not be cost-effective.
- Accuracy of QTL mapping studies not high.
- QTL effects may depend on genetic background or may be influenced by environmental conditions.
- Lack of marker polymorphism in breeding material.
- Poor integration of molecular genetics and conventional breeding.

Future Challenges

- Improved cost-efficiency- Optimization, simplification of methods and future innovation.
- Designing of effective and efficient MAS strategies.
- Greater integration between plant breedingand molecular genetics.
- Data management.

References

- Gasdaska, John R.; Spencer, David; Dickey, Lynn (2003). "Advantages of Therapeutic Protein Production in the Aquatic Plant Lemna". BioProcessing Journal. 2 (2): 49–56. doi:10.12665/j22.gasdaska
- Norris, Sonya (4 July 2005). "Molecular farming". Library of Parliament. Parliament of Canada. PRB 05-09E. Retrieved 2008-09-11
- Kaiser, Jocelyn (25 April 2008). "Is the Drought Over for Pharming?" (PDF). Science. 320 (5875): 473–5. PMID 18436771. doi:10.1126/science.320.5875.473
- Baur, Armin; Reski, Ralf; Gorr, Gilbert (2005). "Enhanced recovery of a secreted recombinant human growth factor using stabilizing additives and by co-expression of human serum albumin in the moss Physcomitrella patens". Plant Biotechnology Journal. 3 (3): 331–40. PMID 17129315. doi:10.1111/j.1467-7652.2005.00127

- Brown, Terry A. (2007). Genomes 3. New York, NY: Garland Science Publishing. ISBN 9780815341383. OCLC 444522997
- "ProdiGene Launches First Large Scale-Up Manufacturing of Recombinant Protein From Plant System" (Press release). ProdiGene. February 13, 2002. Retrieved March 8, 2013
- Dove, Alan (2000). "Milking the genome for profit". Nature Biotechnology. 18 (10): 1045–8. PMID 11017040. doi:10.1038/80231
- Rosyara, U.R. 2006. REQUIREMENT OF ROBUST MOLECULAR MARKER TECHNOLOGY FOR PLANT BREEDING APPLICATIONS.Journal of Plant Breed. Gr. 1: 67 – 72
- "Gene mapping - Glossary Entry". Genetics Home Reference. Bethesda, MD: Lister Hill National Center for Biomedical Communications, an Intramural Research Division of the U.S. National Library of Medicine. 2013-09-03. Retrieved 2013-09-06
- Nowicki, Marcin; et al. (26 October 2013), More than meets the eye: A multi-year expressivity analyses of tomato sterility in ps and ps-2 lines (PDF), Australian Journal of Crop Science, 7(13):2154-2161; Southern Cross Publishing, retrieved 2013-10-29
- Decker, Eva L.; Reski, Ralf (2007). "Current achievements in the production of complex biopharmaceuticals with moss bioreactors". Bioprocess and Biosystems Engineering. 31 (1): 3–9. PMID 17701058. doi:10.1007/s00449-007-0151-y

PERMISSIONS

All chapters in this book are published with permission under the Creative Commons Attribution Share Alike License or equivalent. Every chapter published in this book has been scrutinized by our experts. Their significance has been extensively debated. The topics covered herein carry significant information for a comprehensive understanding. They may even be implemented as practical applications or may be referred to as a beginning point for further studies.

We would like to thank the editorial team for lending their expertise to make the book truly unique. They have played a crucial role in the development of this book. Without their invaluable contributions this book wouldn't have been possible. They have made vital efforts to compile up to date information on the varied aspects of this subject to make this book a valuable addition to the collection of many professionals and students.

This book was conceptualized with the vision of imparting up-to-date and integrated information in this field. To ensure the same, a matchless editorial board was set up. Every individual on the board went through rigorous rounds of assessment to prove their worth. After which they invested a large part of their time researching and compiling the most relevant data for our readers.

The editorial board has been involved in producing this book since its inception. They have spent rigorous hours researching and exploring the diverse topics which have resulted in the successful publishing of this book. They have passed on their knowledge of decades through this book. To expedite this challenging task, the publisher supported the team at every step. A small team of assistant editors was also appointed to further simplify the editing procedure and attain best results for the readers.

Apart from the editorial board, the designing team has also invested a significant amount of their time in understanding the subject and creating the most relevant covers. They scrutinized every image to scout for the most suitable representation of the subject and create an appropriate cover for the book.

The publishing team has been an ardent support to the editorial, designing and production team. Their endless efforts to recruit the best for this project, has resulted in the accomplishment of this book. They are a veteran in the field of academics and their pool of knowledge is as vast as their experience in printing. Their expertise and guidance has proved useful at every step. Their uncompromising quality standards have made this book an exceptional effort. Their encouragement from time to time has been an inspiration for everyone.

The publisher and the editorial board hope that this book will prove to be a valuable piece of knowledge for students, practitioners and scholars across the globe.

Index